# NUCLEAR POWER AND THE ENERGY CRISIS

# Nuclear Power and the Energy Crisis

## Politics and the Atomic Industry

by
DUNCAN BURN

New York · New York University Press · 1978

© Duncan Burn and the Trade Policy Research Centre 1978

Library of Congress Catalog Card Number: 78–412

ISBN: 0–8147–0998–2

Published in the U.S.A. by
New York University Press
Washington Square
New York, NY 10003

# Trade Policy Research Centre

Having general terms of reference, the Centre does not represent any consensus of opinion. Intense international competition, technological advances in industry and agriculture and new and expanding markets, together with large-scale capital flows, are having profound and continuing effects on international production and trading patterns. With the increasing integration and interdependence of the world economy there is thus a growing necessity to increase public understanding of the problems now being posed and of the kind of solutions that will be required to overcome them.

The principal function of the Centre is the sponsorship of research programmes on policy problems of national and international importance. Specialists in universities and private firms are commissioned to carry out the research and the results are published and circulated in academic, business and government circles throughout the European Community and in other countries. Meetings and seminars are also organised from time to time.

Publications are presented as professionally competent studies worthy of public consideration. The interpretations and conclusions in them are those of their authors and do not purport to represent the views of the Council and others associated with the Centre.

The Centre is registered in the United Kingdom as an educational trust under the Charities Act 1960. It and its research programmes are financed by foundation grants, corporate donations and membership subscriptions.

# Contents

Trade Policy Research Centre                                          v
List of Tables                                                      xi
List of Figures                                                    xii
Biographical Note                                                 xiii
Preface                                                             xv
Abbreviations                                                     xvii

1  PROLOGUE AND PLAN                                                 1
     (i) The Development Process                                     1
    (ii) American and British Organisations Contrasted               4
   (iii) The Illusion of British Leadership                          8
    (iv) The Unfolding in Phase II                                  14
         *Notes and References*                                     17

PART I  FIRST COMPETITIVE REACTOR: THE
        LIGHT WATER REACTOR

2  THE PARADOX OF LWR DEVELOPMENT                                   23

3  COMPETITIVENESS ACCEPTED                                         25
     *Note*                                                         28

4  THE RISE OF COSTS – EXTENT AND SOURCES                           30
     (i) The Extent: 1967–76                                        30
         (a) Changes in Costs of Plants                             30
         (b) Changes in Fuel Costs                                  31
         (c) Changes in Generating Costs                            32
    (ii) The Moving Target: Costs of Fossil-fuelled Plants          33
   (iii) Sources of Cost Increases                                  36
         (a) General Economic Circumstances                         36
         (b) Changes in Design and Specification                    41
         (c) Delays and Long Lead Times                             43
         (d) Coincidence of Intense Development and
             Diffusion                                              44

(e) Political, Administrative and Judicial
     Infrastructure                                          45

  *Notes and References*                                     51


5  SAFTEY OF THE LWR                                         55
   (i) 'Overriding Requirement'                              55
  (ii) Bases of Assurance                                    56
 (iii) Probability Assessments                               57
  (iv) Phases in the Campaign against Nuclear Plants         60
   (v) Radioactive Effluents – G and T                       62
  (vi) Risk of Catastrophe                                   64
 (vii) The Layman in Judgement                               68
(viii) The Impact of the Intervenors                         74

  *Notes and References*                                     77


6  PEAK AND COLLAPSE 1974–76                                 82
   (i) The Peak                                              82
  (ii) The Collapse                                          82
       (a) 'Marking Time'                                    82
       (b) Competitiveness                                   83
       (c) Prospects for Lower Real Costs                    84
       (d) Safety and the Anti-nuclear Campaign              85
           LWR Safety                                        85
           Brown's Ferry Fire                                86
           Resignations at GE and NRC                        87
           First LOFT Results                                90
       (e) Anti-nuclear Campaign: Ebb and Flow               91

  *Notes and References*                                     92


7  THE LWR OUTSIDE THE UNITED STATES                         96
   (i) The Flow of Orders                                    96
  (ii) Cost Comparisons                                      96
 (iii) Organisations Compared                                97
       (a) France                                            98
       (b) West Germany                                      98
       (c) Sweden                                           103
       (d) Japan                                            105
  (iv) Delays and Prospects                                 106

  *Notes and References*                                    108


PART II  AGR TO SGHWR

8  PREAMBLE                                                 113
   *Note*                                                   115

# Contents

9  THE COURSE OF AGR DEVELOPMENT                          117
   (i)   Prelude and Prototype                            117
   (ii)  Tenders and Contract                             120
   (iii) Tender Assessment Procedure                      123
   (iv)  First Design Deficiencies                        124
   (v)   The Select Committee Investigates                125
   (vi)  Export Prospects                                 128
   (vii) All AGRs in Trouble: and the CEGB Changes Course 129
   (viii) Costs and Prices                                133
         *Notes and References*                           142

10 SOURCES OF THE AGR DISASTER                            150
   (i)   Whose Responsibility?                            150
   (ii)  The Atomic Energy Authority's Part               152
   (iii) The CEGB: Bond or Free?                          161
   (iv)  The Consortia in the Wings                       164
   (v)   Ministers Decide                                 168
   (vi)  Civil Servants Anonymous                         176
   (vii) The Select Committee Opts Out                    180
         *Notes and References*                           186

11 RESPONSE TO FAILURE: TWO GOVERNMENTS DECIDE            194
         *Notes and References*                           196

12 SINGLE D AND C COMPANY                                 197
   (i)   The Arguments For and Against                    197
   (ii)  The Select Committee Divides                     203
   (iii) Benn Rejects the Single Company                  206
   (iv)  Conservative Formula                             209
   (v)   Nuclear Power Advisory Board                     216
         *Notes and References*                           218

13 LWR VERSUS SGHWR                                       222
   (i)   Low-cost Contenders in 1969                      222
   (ii)  The Atomic Energy Authority has the Best Reactor 223
   (iii) How the Interested Parties Lined up              224
   (iv)  The Balance of Evidence                          228
         (a) Costs                                        228
         (b) Lead Times and Repeatability                 230
         (c) Safety                                       232
         (d) The NPAB Divides                             244
   (v)   The Choice and After                             245
         (a) The Basis of Choice                          245
         (b) SGHWR Development, 1974–76                    247

(c)  Reluctance and Delay                         251
(d)  Economic Impact                              251

   *Notes and References*                         256

14  RETROSPECT                                    264
    (i)  Major Mistakes at Length Acknowledged    264
   (ii)  Responsibility at the Centre             266
  (iii)  The United States Pattern – At Home and
         Elsewhere                                267
   (iv)  British Mistakes Identified Early        269
    (v)  Ministers' Options                       270
   (vi)  The Reason Why                           271
  (vii)  Responses to Information                 277
 (viii)  The Alternative Course                   282
   (ix)  Ministers and Monopoly                   285
         (a)  Working of the System               286
         (b)  A Technical Assessment Unit?        290
         The Scope for Competition                295

   *Notes and References*                         306

Glossary                                          312
Select Bibliography                               319
Index                                             327

# List of Tables

4.1 Forecasts of the Cost of a 1,000 MW(e) LWR Plant in
the United States, 1967–76 ($/KW)                                       31
4.2 Estimates of LWR Fuel Cycle Cost, 1967–82 (mills/KWh)        32
4.3 Estimated Generating Costs in 1,000 MW LWRs in the
United States (mills/KWh)                                               33
4.4 Forecasts if Generating Costs in Coal-fuelled and Nuclear
Plants in the United States (mills/KWh)                                  35
5.1 Risk of Death in the United States from Selected Hazards     57
5.2 Probability of Major Man-caused and Natural Events           60
9.1 Original Cost Estimates for AGR Plants (excluding initial
fuel and IDC)                                                          133
9.2 Estimates of Present-worth Cost of Nuclear Plants (£/KW)
and Generating Costs (p/KWh) in Britain in 1973 (1973
prices)                                                                134
9.3 Generating Costs at 1 January 1972 Money Values
(p/KWh) in Magnox, Coal and Oil-fired Plants in Britain          135
9.4 Estimated Generating Costs at 1 January 1972 Money
Values (p/KWh) in Nuclear, Coal and Oil-fired Plants
under Construction in Britain                                          136
9.5 CEGB Estimate of AGR Capital Costs – November 1975          137
9.6 Utilities' Expenditure on Construction of AGR Plants
(including IDC capitalised)                                            139
9.7 Cost of Electricity from Alternative Sources                 140
9.8 Total Cost of AGR Programme (£m, 1975 prices)                141
10.1 Load Factors in Magnox Plants 1972–73                       184
11.1 (A) Estimated Capital Costs in Britain of Nuclear Plants
of Different Systems 1973–74 (£/KW); (B) Estimated
Generating Costs in Britain 1973–74 (p/KWH)                          229

# List of Figures

3.1 Capacity of Nuclear and Fossil-fuelled Plants Ordered and
    Nuclear Plants Operating in the United States 1963–75      26
3.2 LWRs Ordered and Operating Outside the United States
    1968–75      28
4.1 Fuel and Power Prices in the United States 1960–75      34
4.2 Wholesale and Consumer Prices and Construction Costs in
    the United States 1963–75      37
4.3 Wage Increases in the United States 1960–75      38
4.4 Cost of Money in the United States 1960–75      39
4.5 Nuclear Electric Power Plant Construction and Operating
    Schedule in theUnited States (1973)      50

# Biographical Note

DUNCAN BURN began analysing the contrasts between the rates of industrial and technological development in different countries in 1930 when he was a Lecturer in Economic History at the University of Cambridge. He completed *The Economic History of Steelmaking 1867–1939* just before World War II. From 1939 to 1943 he worked in the British Iron and Steel Control and in 1943 was a member of a Joint Metallurgical Committee, representing the United Kingdom, Canada and the United States, aimed at optimising the use of scarce alloys. In 1944–45 he worked on post-war policy and planning problems for steel in the British Ministry of Supply. From 1946 to 1962 he was a leader writer and industrial correspondent on *The Times* in London, leaving to become Director of the Economic Development Office set up in the United Kingdom by the Heavy Electrical Generator Manufacturers. Duncan Burn was later a consultant to firms in the aircraft and chemical industries, but returned to academic work as Visiting Professor of Economics at Manchester in 1967–69, then at the University of Bombay in 1971.

The second of Duncan Burn's books on steel, *The Steel Industry 1939–1959*, was published in 1961, and the first of his books on atomic power, *The Political Economy of Nuclear Energy*, which was for the Institute of Economic Affairs, London, was published in 1967. His first study for the Trade Policy Research Centre, *Chemicals under Free Trade*, appeared in 1971 and was reissued in 1972 in *Realities of Free Trade: Two Industry Studies*. He edited and contributed to *The Structure of British Industry* (2 vols), for the National Institute of Economic and Social Research, London, published in 1958. He has served on the Executive Committee of the National Institute of Economic and Social Research (1959–69), the Board of Trade Consultative Committee on the Census of Production (1956–67) and the Economic Committee, Department of Scientific and Industrial Research (1962–64).

# Preface

First, I have many people to thank. I am grateful to the Trustees of the Wincott Foundation for a grant when I started, longer ago than I care to think, to write this sequel to *The Political Economy of Nuclear Energy*. Had I realised how the complexities of the subject would mount up, to such an extent that I have been led to put some of them aside for separate treatment, I might have resisted the persuasion of Ralph Harris and John Wood at that time. I am deeply indebted again to the staffs of many government departments and agencies and of many manufacturing firms, electricity supply utilities, architect-engineers and consultants in Britain, the United States, Germany, France and other European countries, for help by discussion, correspondence, provision of unpublished information and facilities to visit plants and laboratories. In the United Kingdom, I am particularly indebted to the Atomic Energy Authority and the Central Electricity Generating Board. Without this help, spread now over two decades (I paid my first visit to Risley and Windscale in 1955 and made my first contacts with the American, German and Swedish activities in 1956) this task would have been impracticable. I am also grateful to all those who permitted me to use copyright material.

Several friends in and out of the industry have read part, or all, of the text as it has evolved, and their comments have been of great help. I would mention especially Neville Beale, of Shell, David Henderson, Professor of Political Economy at University College in the University of London, and David Rapoport, Professor of Political Science at the University of California, Los Angeles. My wife has helped immensely, not merely by what Professor Clapham felicitously called 'domestic industry', but much more by performing the function of lay critic. I should add, too, that a book rewritten several times makes great demands on patience and tolerance. In the final preparation of the book for publication I have been greatly assisted at the Trade Policy Research Centre by Hugh Corbet, the Director, and his staff, particularly Patricia Seligman.

The bibliography which is included lists the main books, reports, papers and articles used and referred to in the text, and is not a bibliography in any wider sense. Many of the reports quoted have a exceptionally long descriptions: I have therefore included the abbreviations of the titles which are used in the notes. Abbreviations used in the text itself, including all acronyms (in which the subject proliferates), are set out in a separate list. One abbreviation I must mention individually here: I refer so much to the Select Committee on Science and Technology of the British House of Commons that I call it simply *the* Select Committee. When other Select Committees of the British Parliament are referred to they are more completely described. I have included a glossary, based on the one in *The Political Economy of Nuclear Energy*, but extended: it primarily covers technical terms, but some unfamiliar non-technical terms are also included.

I intended, in the final rewriting, to cover 1975 fairly fully, but none of 1976. With further delays, however, I felt compelled to refer to some conspicuous events, such as the check to the SGHWR programme, the setback to the anti-nuclear campaign in the Californian vote and the beginning of another round of Select Committee proceedings on atomic energy. The most recent data could not usually be as fully processed as material which has been available longer. Equally it must not be allowed to give the illusion that 1976 is fully covered up to, say, the summer: no attempt has been made to do this. Reviewers will certainly be able to point out that some things, possibly quite important, have been left out. I think the omissions will not disturb the argument of the book.

DUNCAN BURN

*London*
*December 1976*

# List of Abbreviations

| | |
|---|---|
| AAEA | Australian Atomic Energy Authority |
| ACRS | Advisory Committee on Reactor Safeguards (United States) |
| AECL | Atomic Energy of Canada Ltd |
| AEC | Atomic Energy Commission (United States) |
| AEA | Atomic Energy Authority (United Kingdom) |
| AEG | Allgemeine Elektrizitäts Gesellschaft (West Germany) |
| AGR | Advanced Gas-cooled Reactor |
| ANS | American Nuclear Society |
| APC | Atomic Power Construction Limited (United Kingdom) |
| ASEA | Allmänna Svenska Elektriska AB (Sweden) |
| ASLB | Atomic Safety and Licensing Board (United States) |
| ATWT | anticipated transient without trip |
| BASF | Badische Anilin and Soda Fabrik AG (West Germany) |
| BBC | Brown Boveri and Co. |
| BBC–KRUPP | Brown Boveri-Krupp, later HRB, q.v. (West Germany) |
| BBR | Babcock Brown Boveri GmbH (West Germany) |
| BEEN | Babcock-English Electric Nuclear Co. (United Kingdom) |
| BNDC | British Nuclear Design and Construction Ltd |
| BNF | British Nuclear Forum |
| BNFL | British Nuclear Fuels Ltd |
| BNX | British Nuclear Export Executive |
| Btu | British thermal unit |
| BWR | Boiling Water Reactor |
| CBI | Confederation of British Industry |
| CE | Combustion Engineering (United States) |
| CEA | Commissariat à l'énergie atomique (France) |
| CIF | Creative Initiative Foundation |
| CEGB | Central Electricity Generating Board (United Kingdom) |

| | |
|---|---|
| d | penny, pre-decimal coinage (0.417p) |
| D and C | design and construction (companies) |
| DSIR | Department of Scientific and Industrial Research (United Kingdom) |
| DTI | Department of Trade and Industry (United Kingdom) |
| ECCS | emergency core cooling system |
| EDF | Electricité de France |
| EEC | European Economic Community |
| EPA | Environmental Protection Agency (United States) |
| ERDA | Energy Research and Development Administration (United States) |
| EURATOM | European Atomic Energy Community |
| FBI | Federation of British Industries (now *Con*federation of British Industry [CBI]) |
| FBR | Fast Breeder Reactor |
| FOE | Friends of the Earth |
| FRC | Federal Radiation Council (United States) |
| GE | General Electric Company (United States) |
| GEC | General Electric Company (United Kingdom) |
| GW | gigawatt |
| GWh | gigawatt-hour |
| HRB | Hochtemperatur-Reaktorbau GmbH (West Germany) |
| HTR ⎫ HTGR ⎭ | High Temperature (Gas-cooled) Reactor (in some countries HTGR is preferred) |
| ICI | Imperial Chemical Industries Ltd (United Kingdom) |
| ICL | International Combustion Ltd (United Kingdom) |
| ICRP | International Commission on Radiological Protection |
| IDC | interest during construction |
| IPCS | Institution of Professional Civil Servants (United Kingdom) |
| IRC | Industrial Reorganisation Corporation (United Kingdom) |
| JAIF | Japanese Atomic Industries Forum |
| JCAE | Joint Committee on Atomic Energy, Congress of the United States |
| KW, KWh | Kilowatt, Kilowatt-hour |
| KRT | Kernreaktorteile GmbH (West Germany) |
| KWU | Kraftwerk Union (West Germany) |
| LMFBR | Liquid Metal Fast Breeder Reactor |
| LOCA | loss of coolant accident |
| LOFT | loss of fluid test facility |
| LWR | Light Water Reactor |
| MAN | Maschinenfabrik Augsburg-Nuremberg AG |
| MIT | Massachusetts Institute of Technology (United States) |
| MW | megawatt |

| | |
|---|---|
| MW(t) | megawatt thermal |
| MW(e) | megawatt electric |
| MOS | Ministry of Supply (United Kingdom) |
| NCB | National Coal Board (United Kingdom) |
| NEDO | National Economic Development Office (United Kingdom) |
| NEPA | National Environment Protection Act, 1969 (United States) |
| NII | Nuclear Installations Inspectorate (United Kingdom) |
| NNC | National Nuclear Corporation |
| NPAB | Nuclear Power Advisory Board (United Kingdom) |
| NPC | Nuclear Power Company (United Kingdom) |
| NRC | Nuclear Regulatory Commission (United States) |
| NSHEB | North of Scotland Hydro-Electric Board |
| NSSS | Nuclear Steam Supply System |
| O and M | operation and maintenance |
| OEEC | Organisation for European Economic Cooperation, succeeded by the OECD |
| OECD | Organisation for Economic Cooperation and Development |
| ORNL | Oak Ridge Nuclear Laboratory (United States) |
| PDA | preliminary design approval |
| PFR | Prototype Fast Reactor (Dounreay) |
| psi | pounds per square inch |
| Pu | plutonium |
| PWR | Pressurised Water Reactor |
| R and D | research and development |
| RTZ | Rio Tinto-Zinc Corporation Ltd (United Kingdom) |
| RWE | Rheinisch-Westfälisches Elektrizitätswerk (West Germany) |
| SESE | Scientists and Engineers for Secure Energy (United States) |
| SGHWR | Steam Generating Heavy Water Reactor |
| SSEB | South of Scotland Electricity Board |
| TNPG | The Nuclear Power Group (United Kingdom) |
| TVA | Tennessee Valley Authority (United States) |
| UCLA | University of California, Los Angeles |
| UPC | United Power Company (United Kingdom) |
| W | Westinghouse Electric Corporation (United States) |

# CHAPTER 1

# Prologue and Plan

(i) THE DEVELOPMENT PROCESS

This book is a sequel to *The Political Economy of Nuclear Energy* which was published in 1967 and, like its predecessor, is concerned with the comparative efficiency of the organisations adopted in different countries to apply atomic energy to the generation of electricity. Because of inherent technical difficulties and the dangers involved, the development – still in important aspects incomplete – has been both very long and very costly. It has been exceptional in two other respects. First, since the United States, British and Canadian governments owned and controlled the basic technology, developed for war purposes, these governments had to initiate the developments, although the driving force came partly from other sources – in particular from industrial firms in the United States. Second, since atomic energy presented great dangers close government supervision and control were regarded universally as inevitable and essential.

When the first book appeared in 1967 the first large nuclear power plants which were to be competitive with coal-fuelled plants were nearing completion in the United States. This was an economic landmark in the development, the end of the first phase (called here Phase I). Once it was widely accepted by utilities that nuclear plants could be competitive in many areas a rapid diffusion of the new technology was undertaken. At the same time development itself became more intense; and partly because of the greater activity, public reaction to the whole process became more lively. Thus in the new phase, Phase II, with which this book is primarily concerned, new and more complex problems faced the organisers and policy makers.

The likely time scale of development was set out perceptively in the United States in the early 1950s. The Report of the Paley Committee on *Resources for Freedom*, issued in June 1952, forecast that, as a result of the prospective growth of energy consumption and diminishing returns from existing sources, 'some new low cost sources [of energy] should be made ready to pick up some of the load' by 1975 'if we are to avoid serious risk of increase in real unit costs of energy' in the

1

United States. It thought nuclear fuel was likely to be one such source. For technical reasons other sources were needed because atomic power could not supply low-cost energy in all forms: the view was expressed that it possibly could never take more than 20 per cent of the total energy load. Hence other new sources should be researched intensively – solar heat was particularly stressed. Atomic power, the Paley Report emphasised, though now being intensively researched, was still in its infancy and would contribute nothing for ten to fifteen years at least. Most likely the atomic industry would be a net consumer of electric power for twenty-five years, approximately till 1975. Its major contribution to energy would only come, moreover, when breeder reactors, on which the United States experimental programme started in 1945, were developed.[1]

At the end of 1955 reports by experts for the Joint Committee on Atomic Energy (JCAE) of the United States Congress, broadly confirmed the Paley Report forecasts. They had a greater body of analysis and experiment to guide them. The 'consensus' of the experts was that it would be from ten to twenty years before a nuclear plant producing power as cheaply as the best fossil-fuel plant could be operating. This was largely because in order to be competitive a nuclear reactor would need to have a very high capacity – the figure was put at 200 to 300 MW already in early 1953 and quickly increased[2] – otherwise capital costs per KWh would be too high. Costs of generating power in fossil-fuel plants were themselves being reduced impressively by sensational increases in capacities, in which the United States was much ahead of Britain.[3] In developing nuclear plants one could not extrapolate from a laboratory scale to these high capacities in one step: it would be necessary to have two or more intermediate stages, at least an experimental reactor of 20 to 100 MW(t) and a small or pre-commercial scale prototype plant of 40–200 MW(e).[4] Each stage would take upwards of three to four years for constructing the plant and obtaining operating experience to incorporate in the succeeding stage. All this work was essentially development engineering – with laboratory support, but not laboratory control.

Several reactor systems were being developed in the United States in the early 1950s; the first to prove competitive (the experts were not sure which system it would be, although the Light Water Reactor [LWR] looked the most probable) was likely not to be the ultimate best. Development of other systems would thus continue after one type became competitive and the experts took the same view as the Paley Report of the importance of breeders. On the most optimistic assumptions nuclear plant capacity in the United States might reach 12,000 MW by 1970, 48,000 MW by 1975; but substantially lower figures of 7,200 MW and 22,600 MW would result on more pessimistic assumptions.[5]

The general pattern and timing set out in the United States assessments of 1952 and 1955 turned out to be well judged. The scope for new sources of low-cost energy became progressively more clear as the price of domestic fossil fuels rose steadily from the mid-1960s and sharply in 1970, and it was emphasised with unexpected drama after September 1973 by Middle East policies with regard to oil supplies and prices. Development of nuclear power plants had been possibly slightly slower than the forecasts suggested. Although most utilities were persuaded that LWRs were 'competitive' by the early 1970s, costs were not as low as had been forecast, there was much more to be done before LWR design would be stabilised and there was unexpectedly strong resistance from environmental groups who claimed LWRs were unsafe. In 1970 the capacity of nuclear power plants operating in the United States, 6953 MW, was slightly below the lower figure in the long-range forecast in 1955, but capacity operating by the end of 1974 – 33,941 MW – was on a trend midway between the higher and lower figures in the forecast for 1975.

As foreshadowed other reactor types were being intensively developed, for which advantages over the LWR were claimed; greater inherent safety, fuller use of the energy available in uranium and provision of higher temperatures for industrial heat. These had not reached the stage when their competitiveness had been established in practice, and not merely on paper.

The ultimate emergence of a 'competitive' reactor (the term involves ambiguities which will be looked at later) does not prove that the result has been achieved in the shortest possible time by the least costly route. The long and vast development gave great scope for waste and error. It bristled with difficult decisions. First, there were decisions on major strategy: which reactor systems to start developing, which to continue, which to drop, whether and when to start developing yet another system, and so on. Development of each system involved similar crucial choices. It was not just moving along an easily prescribed line once the system was chosen; there were alternative ways of designing any system, different possible parameters of pressure and temperature, different configurations of coolant flows, different possible designs of fuel and control rods, different types of pumps to choose from (or design), different specifications of material to choose or develop, and so on. This work of choosing, designing and researching, must be coordinated to allow the work of manufacture and construction to maintain momentum and keep approximately to a timetable. It was equally necessary that development of ancillary services, fuel supply and reprocessing, should be coordinated with reactor development. I believe the nature of the development process was more fully understood at the outset in the United States than in the United Kingdom.

(ii) AMERICAN AND BRITISH ORGANISATIONS CONTRASTED

The extent to which – and the rate at which – nuclear power will contribute to energy supplies, and therefore lessen what was spoken of by 1973 as the 'energy crisis', will reflect the quality of the development performance, past, present and future, including in performance the establishment of public acceptability. Hence the practical interest in comparing the industrial and political organisations in different countries. The interest extends beyond the nuclear industry, as the first book pointed out,[6] because many of the problems involved arise also in the development of other high-technology industries.

When the United States and the United Kingdom decided to promote the civil applications of atomic energy, they chose sharply contrasting forms of organisation. These reflected differences in their industrial and political skills, traditions and attitudes.

In the United States the initiating legislation required that civil applications should be developed as soon as possible 'within our normal economic and industrial framework' and should 'strengthen free competition in private enterprise'.[7] It was an obligation on the executive agency which had the responsibility for the transfer of the technology, the Atomic Energy Commission (AEC), to implement these principles. Congress through a powerful Joint Committee on Atomic Energy (JCAE) zealously watched that the obligation was observed. In its World War II and early post-war development of the atomic industry for military purposes, the Federal government used large industrial firms as contractors to design and construct major installations (diffusion plants, reactors for making plutonium and the like) and they used universities and engineering and chemical firms to manage their national laboratories. The object of AEC policies in regard to nuclear power plant development from 1950 on was to encourage industrial firms, architect-engineers, and utilities to participate in developments of nuclear power reactors based on AEC knowledge of reactor systems, which was made available for study by interested groups. The object was that as quickly as possible the selection of reactor systems for development and the conduct and management of the developments from the research stage onwards, and the financing of the projects, should be wholly in the hands of competing groups. The decision to buy nuclear plants for use on the electricity supply system when they were believed to be competitive was left entirely to the utilities – many of which were of substantial importance, mostly privately owned, but one of the leaders, the Tennessee Valley Authority (TVA), was publicly owned. These were also encouraged to help finance development from which they would benefit. State aid for developments would be cut off where the work failed, and as work succeeded aid would be reduced: there would be none on plants bought by utilities as 'commercial'. It was intended that competition should be established also in the ancillary industries, uranium ore mining and processing, enrichment, chemical processing of spent fuel, and the like.

The belief was that by promoting the parallel development of different reactor systems by increasingly autonomous groups, a premature decision between systems before their technical or economic values could be assessed was avoided, scope was given for a variety of independent initiatives, one could use more of the available ideas and there were strong incentives for rapid development (thought of in terms of fifteen to twenty years). There would be a possibility of comparing systems and developers, and the final choice of 'competitive' reactors would be on commercial grounds, by the persons most experienced in this kind of judgement and most familiar with the economics of alternative methods of generating electricity. The AEC efforts to secure 'participation' were quickly successful – by the end of 1955 there were over a dozen substantial joint development projects covering nine different reactor systems in which the proportions of government and private enterprise varied, but already some plans were under way for utility-manufacturer plants of almost 200 MW capacity with little or no AEC contribution.[8]

The United Kingdom followed a completely different route. There was no wartime growth of a nuclear industry because it would be too vulnerable in Britain. The early post-war build-up of a nuclear weapons industry was undertaken by a department of the Ministry of Supply (MOS). At the outset they would have liked Imperial Chemical Industries (ICI) to design, build and run a production reactor and a diffusion plant, but for a variety of reasons, including fear of nationalisation, ICI refused. English Electric also refused to undertake the diffusion plant – it was not in their range of business.[9] Atomic research was concentrated at Harwell under Sir John Cockcroft, and the design and architect-engineering of the various plants needed for atomic weapon production was carried out by the Industrial Division of the Atomic Energy Department of the MOS at Risley under Sir Christopher (later Lord) Hinton, who had been Chief Engineer of the Alkali Division of ICI before the war.

By 1947–48 some engineering firms – for example, Parsons, Metro-Vick, and Babcock & Wilcox – were anxious to take part in reactor development on the American pattern.[10] For the first Harwell nuclear power plant programme – proposed tentatively in 1949 – Cockcroft suggested that the 'major design and development work should be carried out through contracts with industrial organisations who would build particular reactors under Harwell guidance. Harwell would do the metallurgical work.'[11] Cockcroft in 1952 envisaged Parsons as major contractor and ultimately operator for the $CO_2$/graphite/natural uranium reactor (then called 'Pippa': later Calder Hall, later Magnox) which was one part of the programme. But the demand for military plutonium suddenly doubled: 'Pippa' must be built to supply this; the job was handed to Risley to organise in 1953, and Hinton decided that only Risley itself could do it quickly, that private industry would be bound to be late.[12] Whether that would have been so we shall never know; Margaret Gowing, official historian of Britain's atomic project, appears to assume it.[13] What is certain is that Calder Hall was only completed on time because all the private subcontractors who did almost all the manufacturing and construction and designed most of the components, were sufficiently on time. For nuclear work it was a simple job. The Americans at this stage, on more complex jobs, were beating exacting targets.

With the Calder Hall decision the nuclear industry remained highly centralised, the research on reactor systems and the choice of research programmes being centralised in government hands. There had been no sign of a vigorous purposeful policy of bringing in private firms to study Harwell's research results and make what use they liked of them – which was the first step in United States policy. (Firms who wanted to use the AEC results in the United States were helped financially in doing so and AEC staff joined them to assist in the good work.) The Harwell approach was, I am informed, often secretive: one of Parsons' chief scientists, for example, told me he was not allowed in the area where the initial work on the high-temperature gas-cooled reactor (HTR) was taking place in the mid-1950s.

By 1954 the Conservative Government had been persuaded to adopt a restrictive centralising policy expressed in the formation of the United Kingdom Atomic Energy Authority (AEA). This was a body whose members would be appointed by the responsible Minister, would be answerable to him and would be subject to directions from him. It would cover all military and civilian atomic activity. It was to be the principal adviser to the Minister on all atomic affairs. In regard to nuclear power plant development, the AEA was given responsibility for research and development of reactor systems up to the point at which a prototype existed, at which time a decision would be taken to build a plant on a commercial scale. At that point industrial firms – whether public or private – would take over and would make 'commercial' plants based on AEA development results which would be handed over to them. They would not, however, research on new systems, have free access to AEA research results or be free to initiate research with state aid. The nationalised utilities would be the buyers. Like the AEA, these were bodies whose 'members', the directors, were appointed by and responsible to a Minister who could give directions, who had to approve their investment plans and could influence their choice of generating plant, including nuclear plant.

One function of the AEA was to avoid wasteful expenditure. Atomic energy was 'the most promising of all the discoveries made by man', according to Sir David (later Lord) Eccles, in presenting the Atomic Energy Authority Bill to the Commons, but the general public had still to realise it was also the most expensive.[14] This has since been the subject of a liberal education. Eccles said there was now a 'risk of excessive spending by scientists ... Explorers are usually enthusiasts. ... There are so many fascinating paths'. So the 'top scientists and engineers' had agreed it was a good idea to bring them round a table where they 'could examine the problem as a whole and make good recommendations which would become the responsibility of all the members of the Authority'.[15] It was essentially a planning exercise. The Minister had a duty to ensure that 'the proper degrees of importance are attached to the various uses of atomic energy'[16] and for this planning they were fortunate to have Sir Edwin (later Lord) Plowden, 'celebrated as ... the chief of all planners ... but his feet never left the ground.'[17]

The idea of having all the experts round a table was fashionable at the time in the United Kingdom. The Herbert Committee, of which Sir Ronald Edwards, who became chairman of the Electricity Council, was a member, recommended in 1956 for example that in the proposed Central Electricity Generating Board (CEGB) 'all the best brains should be assembled with the object of designing future [advanced] stations with maximum efficiency.'[18]

In the AEA the initial expression of this policy was the adoption of the 'narrow front' – the decision to develop on prototype scale only one 'family' of thermal reactors, those using graphite as moderator and a gas, at first $CO_2$, as coolant. It was the one reactor system which in 1954–55 was specifically excluded from American development programmes as having a low material economy. Outside this family prototype development in the United Kingdom would be limited to the fast breeder needed for the 1970s. So in due course the AEA moved from Magnox to the Advanced Gas-cooled Reactor (AGR), also using graphite and $CO_2$ but using enriched uranium as fuel.

There were many dissentient voices within the Authority over this move. In 1954–55 a distinguished group at Harwell believed the best course of development, once enriched uranium fuel was adopted, was to omit the AGR, to adopt for prototype development an LWR for early use and for later use an HTR, using helium as coolant (which presented greater difficulties but offered greater rewards, including high temperatures for industrial processes) or a heavy water type reactor – or even go straight on to the Fast Breeder Reactor (FBR).[19] Members of this group were no doubt some of the explorers whose enthusiasm alarmed Eccles. Two of them went to key positions in the American nuclear industry.[20] Their views did not prevail against the dominant view, which seems to have been growing in the Atomic Energy Division of the MOS since at least the end of 1951. The view of the AEA was of course quite simply the view which prevailed – not necessarily a representative or majority view, still less by definition the best.

In February 1955 British organisation was still further defined and exemplified by the Government's decision to launch a ten-year nuclear plant programme – the 'first' programme it was sometimes called, implying there were to be others. It envisaged a dozen 'commercial' stations with a total capacity of 1500 to 2000 MW to be built by 1965 – carrying out a project conceived in the Atomic Energy Division of the MOS at least as early as the beginning of 1953. The first plants would be of the Calder Hall (Magnox) type; by the end of the period other – liquid-cooled – types might be chosen. The electricity authorities were informed of the plan, not consulted. The plants were to be designed and constructed by groups of firms having complementary activities – 'the consortia' – who would compete with each other in tendering for 'commercial' plants based on AEA protoypes, and would use their commercial experience to build up a great export trade based on Britain's lead in the new technology.[21] This plan was trebled in 1957 – the capacity of stations to be built became 5000 to 6000 MW.

By 1955 the contrasts between British and American organisation and policy were sharply drawn; Britain was planning through government-controlled programmes to save money and resources on development and spend much more lavishly on 'commercial' plants of which the first at least would not be competitive. The United States was planning to devote large sums and resources for development increasingly by private enterprise: commercial plants would be ordered by utilities if and when in their judgement a competitive plant emerged, but the Government would not come into the decision and there would be no subsidy.

(iii) THE ILLUSION OF BRITISH LEADERSHIP

*The Political Economy of Nuclear Energy* was published in May 1967 at the close of Phase I. It seemed clear that within the next two years or so competitive nuclear plants would begin coming into operation in the United States – at least two years before what was expected to be a competitive plant would operate in Britain where the competitive target set by coal-fuelled plants was less severe. It seemed likely, although it was vigorously denied by the British Government and the AEA, that the plants in the United States would cost less and provide electricity more cheaply than the British plant. The American plants were light water reactor (LWR) plants of two types – pressurised water (PWR) and boiling water (BWR). Both use ordinary water as moderator and coolant; in the BWR the coolant, which reaches a pressure of about 1050 psi (pounds per square inch), boils and the steam is used in a direct cycle to drive a turbo generator, whereas in a PWR the coolant is kept in a closed cycle at a pressure of 2250 psi and does not boil, its heat being transferred in a steam generator to a secondary cycle which provides the steam to drive the turbo generator. The British plant would be an advanced gas-cooled reactor plant (AGR).

Realisation of this American lead in reaching the first economic landmark in the development process – in time, and most probably in cost – came as a shock and a source of dismay to many people in Britain where politicians and the atomic establishment had nurtured the belief since the early 1950s that the United Kingdom led the world in the peaceful application of atomic energy. Britain had indeed scored two 'firsts'. In 1956 the AEA had started operating the Calder Hall plant which, although built primarily to make plutonium, was more famous as the world's first nuclear power station to supply power to a utility, not just an experiment. And the British Government had launched the world's first Ten-year Programme of nuclear power plants in 1955 and trebled it in 1957. These seemed to many to be tangible proofs of leadership.

But this was misleading. All nine plants built under the first Nuclear Power Programme were based on the Calder Hall Magnox reactor.[22] This could not be developed into a competitive reactor, as American experts said from the outset.[23] After two initial orders, it proved to be unexportable. Hence the consortia formed in 1955–57, to design and construct the plants in this costly programme, were all engaged in the last development stages of a system with no future. This was disguised by the practice of calling the plants 'commercial'. They did produce more nuclear power for the British grid than was produced for any other electricity supply system. The Americans said, understandably, that they were not interested in producing 'high-cost nuclear KWhs'. But these large outputs were another British 'lead'.

The implication in asserting 'leadership' was that Britain had a lead in the technology. This was not so. A case for 'level pegging' could have been made out in 1956,[24] although it would have been invalid. Nevertheless, there was a further *démarche* in 1965 to perpetuate or restore the myth. In 1957 the AEA decided that the AGR should succeed Magnox. A 30 MW prototype AGR began operating at full power early in 1963. In autumn 1963 two American utilities ordered 600–650 MW BWRs as competitive plants and the estimated costs were attractive. They attracted the CEGB. The British Government decided that the CEGB should compare the estimated costs in AGR and BWR plants before ordering its next nuclear plant – Dungeness B.

The result was, the Government announced, a triumphant vindication of British development. Fred Lee, Minister of Power in 1965, told the House of Commons: 'We have made the greatest breakthrough of all time... We have hit the jackpot this time.' These must rank among the most absurd of ministerial pronouncements.[25] Its author after holding two other Ministerial jobs became a member of the House of Lords. But the news had a good 'press', and detailed substantiation of the claim published in an *Appraisal* by the CEGB[26] was accepted as impressive evidence (except in some professional journals) and people used the figures for pretentious forecasts. Ministry of Power 'experts', including economists and statisticians, used them for estimates of nuclear prospects in the seventies.[27]

The advantage in cost claimed in the *Appraisal* for 'the greatest breakthrough of all times' was 0.01p per KWh,[28] trivial and within the statistical margin of error. As competent observers pointed out quickly (unnoticed in popular discussion) the comparison involved great hazards and some bias. The design of the 600 MW AGR was extrapolated from a 30 MW prototype which involved much greater uncertainty than was involved in extrapolating from a 200 MW plant for the BWR. Recent design improvements in the BWR accepted by American utilities were rejected as unproven by the CEGB, but more recent radical and untested changes for the AGR were accepted. The *Appraisal* showed no recognition of the disadvantages of a permanent graphite core, contained inadequate data on fuel cycle costs and assumed that a 660 MW turbo alternator, of which none had been made in Britain, would stop for maintenance only once in two years, contrary to all experience. 'The evaluating team are at some pains to impress the reader how expert they are, but the reader is left with no doubts about their partiality', an outspoken nuclear consultant, R. F. W. Guard, wrote, saying publicly what several engineers who still had to do business with the CEGB were only prepared to say privately. He added that 'uncommitted observers elsewhere will hardly agree that the decision would be the same in different circumstances'.[29] He proved right: foreign experts' assessments were almost uniformly unfavourable.[30]

This was beginning to strike home in 1966 because efforts to export the AGR failed. There was general agreement that something needed to be explained: pressures which had been building up since 1964 for a reorganisation of the industry increased, and a new Select Committee of the House of Commons – on Science and Technology – decided to make a study of the industry its first task.

My purpose in writing *The Political Economy of Nuclear Energy* was to show why, in Phase I, Britain had not in fact lost a lead, which in the significant sense it never had, but had fallen behind the United States in developing a competitive reactor. It had certainly fallen behind in time. I believed it almost as certain that, while prices of plants would rise in both countries, the cost of generating power in an LWR would be significantly lower than in an AGR.

For the United States to move faster than the United Kingdom in industrial innovation was 'a normal experience', and the Americans had special advantages in nuclear development; a much bigger, faster-growing market for power, for example, a vast demand for nuclear weapons and submarines powered by the PWR, immensely powerful engineering and chemical firms whose nuclear experience started in the war, strong utilities keen on cost cutting innovations.[31] But promising LWR development by German firms showed that American advantages were not insurmountable. The British Government had argued that nuclear power development was more urgent for the United Kingdom than for the United States, and chose its organisation and policies to ensure rapid advance, to offset relative disadvantages, and make the best use of scarce resources. Germany, clearly doing well in developing competitive reactors, had adopted a form of organisation close to the American. Would Britain have done better without her special arrangements? The answer appeared to be yes.

The apologia popular among ministers, civil servants, the atomic 'establishment' and most journalists, rejected this view and urged *more* centralisation. According to this, Britain's weakness lay in the consortia and the way in which the development was divided between them and the AEA. There were still, it was said, too many consortia, this reflected the excessive subdivision of British engineering and consequent ineffectiveness, the consortia structure meant that the partner companies got subcontracts without competing for them, and when the AEA passed on its AGR technology to the consortia it took a year and cost each consortium about £1m. Here was the source of high costs, delays and lack of exports. It could be readily removed by concentrating design and construction round the AEA. Then all would be well since the AEA still led in the HTR and FBR, was developing the best water reactor – the Steam Generating Heavy Water Reactor (SGHWR) – and would soon prove the remarkable development potential of the AGR. Britain's apparent lag and the lack of exports were, one might say, a temporal illusion.

This would not stand serious analysis. There were deficiencies in the consortia system and subdivision was a weakness in engineering. But the British trouble in 1965–67 was the narrowness of prototype development and its slightness and lateness, which was due entirely to ministers and the AEA. The weakness was at the heart of Britain's organisation, in its central organs, not on the periphery. What had to be explained was the inadequacy of the AEA's prototype AGR. This could be due partly to design and management: the AEA had a monopoly and there was no basis of comparison. But the Windscale prototype clearly reflected, one might say symbolised, the three basic

decisions of 1955–57: (i) in 1955 to have a ten-year nuclear plant programme based mainly on Magnox, (ii) in 1957 to treble this and base
it exclusively on Magnox and (iii) to confine prototype development
for thermal reactors to the gas/graphite family. The choice of system
was made with no prototype experience of alternatives. But when it
was decided in 1957 to develop the AGR as a successor to Magnox to
secure lower costs, it would have been as practical in Britain, as it
proved in Germany at the same date, to try alternative systems side by
side. It was certainly known in Britain that many European experts –
some at Harwell – thought the American LWRs more promising than
gas-graphite plants and significantly, as in America, the $CO^2$/graphite
system was not among those chosen for testing in Germany. The outcome of the British decision was that the large volume of engineering
resources committed to design and construction of nuclear plant was
used mainly to make plants to produce power based on a system with
no future. Resources for prototype development of new systems were
restricted in scale, confined to the 'narrow front' (until construction of
the SGHWR prototype started late in 1962), chosen and managed by
a monopoly, and offering no basis for comparisons. One can choose
the wrong horse, and ride a horse badly; Britain appeared to do both.

The Select Committee argued in 1967 that the 'dominant motive' of
United Kingdom policy in adopting the 1955–57 programme was to
'overcome an assumed physical shortage' of coal and oil and that 'cost
was a secondary matter'.[32] It was a comforting rationalisation of a
misjudgement. The increased programme of 1957 was in part a panic
reaction to the Suez crisis,[33] but it would not have occurred without a
grotesque overestimate of the technical and economic significance of
the Calder Hall plant. Most advocates of the 1957 programme appeared to believe in Hinton's forecast that power from Magnox plants
to be completed in 1962–63 would be cheaper than power from new
coal-fired stations.[34]

The panic fears had vanished by the end of 1957[35] and the promise
of early competitive Magnox power was recognised by everyone, including Hinton, as unfounded by 1958. The Ten-year Programme was
deferred for a year in autumn 1957 to help reduce inflation, the early
Magnox plants took longer and cost much more to build than was
forecast; by 1960 the CEGB was pressing for further curtailment of
the programme; and electricity from the first Magnox plant, finished in
1962, cost 80 per cent above the 1957 estimate.[36]

The fateful policy decisions made in Britain would have been impossible under American arrangements, where a narrow limit to the
range of development to be helped by the AEC was ruled out and
neither Congress nor the Executive could impose a plant programme
on the utilities.

The decisions taken were not inevitable under the British arrangements. But these seemed conceived to encourage such decisions, and contained no safeguards to ensure that their probable effects would be examined and their actual effect monitored. Ministers gave a monopoly of selecting and developing new systems to the Authority which they also constituted as their main adviser in atomic affairs. They created a situation in which they could not from indigenous activity compare the progress of major reactor systems, and in which there were no incentives to efficiency and speed in developing such as parallel development by autonomous firms would have given. When the CEGB publicly disagreed with AEA reactor policies, which might have seemed a protection for ministers, this was treated as 'unwise' and to be suppressed. The 'differences of view' must be reconciled in unpublished proceedings of an inter-departmental committee of unnamed officials.[37] Members of Parliament fell for the AEA's 'narrow front' policy as readily as ministers: so when the Authority began belatedly in 1959 to plan to broaden its programme, the Select Committee on Estimates thought this might lead to a waste of resources. No member suggested that resources were already being wasted. The crucial question, which *The Times*[38] spelt out in a leader on the Committee's Report – 'whether one judgement and that the Authority's is necessarily sufficient in questions where there can be reasonable doubt' – was never put.

In retrospect it is clear [I concluded in Spring 1967], first, that to give a monopoly of research and development in an advanced technology to a strongly centralised authority owned, financed, appointed and supervised by the government is not a reliable means of promoting rapid growth. Second, if the object is rapid technical advance, with a hope of leadership, it is dangerous to dispense with variety, the possibility of comparison, the stimulus of competition between autonomous groups and the measure of profit, crude and blunt though these may be. Third, it has to be recognised that a minister in charge of a monopoly cannot make useful judgements or exercise control to protect the public when he has no basis of comparison, and the public is particularly unprotected when the Minister is nominally responsible for a national monopoly's programme. As for risks, the three outstanding ones are that ministers and departments become emotionally involved and become protective of their charges when things go wrong; that even outstanding men, like Lord Hinton, are liable to error; and that authorities and boards become vested interests, eager for more power, for larger staffs and larger empires, anxious to conceal or explain away what goes wrong. The public property becomes a private interest.[39]

This analysis of the reasons why British development lagged behind that of the United States was not popular. Leaders of both main parties in Parliament joined in expressing disapproval of it, without establishing that they understood it. Sir Keith Joseph alone suggested that until the book has been fully studied the forthcoming decision to order another AGR station should be deferred. On the same side of the House, Quintin Hogg (earlier and later Lord Hailsham) implied that the book might be motivated by American industry. Richard Marsh, Minister of Power, did not think 'that one can get to a position where major government policies are held up because someone has written a book.'[40]

Anthony Wedgwood Benn, as Minister of Technology, wrote in a memorandum on the book prepared for the Select Committee on Science and Technology that he 'could not escape the conclusion' that the analysis was 'polemical' (a tribute from a master!) and the picture it presented 'false', and expressed his 'belief that this country had been given good value for the money it has spent in the field of civil nuclear power'. He thought 'the position will be seen more clearly in our favour in a few years' time.'[41]

This was the favourite ploy of the apologists. There was probably a lag, they would allow, in the trivial sense that the American reactors would probably be competitive *first*. This could be attributed to the small British market, the subdivision of production, the split in the development process – all the irrelevances were repeated. Once these minor weaknesses were removed the AGR would prove a much better reactor than the LWR. Sir William (later Lord) Penney, then Chairman of the AEA and later Rector of Imperial College, London University, gave sensational figures to the Select Committee on Science and Technology in 1967.[42] Beyond the AGR Britain still led the world.

(iv) THE UNFOLDING IN PHASE II

Such was the position in 1967, the starting point of this book. During the subsequent nine years the uncertainties on which United Kingdom apologists relied had largely disappeared. The LWR had proved an exceptional success, and was the basis of a significant growth in use of nuclear power in the United States and most industrial countries. No AGR operated until 1976, eleven years after the first order was placed – and it was generally accepted that the system was not as good as was promised. Two reorganisations of the United Kingdom's industry had not revived it; the SGHWR had been adopted as the AGR's successor, but nothing more mature than the 100 MW prototype was there to build upon. The United States and Germany were ahead in development of the HTR, and Britain had withdrawn from the field, at least temporarily. The French had stolen a march on Britain by having a 250 MW FBR – Phénix – operating some months before the Dounreay Prototype Fast Reactor (PFR); but since few people supposed the FBR could be economic before 1990, and the United States and

Germany were doing important work on components and fuel, the significance of this was obscure. A programme of 'commercial' FBRs, which some thought might be a way out of Britain's doldrums, would probably be Magnox and AGR over again – a third premature choice. By contrast the SGHWR seemed likely to prove an anachrosism.

This book falls naturally into two parts, the first focusing on the rapid diffusion of the LWR, the second on the setback to the AGR. Success and failure posed different problems to industries and governments.

The progress of the LWR in the United States, and to a lesser extent in the countries which followed the American lead, was fast but not smooth. This was symbolised by a dramatic ebb and flow in the placing of orders for plants, and a continuous steep rise in their cost, instead of a promised fall. Part of the rise in cost was due to exceptional inflation and general economic circumstances. But a larger part was due to increases in real costs. This had three main sources. Some increases were due to changes of design or manufacture, to add to safety, increase efficiency or cater for particular site conditions or environmental criteria. Some, particularly in the earlier years of Phase II, were the result of manufacturing and constructional delays and defects due to an initial overload when orders rose beyond existing industrial capacities. A growing proportion arose from delays due to the extremely prolonged licensing process which was lengthened by the activities of environmentalist groups who 'intervened' in the process.

When the LWR became competitive, the AEC's promotional role towards it ended – or almost, for it still studied LWR costs intensively and organised safety research. But its role towards it became primarily regulatory and, when a large number of LWRs was ordered, this function increased enormously in scale and importance. This was a fundamental change between Phase I and Phase II. Sensibly the public became more concerned as the number of orders grew. It became progressively clear that whereas the American arrangements worked admirably to encourage development in Phase I, they were much less well adapted to handle the problems of rapid diffusion plus rapid development. It was no longer a simple AEC-industry relation, although this itself became a source of friction. But now that the problems concerned safety and amenity many parts of the government machine, state and local government as well as federal, and the executive, legislative and judical branches of the American government, could participate and the checks and balances in the constitution provided great scope for delay by minorities without providing for what is euphemistically called reconciling opposed interests or for sorting out the experts. The possibilities of delay were brilliantly exploited by the environmentalist groups, who rose to a new peak of influence. The

AEC was attacked for combining promotion and regulation and was replaced early in 1975 by two bodies, one to promote and one to regulate; and this caused more delay without apparently bringing content. This unresolved problem in the politics of safety became the dominant focus of interest in Phase II. By 1975–76 it was not concerned solely with the LWR; it became progressively more concerned with problems of waste, sabotage and proliferation, but the damping effect was over the whole nuclear programme.

In 1975–76 the LWR was still held to be competitive by most of the utilities and it seemed clear that, provided regulative delays could be removed, significant reductions in the real cost of LWR plants would become possible: nevertheless because of the environmentalist resistances, which remained effective, although a large majority of people according to opinion polls and a real poll in California appeared to favour the LWR, the prospects of the LWR in the United States seemed less predictable in 1976 than in 1967.

Part I concludes with a brief survey of the advance of the LWR outside the United States, mainly in Western Europe and Japan. Problems similar to those in the United States occurred in most of these countries in varying degrees. Environmentalist activities were growing in 1972–75, based largely on American precedents and often on the participation of American propagandists. But such activities did not derive the same help outside as within the United States from the structure of the respective governments: and it seemed plausible to suppose the LWR might advance faster on the Continent and in Japan than in the land of its origin.

The failure to bring an AGR plant into low-power operation until February 1976, an extremely costly failure, was as unexpected as the steep rise in LWR prices and, coupled with the recognition by 1969 that it was inherently less economic than water reactors, it was more disturbing. In Part II the causes and consequences of the AGR disaster, as it was widely called, are examined. Since the origins lay primarily in Phase I and because new evidence on them came to light in Phase II the analysis retraces the Phase I stage of development. It became manifest that the AEA had overstated the value of its product and launched it without proper development backing to a greater extent than pessimistic critics had feared.

The atomic 'establishment' had become adept in the art of failing without admitting failure, but it was necessary to decide whether another reactor system should be chosen and who should choose it; and, by the same token, the organisation of the nuclear industry had to be reviewed. It was in fact under constant review during Phase II. The change of organisation adopted in 1972–73 took the form of greater centralisation and a further ban on foreign competition, with more power to the AEA and the Minister responsible and the Minister in summer 1974 chose as the next reactor for the CEGB the SGHWR: the Board's preference for a PWR was disregarded. Both decisions suggested that the source of the disaster had either not been perceived or had been disregarded. It was widely thought that the reactor choice would defer the greater use of nuclear power in Britain and keep energy costs higher than they need be. The Select Committee on Science and Technology might have been expected to probe the failure of the AGR deeply, but in this respect it proved a broken reed. Two years after the SGHWR was chosen, its attractions had lost their appeal; even the AEA recommended it should be dropped. By now Britain's nuclear policy was in a state of complete disarray and indecision, even before the Flowers Report, and the nuclear industry had virtually been destroyed.

I intended to close this book with a short analysis both of the development of the fuel cycle industries which must complement the development of reactors and is a vital factor in determining the rate of use of nuclear power, and of the progress of the HTR, which could have industrial uses not open to the LWR, and of the FBR which will at some point become important for nuclear power – if its use is to continue – as the real costs of obtaining uranium rise. For reasons of space, this intention has been abandoned, and the material put aside for separate publication. While it is essential in assessing the prospects for nuclear power, it is not essential for the study of the government–industry–people relations with which this book is primarily concerned, although the problems involved in these relations are presented in new and interesting forms.

## NOTES AND REFERENCES

1. Presidential Commission on Materials Policy, *Resources for Freedom*, Chairman, William S. Paley (Washington: US Government Printing Office, 1952) vol. III, p. 39; vol. IV, pp. 213, 220. Hereafter cited as the *Paley Report*.
2. *Hearings on Development, Growth and State of the Atomic Energy Industry* (Washington: US Government Printing Office, for the JCAE, 1955) pt 2, p. 408. Hereafter cited as JCAE, *Development, 1955*.
3. *National Power Survey* (Washington: US Government Printing Office, for the Federal Power Commission, 1964) pt 1, pp. 14, 69. The capacity of the largest new plants completed rose from 175 MW in 1950 to 260 MW in 1956, 450 MW in 1960, 650 MW in 1963.
4. The AEC listed 'reactor experiments' as well as 'experimental reactors'; they were often smaller than the experimental reactors and they provided sometimes a stage before the experimental reactor, sometimes an experiment on a variation from a type. Full lists were published annually in *Nuclear Reactors Built, Being Built, or Planned in the United States* (Washington: US Government Printing Office, for the Atomic Energy Commission until 1974, subsequently for ERDA).

5. 'A Forecast of the Growth of Nuclear Fuelled Electric Generating Capacity', in *Report of the Panel on the Impact of Peaceful Uses of Atomic Energy*, chairman, Robert McKinney, (Washington: US Government Printing Office, for the JCAE, 1956) vol. II. (background material for the Report) pp. 8–30. Hereafter cited as the *McKinney Report*.

6. Duncan Burn, *Political Economy of Nuclear Energy* (London: Institute of Economic Affairs, 1967) p. 13.

7. *Civilian Nuclear Power: Report to the President* (Washington: US Government Printing Office, for the AEC, 1962) pp. 1–2. Hereafter cited as AEC, *Civilian Nuclear Power, 1962*.

8. Robert A. Charpie, 'The Technology and Economics of Nuclear Power', in the *McKinney Report, op. cit.*, vol. II, p. 37. A few of the projects failed to get very far. Charpie became Deputy Director of Reactor Development at Oak Ridge in 1958, held a series of important positions in Union Carbide from 1961 to 1968 when he became President of Bell and Howell. It was he who emphasised the 'low material economy' of Magnox.

9. Margaret Gowing, *Independence and Deterrence, Britain and Atomic Energy 1945–1952* (London: Macmillan, 1974) vol. II, pp. 156–61. ICI did a great deal of important work for the Atomic Energy Department.

10. *Ibid.*, vol. II, p. 186.

11. *Ibid.*, vol. II, pp. 186–9.

12. *Ibid.*, vol. II, pp. 189–90.

13. I have an impression Margaret Gowing thinks that if the Risley job was faster than a private enterprise job could have been (we do not know this was so) this would establish that this was the better way of developing nuclear power plants in Britain. This would not be so; architect-engineering, which the Risley set-up could supply well, is not the heart of development. Harwell, which provided the reactor technology for Pippa, was thinking in terms of a broader and richer base. This is discussed later.

14. *Parliamentary Debates* (Hansard), House of Commons (London: HMSO, 1 March 1954) col. 844. Hereafter cited as *Hansard*.

15. *Ibid.*, col. 847.

16. *Ibid.*, col. 845.

17. *Ibid.*, col. 848.

18. *Report of the Committee of Inquiry into the Electricity Supply Industry*, chairman, Sir Edwin Herbert (London: HMSO, 1956) p. 72.

19. I have been informed on this by persons who were involved; there are references to the proposals in the Annual Reports of the AEA, but they do not convey a sense of conflict. It is possible one should refer to groups, rather than to one group – because as indicated not all those who were against the AGR and for the LWR as the immediate next step towards a competitive reactor were equally enthusiastic on the HTR; some favoured heavy water, some thought one would go straight to the FBR from the LWR.

20. The most familiar instance is Peter Fortescue, who found greater scope for his HTR in General Dynamics, now General Atomic, world leaders in HTR development.

21. Sir David Eccles said in the speech quoted above 'we cannot doubt that just over the horizon there are immense orders waiting. British engineering firms must be ready to take the lead in the markets of the world.' *Hansard*, House of Commons, 1 March 1954, c. 844.

22. Above, p. 8.
23. *McKinney Report*, p. 35.
24. An important comparative view was given by 'The Three Wise Men', Louis Armand, Franz Etzel and Francesco Giordani, appointed by Euratom to examine reactor development in the United States and Britain, in *A Target for Euratom*, Report to the Foreign Ministers of the Euratom Countries (Brussels: Euratom, 1957), and a summary was published in *The Times*, London, 8 May, 1957, together with an article dealing with the principles of costing involved.
25. *Hansard*, 25 May 1965.
26. *An Appraisal for the Technical and Economic Aspects of Dungeness B. Nuclear Power Station* (London: CEGB, 1965). Hereafter cited as *CEGB, Appraisal*.
27. Below, p. 170.
28. The BWR cost was estimated as 0.489d per KWh, the AGR as 0.457d; but of the difference (0.032d) import duties accounted for 0.0069d, which is to be deducted for the comparison. This leaves 0.0251d.
29. R. F. W. Guard, 'The Year Since Geneva Part 1', in *Euro Nuclear*, London, September 1965, pp. 440–1 Mr Guard, formerly of the consultant firm of Kennedy & Donkin, took up a new appointment in Canada and was thus free to write frankly. He later became a Vice-President of Canatom.
30. Burn, *op. cit.*, pp. 64–5 (including note 1 on p. 65).
31. *Ibid.*, pp. 72–80.
32. Report of the House of Commons Select Committee on Science and Technology on *UK Nuclear Reactor Programme*, Report Minutes of Evidence and Appendices (London: HMSO, 1967) p. xxxiii. Hereafter cited as *SC Science and Technology, 1967*.
33. *The Times*, 6 February 1957, remarked that 'atomic energy has the power to evoke fantasies. It is not only looked on as an "answer to Suez" [the answer to be deferred apparently for twenty years], but as a fairy godmother source of cheap electricity.' There were forecasts that in twenty years electricity would be laid on to houses like water at a fixed charge however much you used.
34. The classic statement of the prospect was in Sir Christopher Hinton's Axel Ax:son Johnson Lecture, *The Future for Nuclear Power* (Stockholm: Royal Swedish Academy of Engineering Sciences, 1957) especially pp. 19–20. The White Paper on the 1955 Programme forecast generating costs from the plants proposed which would be equal to costs from fossil plants. Arthur Palmer, later to be chairman of the Select Committee, regarded this as a conservative claim. Reginald Maudling, then Paymaster-General, introducing a debate on the expansion of the programme in 1957, said the task had been to consider how much acceleration was possible 'in view of the technical advance within the last two years'. *Hansard*, 25 February 1955, col. 1667, and 5 March 1957, col. 184.
35. This was discussed in the context of fuel policy in a leader in *The Times*, 2 December 1957. 'It cannot be a good thing [embracing oil, coal and nuclear power forecasts] to encourage a badly balanced development, to make bad forecasts, to allow policies to be dominated by fears and fashions or by strong personalities who may be wrong.'

36. The generating cost at Berkeley was 1.2d, at Bradwell 1.07d, per KWh; the 1957 forecast cost for 1960–61 plants was 0.66d per KWh. R. D. Vaughan and J. O. Joss, 'The Current and Future Development of the Magnox Reactor', at the Anglo Spanish Nuclear Power Symposium, Madrid, 1964, issued at the Symposium (paper no. 4) fig. 3.

37. Burn, *op. cit.*, p. 92. It was Lord Hailsham who held that the public disagreement of AEA and CEGB was unwise and who took 'appropriate' action.

38. *The Times*, 18 August 1959.

39. Burn, *op. cit.*, p. 120.

40. *Hansard*, 9 May 1967, col. 1264. Quintin Hogg said, apropos Keith Joseph's comment, that the Minister of Power should be cautious about accepting criticisms of the cost of nuclear power.' This was not of course the relevant point. No doubt he knew the real point which related to the choice of the AGR. Hogg continued: 'such questions are often motivated by interested alternative sources of fuel, and sometimes by American industry.' In the House of Lords (*Hansard*, 4 May 1967, cols. 1071–3), there was a similar unedifying exchange in answer to a question by Lord Bessborough asking the Government for a statement concerning my allegation that Britain no longer leads the world in the development of civil nuclear power.

41. *SC Science and Technology*, 1967, p. 413. The Minister abstained from discussing the organisational problems which were the kernel of the book because the Government were 'now considering what should be done'. See below, pp. 172–5.

42. *Ibid.*, Q. 52. By the end of the six-year programme started in 1965, 'we are going to do 30 per cent better than we did at Dungeness B' [in 1965 money].

# PART I

# First Competitive Reactor: the Light Water Reactor

# CHAPTER 2

# The Paradox of LWR Development

The history of the LWR in Phase II was tinged with paradox. Experience proved, to the satisfaction of most utilities, that LWRs were, as had been promised, competitive and would produce electricity at lower costs than fossil-fuelled plants. They were ordered in great numbers both in the United States and in all other major industrial countries except Britain and Canada. But the cost of constructing the plants, the time taken in building and licensing them and the expected cost of the power they would produce rose to levels remarkably above those confidently forecast in 1963–66, and by 1975 utility purchasing was sharply reduced. The plants continued to have an extremely good safety record, but they were attacked passionately as dangerous sources of radioactivity and of potentially catastrophic disaster. They were widely welcomed as coming 'in the nick of time' to provide the only large new indigenous source of energy which could lessen the increasing dependence of the United States (and most other Western industrial countries) on imported energy from the Middle East in the next ten to fifteen years. But their construction and operation was resisted by environmentalist groups in the United States who wished to have a total ban imposed on their use. Thus while the estimates generally accepted in 1973 of total American energy consumption in 1980 assumed that LWR plants would supply about 9 per cent of the total, Ralph Nader – the consumer champion – was prophesying that within five years none of these plants would be allowed to operate. In 1975–76 capital cost, the cost of capital and price restrictions were having more effect than Nader, but environmental forces were still very influential.

In Part I of this book the five questions which emerge from these apparent contradictions are examined successively. What is the basis on which the competitiveness of the LWR is assessed? What is the relative importance of the various sources of the cost increases? What is the measure of the risk involved in the LWR? If the LWR is so dangerous that it should not be built or operated, how have political institutions and administrative and legislative policies allowed such an immense commitment and dependence to be built up; and if it is not such a risk (which appears to me the true position), why was it possible for an influential campaign against it to be mounted and why did Nader, knight errant of consumer protection, choose it for a major campaign? Finally, to descend from the stratosphere, what is the prospective contribution of the LWR to the energy gap?

CHAPTER 3

# Competitiveness Accepted

Orders for LWRs in the United States rose fast from 1965 to 1967, then fell steadily back in 1968–69 (almost to the 1965 figures) then from 1970 forged ahead again remarkably, reaching their highest in 1974, then dropped again very sharply in 1975. The trends are shown in Figure 3.1, which gives the aggregate capacity of both nuclear and fossil-fuelled plants ordered annually from 1963, together with the number and capacity of large nuclear plants which began operating. The total capacity ordered, fossil and nuclear, rose steeply from 1963 to 1967, a favourable market circumstance. The annual average of orders for all steam-generating plant from 1966 to 1970 was three times the average for 1961–65, and orders in 1967 alone were more than half the aggregate of orders placed in the ten years 1950/51 to 1959/60.

The first flush of orders for nuclear plants, from 1965 to 1967, came before any plant above 200 MW capacity had operated. It was based, to some extent, on promise (although there had been five to six years of experience with 200 MW prototypes). The second and larger flush came as operating experience with large plants of 600–800 MW was accumulating, and orders for nuclear capacity overtook those for fossil capacity. This was based on projections which took account of initial performance of the 500 to 800 MW units. A few of the orders in the early 1970s were for large HTR plants, but all these were cancelled in 1974/75.

The fall in orders for nuclear plants in 1975, much greater than for fossil-fuel plants, was primarily a response to the fall in consumption of electricity in the United States in 1974 – a traumatic experience for an industry accustomed to steady annual increments of 7 per cent. Consumption rose slightly in 1975, but at much below the old rate. This brought to a head the financial difficulties of the utilities due to the enormous rise in their capital costs and aggravated by regulatory authorities' restraints and delays on electricity price increases. Nuclear plants were the most capital-intensive: new orders were few, many old orders were deferred, and some were cancelled.

25

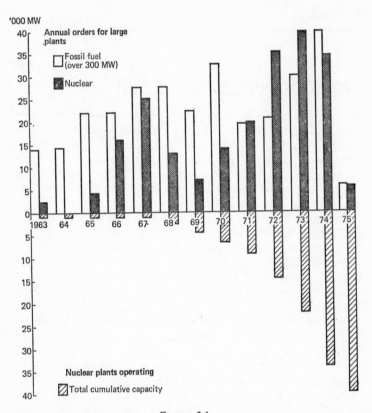

FIGURE 3.1

Capacity of Nuclear and Fossil-fuelled Plants Ordered and Nuclear Plants
Operating in the United States 1963–75

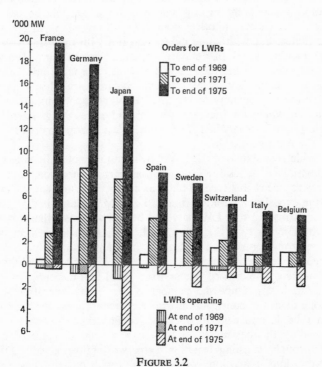

FIGURE 3.2

LWRs Ordered and Operating Outside the United
States 1969–75

The position worsened in 1976, because although votes cast in several states showed that a large majority supported the building of more nuclear plants, which was in line with the results of opinion polls for the United States as a whole, successful stalling actions by intervenors led to increasing delays in obtaining licences to construct or operate plants. This essentially political uncertainty deterred utilities from placing orders. Competitiveness assessed in real costs – and real risks and popular preferences – had little immediate relevance.

Orders for LWRs in other countries followed the pattern in the United States, but with some delay, and not with the same volatility. Several medium-sized plants were ordered by the early 1960s and came into operation between 1964 and 1969. The large flow of orders for big units came from 1969 onwards: it is illustrated in Figure 3.2. which shows the capacity of large LWR plants operating or on order in countries where the increase was particularly striking. There was a falling off in new orders in 1975 in some West European countries, but not on the American scale.

The wide acceptance of the LWR by a majority of utilities in almost all industrial countries for a substantial part of their load was proof that they regarded this type of plant as competitive and safe. The early operating experience with the first 600–800 MW plants had given adequate confirmation that LWRs would have lower costs than fossil-fuel plants in a majority of locations, and, in that crucial respect, would fulfil the promise on which orders for the plants had first been placed, although the competitiveness was on a much higher cost level than had been foreseen. The cost of power from fossil fuel also rose to wholly unexpected levels. Once the nuclear plants began operating – there were long delays in construction whose sources will be analysed – they proved reliable and easy to operate, reached their full power easily (confounding the pessimists in this respect) and in spite of further delays through 'teething troubles' (pipe-weld failures, valve failures, sonic vibrations, corrosion, minor troubles with control rods, failures of small proportions of fuel rods, and so on) they had a good average level of availability. For the first twelve large plants the average reactor availability in their second full year of operation was 76 per cent and the average plant capacity factor 65 per cent.[1] They were promising figures which compared well with those for new fossil-fuel plants, whose records were on average worse. The new, large, high-pressure, fossil-fuelled boilers and turbo generators had a surprising amount of grief in all countries, but most of all, it would appear, in Britain.

## NOTE

1. For the twelve months to May 1974 a Danish analysis showed the average load factor in eighteen Westinghouse plants as 64.9 per cent, in eighteen GE plants 62.5 per cent. In the same period the Magnox average was 65.6 per cent – but the best figures were for low-stressed plants (*Nucleonics Week*, New York, 6 March 1975). In 1975 the Westinghouse average (in the first ten months) for large units (minimum 450 MW, average 720 MW) was 67.5 per cent (based on details in *ibid.*, 27 November 1975).

# CHAPTER 4

# The Rise of Costs—Extent and Sources

## (i) THE EXTENT: 1967–76

### (a) *Changes in Costs of Plants*

Between the end of 1963, when the Oyster Creek plant was ordered, and the end of 1966, the expected cost of obtaining an LWR plant, excluding the fuel, rose by 35 to 40 per cent.[1] This was regarded as substantial. Between spring 1967 and summer 1974 the expected cost rose by a factor of between three and four. This was sensational, and the rise continued in 1975–76.

The broad picture based on AEC figures is given in Table 4.1, which, in addition to total forecast costs at four dates, includes subdivisions giving separately construction cost, interest during construction and escalation, which was the rise of costs expected during construction, mainly from price increases, but partly from design changes. The concept, as described, was not precise, but rates of escalation adopted were not far above the rate of inflation. A forecast by the firm of architect-engineers, Sargent and Lundy, in autumn 1976 for a plant to operate in 1985 is added to illustrate the continued rise, including the further lengthening of the lead time; some features of this are referred to in later chapters.

The AEC figures were seriously based, but had limitations.[2] In the early estimates some figures were significantly too low; the 1969 figures included 'more complete definition of bills of materials than in the original 1967 estimates ... higher than initially estimated unit costs, higher material quantities' and so on. The figures referred to plants of identical capacity but not of identical design and scope. The ranges of cost in each year refer to plants in different areas (construction costs were much higher in the North than in the South) and with different site conditions – some needed cooling towers, or special provision against earthquakes or tornadoes. At the end of the period the scope of plants differed less (all for instance needed cooling towers) but the regional cost variations due to wages were possibly greater.

Over the years designs were substantially changed to give greater efficiency and safety. The costs of interest and escalation were

TABLE 4.1

Forecasts of the Cost of a 1000 MW(e) LWR Plant
in the United States, 1967–76 ($/KW)

| Date of order | 1967 | 1969 | 1971 | 1974 | 1976 |
|---|---|---|---|---|---|
| Projected date of operation | 1972 | 1975 | 1978 | 1982 | 1985 |
| Total cost | 154[a]–(204)[b] | 239–(315) | 355–422 | 600–720 | 1047 |
| Construction cost, without IDC or escalation | 123–(163) | 168–(221) | 215–268 | 327–393 | 515 |
| IDC | 11–(14) | 31–(42) | 50–62 | 106–127 | 180 |
| Escalation | 20[a]–(28) | 40–(52) | 90–112 | 167–200 | 352 |

Sources: *Current Status and Future Potential of Light Water Reactors* (Washington: US Government Printing Office, for the AEC, 1968) (Wash. 1082) ch. 2; *Trends in the Cost of LWR Power Plants* (Washington: US Government Printing Office, for the AEC, 1970); *The Nuclear Industry 1970* (Washington: US Government Printing Office, for the AEC, 1971) p. 89; *ibid.*, 1974, pp. 21–4; and *Electric Light and Power*, Boston, 20 September 1976.
[a] Escalation was not included in the original 1967 figures, but given in *Current Status and Future Potential of Light Water Reactors, op. cit.*, ch. 2, p. 1. It was, I suspect, too low.
[b] Figures in brackets are estimates based on details of extra costs required for special site conditions and given in *ibid.*

naturally uncertain – the lead time, as the table shows, increased steadily and interest and inflation rates, to which we return, rose sharply. The figures must be treated as 'a guide, not a rule'. But because the differences were great they are a useful guide.

## (b) *Changes in Fuel Costs*
Unlike LWR construction costs, the fuel cost fell between the Oyster Creek contract in 1963 and the order for TVA's Brown's Ferry plant in 1966. Manufacturers, consultants and AEC experts thought in 1967 that the fall would continue till possibly 1985–90. Thinking presumably in terms of 1967 dollars, they envisaged a fall of about one-fifth. This did not happen: the cost was stable from 1967 to 1970 in current money terms, which did represent a fall in real costs of possibly 11 per cent, but from 1970 rose steeply – by about 80 per cent (in current prices) by mid-1974. A further rise – in real terms – was now assumed for the future, although the extent was controversial. Table 4.2 gives

the AEC estimates, for 1967, 1971, 1974 and 1982, of the fuel cost per unit of electricity produced, with the estimated costs for the main stages in the cycle. The AEC abstained from giving a breakdown for 1982.[3]

TABLE 4.2

Estimates of LWR Fuel Cycle Cost 1967–82 (mills/KWh)

|  | 1967 | 1971 | 1974 | Percentage change 1967 to 1974 | 1982 |
|---|---|---|---|---|---|
| Mining and milling uranium and conversion to UF6 | 0.41 | 0.45 | 0.82 | +100 | – |
| Enrichment | 0.43 | 0.58 | 1.07 | +150 | – |
| Fabrication | 0.53 | 0.33 | 0.27 | —50 | – |
| Spent fuel reprocessing including transport and waste management | 0.19 | 0.18 | 0.48 | +153 | – |
| Working capital | 0.40 | 0.57 | 0.87 | +117 | – |
| Plutonium (and uranium[a]) *credit* | (0.22) | (0.37) | (0.49) | +123 | – |
| Total | 1.74 | 1.74 | 3.02 | +74 | 5.60 |

[a] This is only referred to in the 1971 figure.

The estimate for 1982 was merely a guess that costs as a whole would rise somewhat faster than the rate of inflation, the costs of the individual stages in the cycle being treated as unpredictable. The AEC's fuel cost figures seemed to become less clearly defined as time passed, unlike the construction cost figures, which became more sophisticated. This was possibly because the terms on which fuel cores were sold by the makers, which included guarantees of the costs at several stages for long periods ahead, were sales gimmicks.[4] The 1967 figure was for the first five years; the 1974 figures were in '1974 prices' – subject therefore to increases due to inflation. By the end of 1975, much higher figures were being given: architect-engineers Sargent and Lundy put the 1984 price at 12 mills a KWh., and raised it to 18.4 mills for 1985–2000, which assumed the continuation of the rapid growth of enrichment and reprocessing costs which occurred between 1971 and 1974, and was 'levelised' for a fifteen-year period.

(c) *Changes in Generating Costs*
By 1966 the prospect held out in 1963 of a privately-owned utility producing power at 4 mills a KWh. had disappeared, although Philip Sporn, doyen of utility leaders, still thought 4.30 mills possible.[5] Esti-

mated generating costs rose through Phase II faster than estimated construction costs. Fixed charges arising from the cost of plant – including interest, profits, depreciation and taxation – were over half the estimated cost of generating power in nuclear plants throughout Phase II and, up to 1974, became a steadily larger part of the total. In 1975–1976 forecasts seemed to reverse the trend – much higher fuel costs accounting for this. Table 4.3 gives the broad picture of what at different dates was expected to happen. The 1967 figure is based on AEC estimates of construction costs and fuel cost, the 1971 and 1974 figures are AEC figures, the 1976 estimate is by Sargent and Lundy and is described as a 'fifteen year levelised cost'. A comparably high figure (42 mills per KWh.) was given by an expert from the General Electric Company (GE) at a public enquiry in Connecticut early in 1976. It was described as a 'ten year levelised cost'.[6]

TABLE 4.3

Estimated Generating Costs in 1000 MW LWRs in the United States (mills/KWh)

| Year of order | 1967 | 1971 | 1974 | 1976 |
|---|---|---|---|---|
| Expected year of operation | 1972–73[a] | 1977–78 | 1982[b] | 1985[c] |
| Fixed charges | 2.8 | 7.7 | 15.5 | 30.7 |
| Fuel costs | 1.7 | 1.7 | 5.6 | 18.4 |
| O and M | 0.3 | 0.5 | 1.5 | 2.5 |
| Total | 4.8 | 9.9 | 22.6 | 51.6 |

[a] 80 per cent load factor, $12\frac{1}{2}$ per cent annual charge.

[b] 75 per cent load factor, 15 per cent annual charge.

[c] 70 per cent load factor, 18 per cent annual charge.

(ii) THE MOVING TARGET: COSTS OF FOSSIL-FUELLED PLANTS

Despite the cost increases shown (and slightly exaggerated) by Table 4.3, LWRs were still judged competitive by most utilities at the end of Phase II. This was because the cost of constructing a fossil-fuel plant also rose steeply, while there was a dramatic rise in the price of fossil fuels, above all coal. Coal was quantitatively the most important and, for simplification, discussion here is limited to the coal case. The rise in the indices of fossil-fuel prices and of electric power are shown in Figure 4.1.

When TVA ordered its first LWRs in 1966, the cost per KW of the nuclear plant and of a coal-based plant priced at the same time were practically identical. But that was exceptional. Coal-fired plants were normally from 15 to 20 per cent cheaper than nuclear in Phase II, and coal-based plants had lower fixed charges as a result; and the extent of this advantage in money terms rose as the cost of the plants rose. Thus in 1968 the difference in annual fixed charges implied by the estimated cost of plants ordered at that time for operation in 1973–74 ($150 a KW for nuclear, $120 for coal) was slightly over 0.5 mills per KWh. on the ground rules then used (80 per cent load factor, 12.5 per

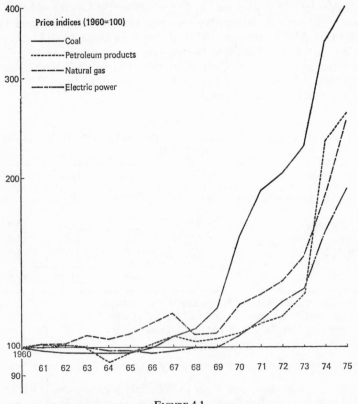

FIGURE 4.1

Fuel and Power Prices in the United States 1960–75

cent annual charge). In 1974, when the comparative costs for a plant to be operating in 1982 were put at 660 and 550 ($ KW) respectively, the annual fixed charge was envisaged as 2.5 mills per KWh. less for the coal plant (on 75 per cent/15 per cent ground rules).

The rise of coal prices made a much greater difference. An efficient coal-fired plant in 1968, using coal at the average price (25.5 cents per million Btu), had a fuel cost of approximately 2.6 mills a KWh.; by 1974, this had risen to 5.6 mills, and it was expected to be 12.2 mills in

1982. The cost of coal which in 1968 was approximately 0.55 mills per KWh. above the cost of nuclear fuel was by 1974 3.6 mills per KWh. higher and by 1982 the margin was expected to be 6.6 mills. Thus despite the big rise in the cost of plants, the margin in favour of buying an LWR had increased substantially between 1968 and 1974. The assessment of the comparison by the new Energy Research and Development Administration (ERDA) in the United States early in 1975 is shown in Table 4.4.

TABLE 4.4

Forecast of Generating Costs in Coal-fuelled and Nuclear Plants in the United States (1000 MW plants ordered in 1974 to operate in 1982) (mills/KWh)

|  | *Nuclear* | *Coal* |
|---|---|---|
| Fixed charges | 15.5 | 13.0 |
| Fuel | 5.6 | 12.2 |
| O and M | 1.5 | 3.7 |
| Total | 22.6 | 28.9 |

Source: *The Nuclear Industry* 1974 (Washington: US Government Printing Office, for ERDA, 1975) p. 20. The fixed charges are calculated on a 15 per cent rate, and a 75 per cent capacity factor. The fuel costs are based on estimates of actual cost in 1974 (3.02 mills for the nuclear plant, 6.6 mills for the coal plant with coal at 69.5 cents a million Btu) all escalated at 8 per cent per year.

Two points are to be noted. First, utilities choosing which type of plant to buy had to guess a wide range of cost changes for many years ahead. In 1967–68, although fossil-fuel prices were rising, coal most of all, no one foresaw that coal prices would almost treble from 1969 to 1974. Nor indeed was the rise in the cost of plants foreseen. The orders for nuclear plants in 1966–68 reflected the impact of the TVA order for two BWRs for Brown's Ferry in 1966, and a third in 1967, and the optimistic view sponsored by the AEC that nuclear plants and fuel would get cheaper. Although the gloss had begun to wear off the optimism by 1968–69, by 1970–71 the prospect of steeply rising coal prices was favouring the nuclear choice. This was discussed at length by the JCAE in March 1971.[7] It was recognised that for several reasons coal prices would rise above the rate of inflation. There were significant differential cost increases – miners' wages rose more than average wages from 1965 (Figure 4.3), legislation required more safety provisions in mines, state and local governments increasingly required mining companies to restore land after open-cast mining. Demand for coal was expanding after a long period of decline and low profits; higher profits were looked for before investment in expansion occurred.

Moreover, the use of coal in generating plants was made more costly by controls, imposed again mainly by state and local governments, requiring elimination of sulphur from the smoke – a costly operation for which no economic solution was known. Several utilities found it necessary to buy low-sulphur coal, whose price was high: sometimes it had also to carry a long transport cost. Commonwealth Edison of Chicago, for instance, brought coal from New Mexico to Illinois. Hence the prospect of continued high costs in buying and using coal became normal in the 1970s.

The second point is that whereas the cost of LWR fuel is broadly uniform for all plants the cost of coal varies greatly according to location. In 1971, for example, when the average cost of coal delivered to utilities was 36 cents per million Btu, there were some generating plants for which it was 17 cents and others, more numerous, for which it was over 70 cents. The differences rose partly from mining costs, but mainly from transport costs. Where the cost was very low, coal-fired plants might still be cheaper than nuclear, even when on an *average* fuel cost basis nuclear plants had a considerable advantage. The break-even point in 1971 was put at from 18 to 21 cents a million Btu.[8]

(iii) SOURCES OF COST INCREASES

The sources of the remarkable increases in the costs of nuclear plants and nuclear power have been extensively analysed, but although much of the analysis has an air of statistical precision it is difficult to quantify the factors in isolation because they interact on each other. They fall into three groups: general economic circumstances, changes in design, specification and scope, and delays and long lead times.

(a) *General Economic Circumstances*
A surprisingly large part of the increase in costs, actual and forecast, resulted from changes in the economic environment. Three factors were significant: an increase in the rate of general price inflation, an above average increase in construction costs and a substantial rise in the cost of money. The first two are illustrated in Figure 4.2.

The increase in prices, retail and wholesale, reflected much stronger inflationary forces in Phase II than operated in Phase I. The average annual increase of retail prices and wholesale prices in the United States from 1967–1973 was $4\frac{1}{4}$ per cent, over three times the rate in the early 1960s. From 1973 to 1975 the rate was almost $7\frac{1}{2}$ per cent. This raised the money cost of inputs into construction and operation of power stations.

Prices of different cost components moved up at differing rates. Construction costs are a particularly important component in all power stations and particularly so for nuclear plants where building construction and on-site labour in erection amounted to over one-third of total cost.[9] The cost of construction for all large plants, for oil refineries and chemical works for example as well as for power stations, rose much faster between 1967 and 1973 than prices in general.

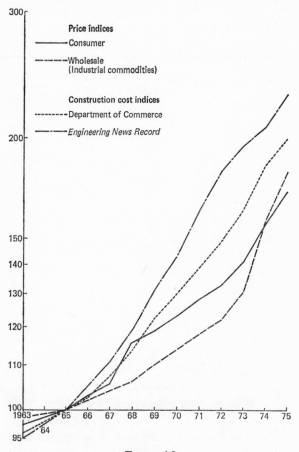

FIGURE 4.2

Wholesale and Consumer Prices and Construction Costs
in the United States 1963–75

The indices of construction costs all show this; of the two illustrated in Figure 4.2, the higher seems the most appropriate. The rise in construction costs was primarily due to above-average increases in construction workers' earnings (Figure 4.3) together with falls in productivity for which there are no comprehensive statistics.[10] The cost of nuclear power plants for this reason would have risen at a higher rate than average prices. Already by 1968 the 'escalation' of costs was taken to be 6 per cent a year; it was raised to $7\frac{1}{2}$ per cent in 1969 and to 8 per cent in 1971. The exceptional rise of construction costs must be seen partly as a result of the increased demand for power plants, not as a wholly external factor, but only partly because,

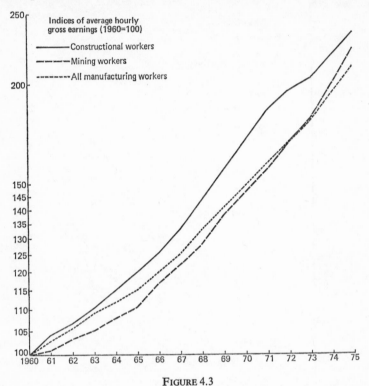

FIGURE 4.3

Wage Increases in the United States 1960–75

as noted, it coincided with high demands from other 'constructional-intensive' industries.

The increase in interest rates on short-term commercial loans brought a further increase in the plant cost because interest during construction is substantial. The movements from 1960 to 1975 are shown on Figure 4.4 together with changes in the return on utility bonds.

The effect of these external factors on construction costs was substantial. If it is assumed that the total cost of identical plants, built according to the same time schedule and completed at yearly intervals, would rise broadly in step with the average rise in prices, then the estimated construction cost net of interest during construction (IDC) would have risen 50 per cent for this reason between 1967 and 1974: and when the exceptional importance of construction costs is also allowed for, the rise would have been broadly 60 per cent. If there had

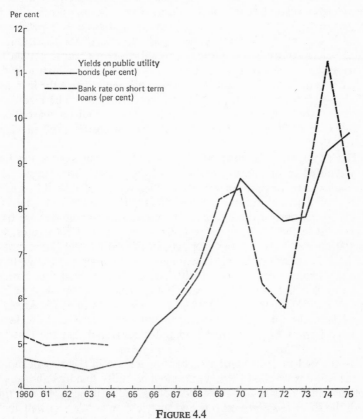

FIGURE 4.4

Cost of Money in the United States 1960–75

been no other source of increase in construction costs net of IDC, the 1974 cost figures would have been in round figures 200–260, compared with the figures of 327–393 in Table 4.1. About two-fifths of the increase in estimated construction costs net of IDC between 1967 and 1974 can thus be attributed to inflation; the remaining three-fifths of the increase, between 35 and 40 per cent of the *total* estimated cost net of IDC and of escalation, must be the result of other changes generic to nuclear plants – mainly changes of design, quality, or materials, additional components and the like, although there were also increases in professional fees for architect-engineers and others and in administrative costs because the jobs were going to take much longer.

The scale of the second and third cost elements (separated out in Table 4.1), IDC and escalation, is derived from the first. IDC is interest paid on money used to finance design and construction, which is spread over several years, as it occurs; hence the amount is directly related to the expenditure on construction: and so, too, is escalation, which arises primarily from increases in the cost of the work due to rising prices. There is no simple functional relation because neither the distribution of the work over time nor the time over which it is distributed is uniform. Naturally the derived elements reflect the increases (35 to 40 per cent) in the basic constructional cost. The significant feature in the changes in these cost elements was their increase (taken together) from 20 per cent of the total forecast cost in 1967 to 45 per cent in 1974.

The explanation is threefold. First, the 1967 figures were too low: they omitted or underestimated some of the constructional costs, used an inadequate method in assessing escalation and possibly underestimated IDC. Second, the changed economic parameters account for a substantial part of the change – the interest rate assumed rose from 5 per cent in the 1967 estimates to 8 per cent in the 1974 figures, and escalation was assumed as 6 per cent in the earlier and 8 per cent in the later forecasts (reflecting, incompletely, the higher rate of inflation). The third source of change was the increase in the 'lead time', which was taken as five and a half years in 1967 (it had been taken as four years in 1963), eight years in 1974 and nine years in 1975. An AEC analysis of the impact of increases in rates of interest and escalation and the impact of lengthened lead times suggested that the first two together had a larger impact than the third.[11]

By the end of 1974 some part of the increase in the cost of money to the utilities resulted from weaknesses in utility finances and so was not just an outcome of general economic circumstances. At this time both the short-term interest rate and the yield on bonds was well above the 8 per cent taken as the basis for IDC in the mid-1974 cost forecasts. Mortgage bonds were used by utilities to provide slightly over half the funds for new investment in the late 1960s and early 1970s.[12] In the past they had raised money for long terms at extremely low rates; the rise in rates had a cumulative effect because as bond issues raised at low rates matured they had to be replaced at the new high rates. Consequently the increased cost of dearer money was not limited to new capital expenditure. Since new capital spending itself grew 'by leaps and bounds'[13] and increases in charges for electricity were delayed by regulating authorities, utilities became progressively embarrassed, their credit-worthiness declined and the average yield on their bonds –

which was almost 10 per cent for most of 1975 – was above the average by $\frac{1}{2}$ to 1 per cent.

To the extent that inflation, high construction costs and dearer money raised the costs of plants, correspondingly they raised the 'fixed charges' arising from depreciation, interest, profits and profit-related taxes which formed the largest component in generating costs. The estimated cost of a plant between 1967 and 1974 rose by a factor of four (Table 4.1) and this would – other things being equal – have led to a fourfold increase in fixed charges, of which one-third would be due to these 'general economic circumstances'. In fact fixed charges rose by a factor of over five (Table 4.3). The difference was due to a change in the ground rules used for the calculation, and the major change in these was a direct effect of one of the 'general economic circumstances', the cost of money. The ground rules used for the 1967 estimates were a $12\frac{1}{2}$ per cent capital charge and an 80 per cent load factor; for 1974 the percentages were 15 and 75. The increase from $12\frac{1}{2}$ to 15 per cent reflected the higher cost of money, and was the major factor in raising the charges. The second change brought the rules more into line with experience; and more in line with prospects, for when nuclear plants were numerous they could not be base-load plants for their whole lifetime.

Estimated fuel costs between 1967 and 1974 (Table 4.2) rose closely in line with inflation plus the higher cost of money. Changes in the cost components show this was a coincidence. The rise in enrichment and reprocessing costs was almost three times the inflation rate; and the rise in financing costs was not much less. These steep rises occurred mainly after 1971. They were partly offset in the 1967/1974 comparison by a fall in fuel fabrication costs, which occurred mainly before 1971, and by a steep increase in the value put on the by-product plutonium. The figures showed there were factors specific to the nuclear industry which had raised the cost of fuel for LWRs since 1971 at a rate significantly above the inflation rate and that this was likely to continue. This was not strongly reflected in the 1974 forecasts for 1982, but forecasts of fuel costs at the end of 1975 and in 1976 were more in line with it. Operation and maintenance (O and M) costs were relatively small; they were raised much above the inflation rate in the estimates – possibly because maintenance was a higher than expected cost.

(b) *Changes in Design and Specification*
When the cost estimates in Table 4.1 have been adjusted to eliminate the effect of inflation and higher interest rates, almost two-thirds of the increase in basic design and construction costs between 1967 and 1974, net of IDC and without escalation, remains to be accounted for.

The major part of this increase is probably due to changes in design and specification. There was a constant upgrading of engineering standards – necessarily far above normal practice – and extensive additions of new systems and components and more redundant equipment, in order to ensure reliable and continuous operation of plants, to make it possible to build plants of increased capacity and run them closer to maximum safe levels and to provide more engineered safeguards against increased risks, and more protection of the environment. The upgrading of standards applied to all major 'pressure-containing parts' – pumps, valves, vessels, tanks, heat exchangers, tubes – throughout the plant, not just in the nuclear steam supply system, and to metals, welding and other processing, and to inspection. The view was accepted that, because of the risk of radioactivity, exceptional precautions should be taken against even minor breakdowns; the steps taken were based partly on progressive experience in large prototypes and the first big LWRs and partly, too, on research results.

The additions of systems included large items such as cooling towers or ponds,[14] but the addition to costs was made up of a surprising number of 'significant' but mostly less conspicuous changes – additional emergency core cooling systems (so that there were three, independent and each sufficient on its own); more extensive monitoring of radioactivity, temperature distribution and variation and fuel performance; more automation and computerisation of controls; new systems of containment; added processes for reducing radioactivity in liquid and gaseous effluents; much taller stacks for gaseous effluents; more diesel-generators to provide emergency power; more redundancy of other electrical equipment; additions to give protection against earthquakes, tornadoes, tsunamis[15] and 'missiles', a euphemism which covers aircraft and, for example, fragments from an exploded oil tanker; provisions to safeguard control rooms against missiles and radioactivity; and so on. The number of changes of design termed 'significant' in each of TVA's first two LWR plants exceeded one hundred and this seemed normal. There were changes in fuel design and manufacture also which raised cost: PWR fuel rods were internally pressurised, BWR rods were hot-vacuum dried. Both types, in their most advanced forms in 1973, were reduced in diameter.[16]

The AEC study of technical and economic prospects for the LWR in 1967 – published early in 1968 – recognised that costs had been raised during 1967 by the addition of engineered safeguards by possibly 10 per cent, and pointed to the volume of further work on which firms were engaged in this area. But they gave more emphasis to the reductions of cost which would come from higher capacities – up to 3000 MW(e) – with higher power densities, and from experience in design manufacture and operation. They thought that a 90 per cent 'learning curve' would probably apply – costs would be reduced by 10 per cent every time the output of plants was doubled, simply because of experience in manufacture, more repetition, mass production and so on, without major changes in unit size.

There were gains from the increases in capacity and power densities in Phase II. But the emphasis of the 1968 study was misjudged. There was a lot of learning, but much of it was learning to do things more expensively. Additions to costs outweighed reductions. The large volume of orders contributed to delays in production so that 'learning' (in the 'learning curve' sense) was slowed down and the flood of orders fostered a subdivision of production which was not good for 'learning'. It seemed likely in 1972–73, starting with the launching by GE of the BWR/6, which they planned as an advanced standard design, that a stage might have been reached in which the 'learning' on which the AEC study placed hopes would become a significant source of cost reduction.

How much the various types of changes of design and specification contributed to the estimated cost rise of a 1974/1982 plant compared with a 1967/1972–73 plant cannot be quantified. The AEC's forecasts of costs of a plant ordered in 1973 to operate in 1981 gave additions to direct construction costs for 'safety and environment items' as one-quarter of total direct costs.[17] This proportion may reasonably be transferred to 1974. If the same proportion of indirect construction costs was also additional for 'safety and environment' items, the two together totalled from $82–$98 per KW of the $327–$393 in Table 4.1 This is slightly less than two-thirds of the total increase in construction costs net of IDC which could not be explained by inflation. The remaining third could be explained by cost increases due to design and specification changes not arising from safety or environmental items.

## (c) Delays and Long Lead Times

Construction costs are raised when the time between ordering and operating a plant is increased. The 1967 and 1974 estimates in Table 3.1 envisaged lead times of five and a half years and eight years respectively. Lengthening increases IDC and general expenses, which are increases in 'real' costs,[18] and escalation which, when the rate corresponds to inflation, is *not* an increase in real costs. The AEC put the minimum increase of IDC, when interest is at 7 per cent and over, for an additional three years lead time at $8\frac{1}{2}$ per cent of total cost, based on an 'ideal stretchout of the construction schedule'.[19] The stretchout may not be ideal. The best situation is when most expenditure occurs in the later part of the schedule. If a substantial time elapses between completion and operation, interest charges may exceed $\frac{1}{2}$ per cent per month. Moreover, if the long lead time was not foreseen the utility may be short of power. Power may be bought expensively from other utilities and old high cost plant run longer than was planned. Even if long lead times are foreseen, as they were by 1974, quick expansion of capacity or replacement of old by using LWRs is ruled out; other means must be found (the use of gas turbines flourished) and large orders have been bunched for years ahead.

The average time between ordering a plant and its commercial operation, which utilities assumed when they announced orders, rose from five years in 1966 – and even less earlier – to 6.5 years in 1970, 7.0 years in 1971 and 9.3 years in 1974. The average time taken for twenty-seven out of twenty-nine plants ordered in the years 1963 to 1966 was 6.3 years. Of twenty-five plants ordered in 1967, the sixteen which were in operation by 1975 had taken on average 6.4 years, but the average would clearly be significantly higher when the remainder were in commercial operation. The first 1000 MW plants (Brown's Ferry I and Zion I) to come into commercial operation took respectively just over and just under seven years: Brown's Ferry suffered from an eighteen months' delay in delivering the pressure vessel.

The lengthening of the lead time during Phase II had two main sources: (i) the intensity of the development and diffusion of the LWR, which was beyond the capacity of all the parties concerned to execute on schedule, and (ii) the political, administrative and judicial infrastructure. The institutions, as President Nixon put it in June 1971 in the first-ever Presidential Energy Message – an able document – were 'not adequate for the job'[20] of reconciling the conflicts of interests involved!

(d) *Coincidence of intense development and diffusion*
Diffusion of a new technology is often conceived as following development, but with the LWR it coincided with the last stage of prototype development. Both were on an astonishingly large scale. The number of orders showed this for diffusion. The quick rise in the capacity of the largest units ordered in 1963–66[21] and the mass of design changes during Phase II showed it for development. The development programme aimed at exploiting quickly the scope for reducing costs by increasing the volume of reactor vessels and the output of steam per unit of volume, the 'power density'. This depended upon using fuel more intensively at higher temperatures, involved more risk and needed more refined controls and more engineered safeguards.

The coincidence of intense diffusion and development created demands which exceeded existing capacity for design, manufacture, construction, management, research, inspection and licensing. Bottlenecks occurred in all sectors. Late delivery of pressure vessels was the classic instance – many were upwards of eighteen months late, a considerable number being sub-contracted to European or Japanese firms because of the disarray of part of American domestic production.[22] This was the most dramatic among a mass of delays. The load of work led to defective quality which had to be rectified and to labour conflicts.

The AEC licensing staff, faced with applications unexpected in number and in extent of innovation, became another source of delay. Leading members of the JCAE thought much was due to bad AEC organisation and bad use of staff.[23] But the AEC's emphasis in explaining licensing delays up to 1970, mainly by shortage of staff and volume of innovation, seems justified. As a measure of the licensing delays, the AEC stated that the average time between the application for and granting of a construction licence was nine and a half months in 1967 and eighteen to nineteen months in 1969–70.[24] The time taken in securing an operating licence also rose, especially from 1970 on. This was not the whole story, because the regulators revised their judgements on some designs they had licensed and required modifications – it was called 'back fitting', and meant further delay.

By its nature the coincidence of diffusion and development must be a transitory source of delay. Capacity scarcities disappear, major developments are completed, overloading of licensing will end and licensing may benefit from 'learning'. The peak influence of the coincidence was probably reached in 1970; and possibly also the peak of fragmentation in the design and construction of plants and fabrication of fuel, although this influence was more prolonged.[25]

*(e) Political, administrative and judicial infrastructure*
By contrast, the influence of political, legal and administrative factors in increasing delays and lead times continued to grow after 1970. The AEC regulatory branch believed the nadir had been passed by spring 1973 and that, in so far as licensing was a governing factor, lead times could within the next few years be dramatically reduced. Nader still hoped to make things worse.

The leading members of the JCAE thought in 1970 that there were 'serious deficiencies in the AEC's procedural and administrative mechanism for licensing nuclear power plants', which resulted in delays that 'contribute nothing to health and safety' or the environment.[26] They saw typical bureaucratic weaknesses – demands for immense amounts of information which could not be digested, needless duplication of investigation, wasteful use of experts and failure to develop quickly norms and guidelines for utilities and manufacturers who were seeking licences. The AEC's replies to these comments implied they had some validity, that there could be improvements, but that these faults were not a major source of delay so far.

But in two ways, both the AEC and JCAE agreed, the licensing procedures became important sources of increased delay from 1970. First, public hearings required by the Atomic Energy Act before a licence was given by the AEC, either to construct or operate[27] a nuclear plant, were found by opponents of nuclear power to be a splendid instrument of delay and propaganda. Second, legislation increased the AEC's responsibilities in licensing, which lengthened the proceedings.

A public hearing was held in the locality where it was proposed to build a plant after the AEC regulatory branch had decided that the plant would involve no unreasonable risk to health and safety and that a licence should be granted. For a construction licence, this followed a detailed review by the regulatory staff – composed of highly qualified people – of the designs of the proposed plant; and for an operating licence, it followed intensive inspection of the construction and manufacture of a plant and pre-operational testing. In both cases the AEC submitted the problems to detailed review by the Advisory Committee on Reactor Safeguards (ACRS), an independent statutory committee set up in 1957 on the initiative of the JCAE to advise the AEC on all safety matters. Its fifteen members, appointed for four years, but open to re-election, appeared by 1971 to be a self-selecting body of distinguished scientists and engineers, each expert in some aspect of nuclear reactor safety and forming a balanced group with complementary disciplines.[28] Membership was onerous: in addition to meeting several days a month in full committee, members served on sub-committees, which called in other experts as consultants (there were sixty-nine consultants in 1971). It was a very intensive high-powered institution.

Modifications to a proposal were negotiated between applicants for a licence, the ACRS and the AEC. Review and negotiation were a long operation. In 1970 it averaged 460 days for a construction licence.[29] The hearing was quasi-judicial by an *ad hoc* Atomic Safety and Licensing Board (ASLB) of three members from a panel selected by the AEC; the chairman was a lawyer, and there were two experts, for whom work on boards was not their main job. The Board's task, which was set out in very general propositions, was to ensure that the regulators had had all the data they needed and that their recommendation was justified. The recommendation of the AEC regulatory branch was submitted by AEC counsel.

People whose interest might be affected by the proposed plant had a right to intervene, to ask questions and to submit evidence. Initially there was little intervention. By 1970 national groups of environmentalists (who came to be called 'public interest groups' and often had names redolent of exclusive virtue or wisdom[30]) were promoting and financing local intervention, providing lawyers and organising effective opposition. What had seemed a charade in support of nuclear energy became a charade against it. Since the AEC defined the objects of the ASLB's proceedings in extremely broad terms intervenors could ask for the production of detailed documentary evidence of all the review process and the negotiations, subpoena any participants as witnesses, cross-examine witnesses and present their own witnesses, and bring in outside experts to participate in cross-examinations – and the Boards were unable to curtail proceedings. 'Almost anybody, for almost any reason, can hold up almost any reactor licence for almost any length of time.'[31] The intervenors would challenge the impartiality of the Board's experts and make scheduling of meetings difficult; they were prompted to manage their case 'as a newsworthy event, constantly alert to the rhythms of media coverage.'[32] The average time in hearings for a construction licence rose from two days in 1966 and eleven in 1969 to fifty-four days in 1970.[33] It continued to rise. Hearings could last over a year and one took three. The average greatly understated the delay because (i) there were some uncontested cases, (ii) there was a gap of thirty days at least between calling a hearing (when all documents must be ready) and holding it and (iii) sometimes there were prolonged manoeuvres to delay the start which should be added to the hearing time. The same problems arose with hearings for operational licences, where delay was more serious. Time taken by AEC staff in prolonged licensing proceedings delayed their other work.

The 'cost' of the long hearings was not limited to delay, since the 'manipulation' of procedural possibilities forced utilities 'to make nearly any concession just to get the plant authorised'.[34]

In 1971 the intervenors became active in a new kind of hearing introduced by the AEC to allow public discussion of rules to set standards in respect of gaseous effluents, emergency core-cooling systems and other design features, whose adoption would make further discussion of these features in respect of individual plants unnecessary. Although intervention here could be factious, a once-only discussion was expected to be preferable to plant-by-plant litigation with no respect for precedents.

The greatest prolongation of licensing proceedings by new legislation resulted from the National Environment Protection Act of 1969 (NEPA), which provided that government agencies should assess the impact of their decisions on the environment and protect the environment 'to the fullest extent possible'. This generalisation obviously had no precise meaning.

The AEC's licensing responsibility had been limited to ensuring the safety and health of the public from radio-activity. In this field it was a competent body. Non-radiological environmental impacts involved in proposals to build power station and associated transmission lines were the responsibility of a host of other federal, state and local government agencies from whom licences, permits or agreements had to be obtained – in respect of land use, building regulations, clean air, clean water, refuse disposal, sewage, health, education, archaeological interest, aircraft routes, railway lines, road, bridge and wharf building, flora, fauna and fishes.[35] Agencies had to be negotiated with separately and there was no coordinating body, no 'one stop regulation'.

As a result of NEPA and an unfavourable court decision (the Calvert Cliffs[36] case, which the AEC decided not to take to appeal, partly perhaps to establish the Administration's commitment to the environmental cause) the AEC was committed in 1971 to submit to the ASLB a detailed cost benefit analysis covering *all* environmental aspects of every application for a licence.[37] Over this wide area the AEC had no claim to be an expert body. The starting point was an immensely detailed Environmental Impact Statement by the applicant, comparing the impact of different ways of dealing with the whole problem and particular stages. The AEC regulatory staff would review and amend this, ask for additions and modifications, submit it for comment to all other relevant agencies and make it available for comment by private persons and groups. On receipt of comments (for which two to three months were allowed) a final statement was prepared for the ASLB, which could reject or alter it. The AEC regulatory officials would take note of comments and must accept federal or Environmental Protection Agency (EPA) standards on, say air or water quality, but their assessments of costs and benefits had to be independent. The new procedure did not institute one stop regulation.

It was a long burdensome process. The average time taken by the AEC to review and process a report was, initially, thirty-one months. It still took fifteen months for an application made in 1973. But the applicant took at least a year, sometimes over two, in drawing up the Environmental Impact Statement. Immensely detailed information was called for – a JCAE official suggested to me in 1973 that the exercise was set up to ensure permanent full employment for all American ecologists and entomologists – and the requirements were frequently revised and increased.[38] This was an important source of long lead times. There were transitional difficulties in dealing with plants for which licences had been given before an Environmental Impact Statement was needed; intervenors held up the operation of several completed plants.[39]

An amendment to the Atomic Energy Act requiring the AEC to be satisfied that applications for construction licences were in compliance with anti-trust laws was expected in 1970–71 to be another source of delay. The AEC was to submit applications to the Attorney-General for review. His review would be accepted by the AEC, but would be submitted to the ASLB, and intervenors could contest it. It was expected some would. There were plenty of critics of private utilities as monopolies (albeit they were subject to public regulation and the monopoly was not complete) who advocated more public ownership. Building enormous nuclear plants having low unit costs was thought likely to give large private utilities an unfair competitive advantage over small municipal utilities in seeking new business. To avoid this, it was claimed the small publicly-owned utilities should either have power from a nuclear plant at the cost to the utility which owned the plant, or should be allowed to be part owner. The Attorney-General seemed sympathetic to these views and there was clearly a risk of hard bargaining. Publicly-owned utilities had substantial tax concessions and exaggerated claims could have been advanced to induce delay.[40] The fears over this had not been justified by 1974, probably because nuclear power was not cheap, but risks remained.[41]

Conflict between federal and state agencies over the right to establish standards of pure air and water provided another source of delay in the licensing. Some states, Minnesota, Vermont and Illinois for instance,[42] claimed the right for their agencies to treat the releases of radioactivity from nuclear plants permitted by the AEC (but determined by scientific bodies and sponsored from 1971 by the EPA) as maxima and to set more severe lower standards. Similarly, some adopted more exacting standards of thermal pollution, requiring cooling towers for example where the AEC thought them unnecessary.[43] This conflict was taken to the courts and while cases were being litigated there was, in effect, an injunction on the use of licences already granted: 'some issues can now', as it was put, 'be litigated before one tribunal and relitigated before another'.[44] The driving force, and finance, behind the 'state rights' claims came from the environmentalist movement.

Environmentalists taking cases to the courts often found the judges friendly, even cooperative. 'The now fashionable theory', according to one observer, 'is that administrative agencies cannot be trusted to take important initiatives altering the *status quo*, whereas courts are more dependably "with it"'.[45] The judge in the Calvert Cliffs case, in delivering his decision, showed at the outset that he was 'with it'. Several recent statements, he said, 'attest to the commitment of the Government to control at long last ... the destructive engine of material "progress". But it remains to be seen whether the promise of this legislation will become a reality. Therein lies the judicial role ... Our duty is to see that important legislative purposes heralded in the halls of Congress are not lost or misdirected in the vast hallways of federal bureaucracy.' In making his judgment he assumed 'it is very likely the planned facility will include some features which do significant damage to the environment', a presumption for which (unless it merely meant that a power plant would *change* the environment) evidence was presumably in the imagination, but not on the record. The AEC's interpretation of NEPA led the judge to comment 'it seems an unfortunate affliction of large organisations to resist new procedures and to envision massive road blocks to their adoption.' The readiness of the AEC to 'defer totally' to air and water quality standards set by other federal agencies charged with responsibility for them, he said, deprived 'concerned members of the public' from raising these as issues at the hearings. 'NEPA mandates a case-by-case balancing judgment' – a 'rather finely tuned and systematic balancing' – in each instance.[46] The competence of the AEC to do such fine balancing was not referred to – not, presumably, considered relevant. Such judicial attitudes were clearly important in guessing the future of delays due to licensing.

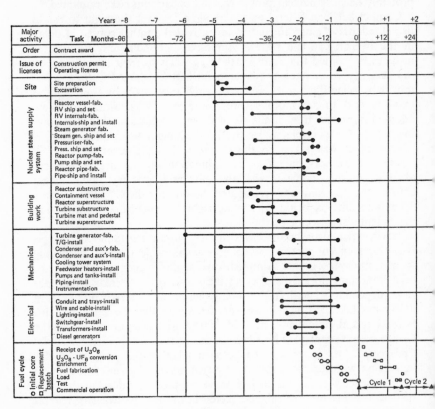

FIGURE 4.5
Nuclear Electric Power Plant Construction and Operating Schedule
in the United States (1973)

Prolonged delays due to concerted exploitation of the licensing processes would only continue if the exploitation continued successfully and there were no changes in law or procedure. An LWR plant could be built in four and a half to five and a half years – production schedules published by the AEC and ERDA (one is reproduced in Figure 4.5) illustrated this forcibly. So this addition to cost was another transitional cost. In summer 1973 the AEC's Director of Regulation,

L. Manning Muntzing, asserted that he had devised means of streamlining regulation so that the lead time, which he put as nine to ten years, could be reduced within two or three years by three years. Time taken by the regulators in reviewing applications could be cut, if makers designed standardised plants there would be no need to duplicate reviews; if utilities chose sites and secured their approval some years before they wanted to start construction that would save time; and if a hearing for an operating licence were turned into a legislative-type instead of a judicial process, that would save time. Much of this required Congressional support which was not so easy to come by now. Muntzing's assertion was received with derision by many. But even if a little of what was promised happened, the deterioration which Calvert Cliffs brought might be reversed.

Exploitation would only continue if a sufficient number of people (possibly a small minority of the population) was prepared to finance it, believing that the object still deserved support. This depended on the effectiveness of the campaign on safety to which we now turn.

## NOTES AND REFERENCES

1. Burn, *op. cit.*, p. 41.
2. *Trends in the Cost of Light Water Reactor Power Plants for Utilities*, Wash. 1150 (Washington: US Government Printing Office, for the AEC, 1970) p. 14. As Table 4.1 shows, estimated construction costs rose 36 per cent, total costs 55 per cent, between spring 1967 and June 1969. It would not be surprising if the construction cost had been 10 per cent or more too low originally; the IDC and escalation deficiencies were greater.
3. Burn, *op. cit.*, pp. 38, 46–50; *Current Status and Future Technical and Economic Potential of Light Water Reactors*, Wash. 1082 (Washington: US Government Printing Office, for the AEC, 1968) ch. 5, p. 166; *The Nuclear Industry for 1971* (Washington: US Government Printing Office, for the AEC, 1972) pp. 91–2; *The Nuclear Industry for 1974* (Washington: US Government Printing Office, for ERDA, 1975) pp. 19–20.
4. A leading New York consultant told me in 1973 for instance that a utility whom he was advising was basing its choice exclusively on differences in fuel-cost guarantees, because estimated capital costs were identical. One plant manufacturer took a particularly optimistic view of the level of future uranium ore prices, and of the reduction possible in labour input into fuel fabrication and based his fuel cost guarantees on these assumptions.
5. This figure was given by Philip Sporn, in *Nuclear Power Economics, 1962 through 1967* (Washington: US Government Printing Office, for the JCAE, 1968) p. 10. He based it on a construction cost of $137.50 (which did not include escalation) and on a fuel cost 0.20 mills/KW lower than the AEC figure quoted in Table 4.3.
6. *Info* (Washington: American Industrial Forum, 1976) p. 9.
7. *Hearings on AEC Authorizing Legislation* (Washington: US Government

Printing Office, for the JCAE, 1972) pt 2, pp. 635–53. Hereafter cited as JCAE, *Authorizing.*

8. *Ibid.*, p. 626.

9. Based on details in *1000 MWe Central Station Power Plants Investment Cost Study*, Wash. 1230 by United Engineers and Construction Inc. for the AEC (Washington: US Government Printing Office, 1972) vol. I.

10. *Concept: Computer Code for Conceptual Cost Estimates of Steam-Electric Power Plants*, Wash. 1180 (Washington: US Government Printing Office, for the AEC, 1971). Hereafter cited as AEC, *Concept*, 1971. In this report productivity factors are given on p. 21 which 'indicate a decrease in site-labour productivity of approximately 6 per cent a year from 1967–69 and of about 4 per cent from 1969–70'. The data on which conclusions have to be based regarding construction costs are not presented with a critical discussion of sources. The evidence of a significant drop in productivity at individual sites seems conclusive.

11. AEC, *Concept, 1971*, pp. 29–34.

12. The sources from which the utilities financed their investments are analysed in the Federal Power Commission's annual *Statistics of Privately Owned Electric Utilities in the US* (Washington: US Government Printing Office).

13. Their expenditures on new plant and equipment rose almost uninterruptedly from $6750m in 1967 to approaching $18,500m in 1975.

14. These were treated as a source of cost above normal estimates in 1967, but not in 1974. Their cost was upwards of $10 per KW by 1970.

15. Tidal waves resulting from earthquakes under the sea.

16. See below, p. 82.

17. I omitted the 1973/81 figures from the Table 4.1 for brevity.

18. They reflect a use of more resources and the keeping of resources unproductive for longer periods before their use on production is possible or permitted.

19. AEC, *Concept, 1971*, pp. 29–34.

20. *Hearings on AEC Licensing Procedure and Related Legislation* (Washington: US Government Printing Office, for the JCAE, 1971) pt 2, p. 526. Hereafter cited as JCAE, *Licensing, 1971.*

21. Oyster Creek, 640 MW (1963); Dresden II, 800 MW (1965); Brown's Ferry, 1050 MW (1966).

22. For about two years a new plant put up by Babcock & Wilcox, which was intended to deliver within little over a year vessels at about one per month, failed to operate satisfactorily.

23. JCAE, *Licensing 1971*, pt 2, p. 479.

24. *Ibid.*, pt 2, p. 481.

25. In 1966 Combustion Engineering and Babcock & Wilcox secured orders for PWRs so that with GE and Westinghouse much in the lead there were four firms offering LWR nuclear steam supply systems – each by 1970 in two or three capacity groups. The AEC said there were eight different designs of containments incorporating over thirty different combinations of engineered safeguards – 'each requiring different and sophisticated analysis'. This reflected the large number of architect-engineers. There were at the same date five firms, in addition to the reactor manufacturers, listed as offering to design and manufacture LWR fuel to provide reload cores in competition with the plant makers, and some other groups were con-

sidering doing this. Three such firms had orders from utilities, and one had already supplied reload fuel.

26. JCAE, *Licensing, 1971*, pt 2, p. 429.
27. A public hearing was obligatory for a construction licence but optional for an operating licence; it must be held only if someone requested it. As environmental interest increased someone usually did request it.
28. Membership and activities are described and discussed in JCAE, *Licensing, 1971*, pt 1, pp. 93–133. Of the 1971 members five were academics (professors of nuclear, chemical, civil and sanitary – it would now be environmental – engineering); three were engineers or physicists from AEC research laboratories, (Los Alamos, Argonne and Brookhaven); the remaining six were consultant metallurgists, engineers and chemists, of whom one, Dr Spencer Bush of Batelle Memorial Laboratories, was chairman.
29. JCAE, *Licensing, 1971*, pt 2, p. 482.
30. Union of Concerned Scientists, Citizens for Survival, Citizens' Committee for Environmental Concern, Friends of the Earth, Business Men for the Public Interest.
31. JCAE, *Licensing, 1971*, pt 2, p. 927.
32. *Ibid.*, pt 3, p. 1402 *et seq.*
33. *Ibid.*, *pt* 2, p. 549.
34. *Ibid.*, pt 3, p. 1448. Leading lawyers for intervenors openly announced their readiness to make deals with utilities, making it plain that otherwise there was no limit to the time they would spin out the proceedings.
35. *Hearings on Environmental Effects of Producing Electric Power* (Washington: US Government Printing Office, for the JCAE, 1970) pt 2, vol I, p. 1883, Hereafter cited as JCAE, *Environmental* (pt 1 1969, pt 2 1970). *Hearings on Prelicensing Antitrust Review of Nuclear Power Plants* (Washington: US Government Printing Office, for the JCAE, 1969) pt 1, pp. 61–3. Hereafter cited as JCAE, *Antitrust*, (p. 1 1969, pt 2 1970).
36. JCAE, *Licensing, 1971*, pt 2, pp. 593–9.
37. For the first reactions of the AEC see *ibid.*, pt 2, p. 620.
38. For example, complete lists were required for all species of vegetation, animals, birds and aquatic life at different seasons in a wide area round a site, not only specially rare types and types to be affected, and diagrams of life cycles of particular species were called for.
39. The most conspicuous case was Quad Cities I and II; intervenors brought a case against the AEC and the first court supported them, though the Environment Protection Agency were on the side of the AEC in granting a licence to operate at 50 per cent prior to the preparation of an Environment Impact Statement.
40. There were long hearings on this topic before JCAE, *Antitrust*, pt 1 1969, pt 2 1970.
41. According to *Nucleonics Week*, a few utilities had anti-trust problems by 1974, but these caused no delay since they had obtained licences before it was necessary to have clearance: one utility, however, was being delayed, Louisiana Power and Light Company's Waterford plant. *Nucleonics Week*, New York, 30 May 1974.
42. JCAE, *Environmental*, 1970, pt 2. vol. I, pp. 1111 and 1198.
43. *Ibid.*, p. 1198.
44. *Selected Material on the Calvert Cliffs Decision, its Origin and Aftermath,*

Report of the JCAE (Washington: US Government Printing Office, 1972) p. 457. Hereafter cited as JCAE, *Calvert Cliffs, 1972.*

45. Professor L. Jaffé (Professor of Administrative Law at Harvard) in a review of a book by J. L. Sax (Professor of Law at Michigan) which espouses the criticised dictum. See JCAE, *Licensing, 1971,* pt 4, p. 1645 (quoted also by Professor Leo Wolman, *ibid.,* p. 1639).

46. *Ibid.,* pt 2, pp. 574 and 899 gives the judgment.

CHAPTER 5

# Safety of the LWR

## (i) 'OVERRIDING REQUIREMENT'

The 'overriding requirement for safety precautions' was recognised from the outset by all concerned with the development of nuclear power plants – in all countries – because the vast quantities of intensely radioactive materials that are generated in the fuel of a reactor during its operation would be immensely dangerous unless they were 'excluded from the environment'.

Hence designers, manufacturers, utilities, the Administration and Congress in the United States were all committed to avoiding any accident or malfunctioning that could result in significant escape of radioactive materials. As Robert A. Charpie, who at the time was Assistant Research Director at the AEC's Oak Ridge Laboratory, put it in 1955, 'We now insist that every reactor be enclosed in a blast-proof envelope which will guard against the spread of radioactive contamination in the event of a full scale chemical or steam explosion occurring in the reactor; ... conservatism in safety requirements' was the main element in capital costs, which 'tend to make estimates high'. As experience is gained Charpie was sure the 'restriction will be relaxed in part.'[1]

The trend was in the reverse direction. For the larger and more highly stressed plants, working at higher temperatures and having larger fuel inventories, containing larger quantities of radioactive materials, there has been a progressive elaboration of safety measures.

Has it been adequate? The great majority of engineers and scientists closely connected with the development, in designing, constructing, buying and operating plants, in regulating these activities in the interests of public health and safety, or in giving independent expert advice to the regulators, have consistently had immense confidence in the safety of the plants as designed, constructed and licensed. But as the environmentalist movement became a major political force, the plants have been attacked as continuous sources of routine, lethal, radioactive effluents and as susceptible to major catastrophic accidents which would release radioactive materials potentially more devastating

55

than a bomb. This became a source of delay and higher costs and a threat to the continuance of nuclear plant building. Was there reason to believe that the developers, regulators and assessors had been wrong? If not, what accounts for the strength of the criticism?

(ii) BASES OF ASSURANCE

The engineers and scientists – the range will be examined later – could give good reasons for their confidence.

The safety record over the twenty years of LWR development was (and still is) impressive. It is a record without parallel in any other major technological development – for as the Rasmussen safety study puts it, echoing all earlier reports, 'risks had to be estimated and not measured because although there are about 50 [nuclear] plants now working, there have been no nuclear accidents to date'.[2] It looks equally as though there have been no near-misses. Development had an exceptional research backing within the firms and the AEC which the engineers and scientists could assess. Development engineering by its nature must involve judging whether adequate precautions were taken to allow for unavoidable uncertainties in innovation: the firms developing the LWR had impressive records in this work. Because a serious accident in a nuclear plant could be disastrous, the precautions taken – in conservatism of design and operation, in materials, specification and quality assurance, in the extent of inspection, monitoring and surveillance, and in training operators – were exceptional.[3] The plants were 'the most thoroughly studied, engineered, constructed and checked out plants in the history of mankind', one of the makers put it.[4] This was confirmed by the chairman of the ACRS.

> The record in the chemical industries, certainly in specific aspects, such as the petroleum refineries, particularly, is good to excellent. But I would say compared with this the record in the reactor industry would be outstanding. The efforts that have gone into reactors, such as quality assurance and improved overall quality, play a major safety role . . . many are being picked up by the petroleum industry now . . . I really believe that there is a substantial jump between the safety and the safety records of the reactor industry as compared even to some of the better ones in the chemical industry.[5]

Designs by firms and architect-engineers had to satisfy stringent conditions set by the AEC and were subjected to long examination in every detail by the regulatory staff and, independently, by the ACRS before being considered by the Licensing Board. Possible malfunctioning or breakdown of components and systems were as far as possible foreseen, their probability was reduced by design, engineered safeguards were provided to counteract their effects should they nevertheless occur, by instantaneous automatic responses, and should the first line of safeguards break down alternative (redundant) safeguards were provided. Beyond the design stage there was an exhaustive surveillance of construction by the regulators, and plants, when completed, were exhaustively tested, and when any signs of malfunctioning, defective work or material occurred before during or after testing they were followed up not merely where they occurred but also where they might occur in other plants. When development experience cast doubt on the validity of design decisions, modifications were built in to existing plants. None of this was a formality; replacing work falling below the extremely high levels called for, and components which failed under test, was, as Milton Shaw, Director of the AEC Division of Reactor Development and Technology, emphasised,[6] a frequent source of delay and added cost in the early 1970s.

(iii) PROBABILITY ASSESSMENTS

By these measures the 'atomic establishment', as its critics called it, believed the risks from nuclear plants were exceptionally low. From the late 1960s systematic methods have been applied and developed[7] for estimating the probability of accidents of different kinds occurring in nuclear plants, their probable scale and the deaths, injuries and property destruction they would involve. The results supported the Establishment's confidence. They were not its source, but they reflect the sources. In an impressively thorough analysis of *The Safety of Nuclear Power Reactors* (*Light Water Cooled*), published in July 1973, the AEC summed up the outcome of the probability studies at that date in a table here reproduced (Table 5.1).[8] They also indicated that

TABLE 5.1

Risk of Death in the United States from Selected Hazards

| Hazard | Annual chance of death: one person out of | Probability of death per person/year |
|---|---|---|
| 1. Cancer (all types) | 625 | $1.6 \times 10^{-3}$ |
| 2. Auto accident | 3,600 | $2.8 \times 10^{-4}$ |
| 3. Drowning | 27,000 | $3.7 \times 10^{-5}$ |
| 4. Choking on food | 100,000 | $\sim 10^{-5}$ |
| 5. Cancer from medical X-rays | >100,000 | $<1 \times 10^{-5}$ |
| 6. Lightning | 1,000,000 | $0.8 \times 10^{-6}$ |
| 7. Cancer from nuclear effluents | 25,000,000 | $<4 \times 10^{-8}$ |
| 8. Nuclear power plant accidents leading to releases of radioactivity | $\sim 10,000,000,000,000$ | $\sim 10^{-10}$ |

they had under way a much more exhaustive high-powered study. In Table 5.1 the degree of safety according to current estimates was expressed in a comparison of two major risks from using LWRs with risks from other hazards. The risks were measured in probabilities of deaths. Risks of disease, injury and disablement and damage to property are also significant, but death rates gave an adequate guide to relativities.

The estimated risks from nuclear plant were negligible compared with the others chosen. The probability of dying from cancer from radioactivity due to a nuclear accident was put at one ten thousandth of the probability of being killed by lightning. The risk from routine effluents was greater – one twenty-fifth of the risk from lightning. But by any standards this too is negligible. The estimated low probability of death due to a major catastrophic accident reflected mainly an extremely low probability of such an event happening, (1 in $10^7$ years); but in addition the severity of a major accident was judged to be far from the catastrophic proportions presented by critics, who suggested that one disaster might lay waste half Pennsylvania and have a greater impact than an atomic bomb. An estimate in a report for the Resources Agency of California, by the University of California, Los Angeles, (UCLA) School of Engineering and Applied Sciences, put the probable deaths resulting from a major (1 in 100 million years) accident at 5000 spread over 'very many years' in a population of ten million – so that 'the public impact of such a nuclear plant accident is unlikely to have much general visibility'. The 'public health impact of a maximum disaster would not be significant compared to the normal incidence of disease.'[9]

The Californian report also compared risks from a nuclear and an oil-fired generating plant, much to the advantage of the nuclear plant.[10]

The probability statistics which gave these amazingly small risks gave a sense of precision which no experts, least of all those who calculated them, would have claimed. They were based on experience, but experience in operating big nuclear plants, of 600 MW upwards, was short, and no 1000+ MW unit had started up. Direct experience with plants was adequate in regard to routine effluents; but for malfunctioning and breakdown of components it was limited. Many malfunctionings and breakdowns (numbered in hundreds), whether in the course of plant construction and testing or during operation, had been reported as 'unusual incidents' to the regulatory authority. Most were trivial and none of a major category. No complete sudden rupture of a main coolant pipe or of a vessel, the sort of incident which might initiate a major accident, had occurred, nor had there been any approach to it. Risks of such ruptures had to be measured from experience in other industries adapted to the different requirements: there was no plant experience of engineered safeguards for these conditions being called into operation in a real accident.

The probability statistics were thus in no sense a proof of safety and at many points they depended upon engineering judgements. Such judgements, for example, were required in translating statistics of vessel and pipe behaviour in chemical plants and fossil-fuelled power plants into figures for the nuclear industry, where some vessels were larger and all were subject to continuous radiation, but where conditions of temperature, pressure and corrosion were less severe. It was the judgement of many experienced engineers exercised on a great mass of design and operating data. The element of judgement in the probability calculations was marginal, but it was there. This is normal in all engineering development.

In August 1974, the comprehensive Reactor Safety Study which the 1973 Report foreshadowed was published – known as the Rasmussen Report, after its director, Norman C. Rasmussen, Professor of Nuclear Engineering at Massachusetts Institute of Technology (MIT). All concerned were aware this was a sensitive subject and, while the AEC had commissioned the study and provided office accommodation, it was stressed in the Report that its direction was wholly independent. Of the sixty qualified staff who worked on the two-year job, only ten, having specially appropriate experience, came from the Commission, the rest from national and private laboratories (Oak Ridge and Sandia, for instance) and from corporations including Boeing (who had great experience in 'fault tree' analysis), and Aerojet Nuclear (with much experience in safety research at Idaho), while research institutes such as Batelle and Stanford, and universities (for example UCLA) as well as national laboratories and a few corporations worked for the project as consultants. It was limited in scope to the LWR, based on the latest BWR and PWR just entering service, and concerned with conditions when the first 100 LWRs would be operating, about 1980. Within its limits, it was far more searching than any earlier work, basing its results on computerised examinations of thousands of potential accident routes, determined by methods, 'event trees' and 'fault trees', developed within the Department of Defense and National Aeronautic and Space Administration and further advanced by the Rasmussen Group study.[11] There were some critics of the methodology and its appropriateness in this application: a small minority of those were competent to have a view, but of course small minorities can at times be right!

The results gave even more support than the 1973 report to the confidence of the majority of scientists and engineers in the LWR. 'If we consider' the Report concluded 'a group of 100 similar plants [LWRs of 1973–74 vintage] the chance of an accident causing ten or more fatalities is 1 in 2,500 per year or, on this average, one such accident every twenty-five centuries. For accidents involving 1000 or more fatalities the number is 1 in a million years.'[12] The following table was given to compare the 'likelihood of a nuclear accident to non-nuclear accidents that could cause the same consequences'.

TABLE 5.2

Probability of Major Man-caused and Natural Events

| Type of events | Probability of 100 or more fatalities | Probability of 1000 or more fatalities |
|---|---|---|
| Man-caused | | |
| Airplane crash | 1 in 2 years | 1 in 2000 years |
| Fire | 1 in 7 years | 1 in 200 years |
| Explosion | 1 in 16 years | 1 in 120 years |
| Toxic gas | 1 in 100 years | 1 in 1000 years |
| Natural | | |
| Tornado | 1 in 5 years | very small |
| Hurricanes | 1 in 5 years | 1 in 25 years |
| Earthquake | 1 in 20 years | 1 in 50 years |
| Meteorite impact | 1 in 100,000 years | 1 in 1,000,000 years |
| Reactors | | |
| 100 plants | 1 in 100,000 years | 1 in 1,000,000 years |

(iv) PHASES IN THE CAMPAIGN AGAINST NUCLEAR PLANTS

Criticisms of the LWR as dangerous came in phases. When plans were first made to develop atomic power, divergent views on safety emerged. In 1946 Glenn Seaborg, Robert Oppenheimer and Enrico Fermi, among the leading figures, favoured developing more new reactors at Argonne near Chicago (where the first 'pile' had been built). Edward Teller favoured isolating new work in the desert at Idaho Falls and was successful. He became progressively troglodytic, recommending by 1960 that underground shelters should be provided for all against a nuclear bomb attack[13] and later suggesting that LWR plants should be built underground. Under pressure from Seaborg, Chairman of the AEC, Teller conceded in 1969 that if it was a question of having LWRs built above ground or not having them at all he would have them.[14] They were not after all *so* dangerous! In 1975 he was reported 'strongly pro-nuclear' in a new report for the Committee on Critical Choices for America.[15]

In 1957, when LWRs of 200 MW capacity were being built, reactor safety was examined for the AEC at its Brookhaven laboratories and all the experts consulted concluded that the danger of a major accident resulting in a substantial release of radioactive nuclides outside the plant was 'exceedingly low'. Those prepared to give a figure put the risk for a reactor as between 1 in 100,000 years and one in a billion years, and only 1 in 10 of these accidents would coincide with the most unfavourable weather conditions – which would result in 3400 deaths.[16] It was analogous to the later Californian study – the reactors considered were smaller, but they lacked the sophistication of safeguards and containments used by 1970. The authors of the report feared they might be accused of complacency. Strangely, however, passages in the report were largely quoted by environmentalists in support of their case. Even *Nucleonics Week* referred to it in 1971 as 'notorious', implying pro-environment! The Rasmussen Study found that it greatly exaggerated the risks – that it estimated 'hazards conceivable in highly improbable reactor accidents', but 'made no serious attempt to assess the probabilities'.[17]

In 1963 David Lilienthal a lawyer administrator who had been the first chairman of the AEC, attacked the AEC's *Report to the President* on civil nuclear development in 1962[18] for its support of several 500 MW LWRs. As former chairman also of TVA, he regarded himself as an expert in experts. But in 1946 he had apparently swallowed the politically convenient doctrine that the atom would create a new world in civil sectors as rapidly as it had in defence – although he later admitted that good engineers had warned him the development could not be fast.[19] When the United States really began developing power reactors for civilian electrical supply from 1950 to 1955, the engineers thought fifteen to twenty years' development work would be needed before any type would be competitive; the Paley Report thought nuclear power might be ready to fill an energy gap foreseen for 1975.[20] Nevertheless, by 1963 Lilienthal thought civil atomic development a failure. Atomic power was not yet competitive, but supported by an enormous and growing government budget, it promised power only as cheap as fossil plants. No scarcity of fossil fuels could be envisaged; they were cheap and coal-mining was revitalised. Nuclear plants would involve more serious dangers than fossil plants.

Lilienthal wrote attractively, but (as in 1946) read the trends wrongly: almost as he wrote, 'competitive' reactors were being bought without state aid, fossil-fuel supplies were seen not to be securely available for the current rate of increase in demands, and the serious environmental costs of fossil plants were more fully recognised. Nevertheless, he helped to set a pattern for opposition; he attacked the administrative arrangements whereby the AEC had the conflicting duties, as he thought (and this was a potential risk), of encouraging nuclear development and of regulating it in the interests of public health and safety. He attacked the JCAE for becoming too interested in a programme of nuclear power expansion (involving enormous state patronage which distorted the pattern of scientific research) to act effectively as a watchdog for the public interest. He almost welcomed what he thought the atomic failure as a rebuff to the 'scientific establishment'.[21]

Although proposals to build nuclear plants in the Bronx and near an earthquake zone in California were blocked at this time, and dropped, distrust and fear of nuclear plants did not conspicuously increase until the late 1960s, when it became an important element through intervenor activities.

## (v) RADIOACTIVE EFFLUENTS – G AND T

Between 1969 and 1974, the emphasis rested at different periods on different phases of the safety problem. First, from 1969 to 1970, great concern was stirred up over the 'routine' liquid and gaseous radio-active effluents which are produced in operating LWRs. Behind this lay the alarms over fallout from bomb tests in the 1950s. The AEC laid down limits to effluents based on guidelines issued by the Federal Radiation Council (FRC), a body of senior administrators appointed by the President which relied upon national and international scientific committees,[22] and whose members included the world's leading experts in measuring radiation effects.[23] The FRC guidelines laid down that individual members of the general public should not receive more than 500 millirems a year of radiation from man-made sources, nor over five rems in thirty years, and the average dose for a 'suitable sample of the exposed population' should not exceed 170 millirems a year. These were outside limits – exposures should be kept 'as low as practicable'.

In 1969 two AEC scientists, John Gofman and Arthur Tamplin, referred to frequently as G and T, and nicknamed irreverently by a leading AEA expert on safety as 'gin and tonic', attacked these standards adopted by the AEC from the FRC as dangerously high, and asserted that if the standards were not reduced at least tenfold then as the number of reactors increased the number of deaths from cancer would be greatly increased. The increase was put in succeeding calculations as 17,000, 36,000 and 104,000 deaths a year, notably higher than the probability figures in Table 5.1 (ten to twenty deaths a year).[24] G and T had good academic credentials. Both were experts in the field of medical physics, and radiation in particular, and were leaders of research teams in the AEC's Lawrence Radiation Laboratory, managed for the AEC by the University of California, Berkeley: qualifications which would persuade many people these were experts who were liable to be right.

Many leading radiologists held that the methods whereby Gofman and Tamplin calculated the additional deaths from cancer which would occur if the whole United States population were exposed to a dose of 170 millirems a year were incorrect and exaggerated the number. There were no means of basing it on direct experience, because there is no clinical evidence of the effects of such low rates of radiation on human beings or animals, so that guesses must be based on hypotheses that, for example, the effect of small doses will be proportionate to that of large doses, and that an accumulation of doses over a long period will have the same effect as one large dose of the same amount. So their figures, based, they claimed, on 'hard incontrovertible data', involved extrapolation from an exposure 500 times greater and could not be regarded as 'firm scientific fact' as G and T claimed.[25] The critics disputed other aspects of their conversion of radiation dose into cancer deaths.[26] It is not necessary to probe these disputes – though the critics make a strong case against Gofman and Tamplin's statistical methods – because their calculations were completely irrelevant in the nuclear controversy.

This was so for two reasons. First, the known and projected radioactive releases from LWRs were far below the maximum permitted limits and, second, even if the maximum release did occur at some point on some occasion on the perimeter of a plant the average dose for a 'suitable sample of the exposed population' would fall far below 170 millirems.

Plants being completed in 1969–70 (hence planned in 1964–65) were designed to have releases of between 1 and 5 per cent of the permitted level.[27] GE said in 1970 they would design for 1 per cent, and Westinghouse offered zero release plants; something close to this was to be achieved for both PWRs and BWRs. BWRs have potentially larger gaseous radioactive releases than PWRs. Actual gaseous releases from BWRs which started operating in 1969–70 ranged from 1 to 6 per cent of the permitted level of 500 millirems at the perimeter; the lower level could be made general and could be improved on.

The average exposure over a 'suitable sample of the exposed population' was far lower than the perimeter level, because the gases become progressively diluted and because some radioactive nuclides which form a large proportion of the total have a very short 'half life' – Krypton 87 and 88, Xenon 135–138, range from seventeen minutes to nine hours.[28] Average annual doses within fifty miles of the large BWR plants operating in 1971 ranged from 0.006 to 0.06 millirems per person – negligible in themselves and in relation to the 170 millirems premised by Gofman and Tamplin for the whole population (not merely those within range of a plant). The number of LWRs (and other reactors) used would greatly increase by the year 2000, but as the effluent would be kept down to the better levels reached in the 1970–71 design there was no reason why averages in areas round power stations should rise above these lowest levels – but there would be more such areas, and some would overlap. All this was taken into account in the probability statistics and other studies of prospects over the next thirty years.

When it was pointed out to Gofman and Tamplin that actual emissions were already much lower than the new 'maximum' they proposed, their reply was: splendid, then why resist what we propose?[29] This was essentially a barrow-boy response. Put temperately, however, it was an apposite question. The answer given by manufacturers, utilities, administrators and the JCAE was: something *might* go wrong which would temporarily involve effluents larger than normal. But a margin of one hundred-fold above the design level was hardly reasonable: only one of all the plants, old or new (it was of course an old one), needed as much as a tenfold margin in 1970–71. Shortly after James Schlesinger – an economist who had worked for the Rand Corporation and was drawn into administration by President Nixon, first as Assistant Director of the Office of Management and Budget – succeeded Glenn Seaborg, a professional scientist, as Chairman, the AEC did more than Gofman and Tamplin had asked, reducing the maximum permitted dose one hundred-fold to 5 millirems instead of 500 at the perimeter. This was part of a deliberate policy of improving the public image of the AEC. The new limit – which was at the design level adopted by GE – would probably require slightly more equipment to reduce BWR radioactive gaseous effluent and might slightly reduce future health risks from power stations, but even this was not certain from the figures.[30] The level was clearly already going to be negligible. Milton Shaw was known to say that LWRs 'could be made safe to any degree of *concern*'. This was a good illustration. When risks had become so minimal the point was being reached when the cost benefit balance might be negative. The focus of effort in regard to effluent risk had now passed to fuel reprocessing; there was no reason to doubt it was an equally manageable problem.

## (vi) RISK OF CATASTROPHE

In 1971, attention reverted to the risk of catastrophic disaster. Here it was not possible, as with routine effluents, to establish the risk completely on experience in existing plants. There had been no major disaster, and no approach to one – but equally no plant of 1000 MW or more had so far operated and only one 600 MW plant had run for four years, whereas the expected life was over thirty years. Moreover, the flood of orders for large plants, which were more highly stressed and had larger inventories of radioactive materials, had raised massive new problems for the regulatory authorities and increased greatly the uncertainties which had to be either resolved or temporarily offset by wide margins of safety, on the principles expressed by Charpie in 1955.

Determining what these margins should be, what degree of conservatism in design was called for, was a matter which involved judgement – and experts did not all form exactly the same opinion. This was implicit in the fact that the regulatory authorities required design modifications, for plant makers clearly thought their designs were safe, and included sufficient margins. There were also differences – relatively narrow – among the AEC regulators, the AEC safety research workers and the ACRS over the precise degree of conservatism required. These differences came to light in discussions concerning the efficiency of the Emergency Core Cooling Systems (ECCS) after the AEC issued Interim Criteria for such systems in June 1971, and submitted them to its laboratories for comment – and to a public rule-making hearing to be held in 1972, which turned into a marathon. The 'criteria' required more conservatism in operating plants, establishing for instance lower maximum core temperatures. The changes, in response to new methods of calculating fuel-cladding performance in a loss of coolant situation (which had never occurred), would involve some changes in operating new plants, might involve some equipment modifications in some older plants, but would permit normal electrical output in all, or almost all.[31]

The intervenors decided to stage a major effort at the public (rule-making) hearing to secure drastic revisions of the criteria proposed by the AEC. The data used by the Commission in drawing up the proposed criteria and submitted to their laboratories, the manufacturers and the ACRS for comment and to the JCAE, were combed by the intervenors, or more accurately by sympathetic scientists for the intervenors, to find a case. The AEC did not formally release the documentation to the press for the excellent reason that lay comment was of no help in its technical evaluation. But it was available for anyone who asked for it. This provided a double opportunity to the intervenors. They could say (without justification) that the Commission was suppressing inconvenient evidence, and they could mass titbits of supposedly suppressed data, isolated, out of context and not carefully evaluated. For this analysis the method is more important than the content: a full study of content would be impossible on a small scale, but also wasteful of time.

The intervenors did not appear to contribute any new data to these discussions: their participation was not based on independent research. They were helped by some scientists, including nuclear physicists who sympathised with their general attitudes and who examined some of the voluminous AEC documentation. Possibly some of the sympathetic scientists had friends on the AEC side of the fence; certainly the intervenors tried to stimulate members of the AEC staff to reveal disagreements to them.[32] The intervenors' case was mainly presented by leaders who were not scientists; the stars in the exercise were Myron Cherry, Anthony Roisman (both lawyers) and David Comey, an expert in Soviet affairs, and it was presented in the spirit of 'adversary' legal proceedings, exclusively to emphasise one side only. [33]

Intervenors latched on, with conspicuous skill, to anything which could be made to spread alarm about the nuclear plants. Thus the large number of 'abnormal incidents', referred to above, in nuclear plants operating or being built, which were reported to the AEC, was emphasised.[34] None had been serious ('none' the AEC reports also stated 'posed any threat to health and safety'), but this the intervenors claimed was only luck – the implication being that a major disaster could easily have occurred. The AEC experts found that analysis of the record 'tends to confirm the view that the principal safety concerns have been identified and that the measures employed to cope with them have been effective'.[35] This conclusion was kept in obscurity by the intervenors.[36]

Many AEC scientists and engineers said (i) that the effectiveness of the ECC systems had not been completely proved (it had not been disproved!), (ii) that full-scale realistic tests needed to establish it had fallen behind schedule and should be hurried up, and (iii) that the computer codes used by the plant makers as the basis of ECCS design were not sufficiently sophisticated.

The intervenors quoted such scientists and administrators,[37] but did not point out (it was characteristic of this method, and quite deliberate) that many, probably most, of the AEC experts and advisers whom they quoted, and certainly those who held the most distinguished positions, believed that the uncertainties about the ECCS were adequately offset by the conservatism in design and operation imposed by the Interim Criteria and that the LWRs, as designed and operated, were safe. This was the conclusion of the ACRS, precisely spelled out by its Chairman: 'The committee believes', Spencer H. Bush said, 'the health and safety of the public has been adequately protected. . . . This is believed to be due to the high degree of conservatism used in both nuclear plant designs and in safety reviews'. His predecessor, J. M. Hendrie,[38] confirmed that 'we are not running a reckless course' in deciding 'that we do not have to shut the reactors down,' or 'stop building more', although the research work on ECCS was not complete; but the 'very conservative point of view', which the Interim Criteria incorporated, protected the safety of the public. When ECC systems were shown to be fully effective, the ACRS concluded, some existing licensing restrictions could be relaxed. This was the view also of Milton Shaw, colourful Director of the AEC Department of Reactor Development and Safety Research till June 1973, who certainly made acrid comments on the computer codes and the lack of complete proof of the systems, but also stated that LWRs were safe, and could be made safe 'to any degree of concern'. Shaw and his staff were convinced that the 'usual techniques used . . . for achieving adequate assurance by use of conservatism and bounding techniques' were 'applicable to ECCS calculations', and that the ECCS criteria would 'provide the requisite degree of confidence in ECCS performance'.[39]

The intervenors gave great prominence to results of a few experiments – by AEC teams – which they argued showed that ECC systems would almost certainly be ineffective. The most sensationalised was a so-called 'semi-scale' test in a model supposedly analogous to a PWR vessel.[40] When loss of coolant conditions were simulated in the model the emergency coolant was kept out of the core by steam pressure built up inside it. When the experiment was repeated with appropriate controls the original research team agreed the result was entirely an 'artifact of the smallness of the test vessel'. The 'semi-scale' test had been made in a model one foot in diameter; a commercial vessel was twenty feet in diameter: and the layout of pipes was entirely different in the two vessels. In a vessel larger than the 'model' the effect recorded could not, it was established, occur or be significant. Presumably it was a badly designed experiment: the vessel, Shaw had told the JCAE in June 1971, had not been designed for this test. Westinghouse had said immediately the experimental results were described that they were irrelevant, and the firm was right. The intervenors abstained throughout from speaking of the actual scale of the test, which might have put even laymen on their guard.[41]

To give one further illustration of the 'case', the intervenors gave a forecast of the results of a major accident following a melt-down of the fuel, which was far more devastating than the forecasts given by the Establishment experts. David Comey said that 'everyone down-wind for 200 miles' from a disaster would 'receive a lethal dose' of radioactive nuclides, and many others up to 1000 miles down-wind would suffer a similar fate; Gofman more conservatively envisaged 50,000 deaths immediately, and many others later. The Californian study, on the other hand, assessed the outcome of the very worst 'one-in-ten-million' accident risk as 5000 deaths spread over many (twenty or thirty) years.

The intervenors assumed that half the maximum of radioactive nuclides in a large reactor would be released in weather conditions which would prevent rapid dissipation of the releases in the upper atmosphere. There is a ten to one chance against such conditions occurring. The Establishment calculations took into account this and the sophisticated steel and concrete containments which had been developed, with pressure suppression systems and multiple barriers, capable of resisting the highest pressures expected, so that a great part, if not all, of the heat generated would be absorbed. It was the intention in licensing that containment should wholly contain the results of a fuel melt: the Establishment did not, however, assume complete success. They took into account the probable effect of measures to lessen the impact of a disaster by moving people temporarily out of the danger area.

The Rasmussen Study gives a similar maximum figure for the most serious accident – 3200 immediate deaths and 3200 deaths from latent cancer spread over twenty years, and 90 per cent within seventy miles of the accident. Only two out of 100,000 core-melts would have this effect – the annual chance per reactor was 1 in a billion, or 1 in 10 millions for 100 reactors. The annual chance of a core-melt for a single reactor was put at 1 in 20,000. Many core-melts would have virtually no public consequence at all: the worst accident would require extremely improbable coincidences. The scare figures were dismissed as based on assumptions unrelated either to nuclear plants or weather probabilities or geography.[42]

### (vii) THE LAYMAN IN JUDGEMENT

The layman faced with this kind of discussion cannot assess the technical value of the data, arguments and judgements. If a designer or a licensing authority says that a plant is safe in spite of some uncertainties, and that the degree of conservatism is adequate, the layman cannot dismiss this on technical grounds. If designers say computer codes are adequate for a purpose and a licensing authority demurs, the layman cannot choose between them on technical grounds. When results of experiments are presented, the layman cannot say whether the evidence presented is all that is available or specifically relevant, how conclusive it is and whether it is presented in such a way that all the circumstances which should influence its interpretation are known. These limitations are not invariably recognised by politicians, administrators, lawyers, journalists, telecasters and others who are addicted to passing judgements or find themselves in positions where they seem committed to doing so – as politicians often are as a result of their collective folly (or is it ours?) and lawyers by their briefs.

All the layman can do is to compare the experts. He can observe their formal eminence in their professions. He can try to establish the closeness and recentness of their familiarity with the engineering and related sciences concerned. He can try and find out, which is usually still more difficult, their record as judgement-makers – which in the present context involves their record in the cumulative judgements on which success in engineering development depends. When there are disputes between experts who all seem to have some claims to respect, he may consider why they do not persuade each other. He may try to compare their qualities, as revealed in the ways in which they handle evidence in the small part of the enormous technical literature and discussion which he may find time to look at and which is in a form which he can understand. Most of the technical literature is of course *not* in such a form: it assumes basic technologies and relies on a mass of cross-references to similar literature. Where results are based on computer calculations, the layman will not understand the programme or know whether the inputs are sufficiently accurate or comprehensive for the weight given to the conclusions. Finally the layman can examine whether there are factors which may bias some experts in favour of one or other conclusion.

This final basis of comparison was of key importance in this instance: because the intervenors were asking the layman to dismiss the judgements of virtually all the leading engineers and scientists closely familiar with the contemporary development of the LWR who said they believed, and acted as though they believed, the LWRs were safe. This included the engineers and scientists in the firms who designed, manufactured and constructed the LWRs, those in the utilities who bought them and in the firms of architect-engineers and consultants, and those concerned in the AEC with licensing them (including the independent outside scientists in the ACRS upon whom the AEC partly relied in this work) and with organising safety research. Some able scientists took part in the intervenors' campaigns, but none appeared to have close participation in and detailed knowledge of recent LWR development. This was tacitly admitted by the intervenors who complained that they lacked resources wherewith to obtain the services of appropriate experts and sought support from public funds to do this – a proposal which Edward Kennedy, among other politicians, was prepared to support.[43]

It could not be claimed that the intervenors presented their case more temperately or with better balance than the Establishment: their presentation was, and was meant to be, tendentious, effective not only by what it included but by what it left out, by its lack of perspective and lack of balance.[44]

This was justified on the grounds that the Atomic Establishment engineers and scientists were all heavily biased by their long personal involvements in nuclear development. They had all been imbued with enthusiasm for the glamorous new technology which was to provide a new source of energy just in time to satisfy otherwise insatiable demands – whose price elasticity was certainly underestimated. They had persuaded Congress and the Executive, plant manufacturers and utilities, to make enormous and progressively increasing investments in the LWR over nearly twenty years on the understanding that all difficulties in making it safe and economic could be overcome. University scientists and administrators had been drawn into the net by research contracts and jobs as advisers; their careers and reputations were thus also committed, and the consultant nuclear engineers of course also thrived best when business was booming. It was the Lilienthal theme writ larger.

By their long commitment it was argued the Establishment experts had all, or almost all – all save those who agreed with the intervenors – become incapable of detached and objective judgements on the

LWR. The regulatory activity of the AEC was overshadowed by the commitment to development, so that no restraint which would make LWRs uncompetitive would be adopted. The JCAE had become protector and sponsor of the AEC. The Establishment were ready to design, manufacture, operate, approve and licence plants 'on speculation and the hope that an accident will never occur' – but were well aware there was a serious risk of one occurring; and all concealed information which was unfavourable to the LWR. The 'limited resource Intervenor litigant' thus faced a 'well orchestrated chorus' – in 'a David and Goliath confrontation'.[45]

This picture of a monolithic Atomic Establishment was mere fantasy. The scientists, engineers, administrators and managers in the so-called 'Establishment' did not present a common front on safety matters. The conflict between those who designed and manufactured plants and those who licensed was plain for all to see and within the ranks of the AEC and its advisers there were conflicts of opinion which the intervenors documented usefully. It was not really credible that all these experts, whose views were not uniform, were either innocently underestimating or deliberately understanding the dangers in the LWR plants as licensed, whether they were doing so because of zeal for nuclear development, or zeal for energy and 'economic progress', or for business profits or personal advancement and prestige, or to conceal past misjudgements.

This was not a matter on which their judgements could be casual – for many of the persons concerned, safety was their primary preoccupation, the plant manufacturers had large groups concerned with safety, and the whole regulatory organisation with its industrial advisers was concentrated on it. Although it was widely accepted that – to quote from a 'Tentative Staff Report' on the whole problem of regulatory organisation, drawn up by two lawyers for the Administration of the United States – 'The AEC's commitment to the development of nuclear energy is likely to lead it to be unwilling to impose requirements that will make nuclear energy non-competitive with other energy sources', the same report states only fifteen lines further down the same page: 'There is no reason to suppose that the procedure fails to provide a valid safety analysis or does not result in a reasonable assessment of risk. The AEC staff and the ACRS contain dedicated people well qualified to make the technical judgements.'[46] The authors of this report were not hardline environmentalists or intervenors, and while there is clearly some ambivalence in their approach (to which we return) it is clear they found no evidence of the regulators underestimating or understating the risk.

Intervenors asserted, as seen above, that the AEC suppressed information unfavourable to the LWR, but in instances quoted to justify this it was plain the AEC had circulated the data immediately on receiving it to all who in their view could react to it usefully; regulators, manufacturers, safety research laboratories, the AEC development division, the ACRS, and the JCAE; it was also available freely to anyone who asked for it. In the case of the experiment supposed to show that water from the ECCS would be excluded from the core, the regulators had increased conservatism in design and operation temporarily[47] on account of this, although it was believed by all competent to judge that the results were irrelevant, as in due course they were proved to be. The AEC stance was on the assumption that wider circulation of such data would be fruitless: the judgement was ultimately shown in the straightforward sense to have been right, and the sensational spreading of incorrect deductions by the intervenors, based upon results they were incompetent to assess, generated heat but not light.

It is seductive to argue that since manufacturers' designs were normally modified in the licensing process to increase safety, the manufacturers were at least ready deliberately to put forward designs which were not safe. This would destroy the intervenors' picture of a monolithic Establishment. But even the limited assertion would be erroneous. Designers and licensers were involved in making judgements of the degree of conservatism to include, and some of the relevant data, rates of heat transfer in the core, for example, were subjects of controversy. It was not surprising that on some matters – a small proportion of the whole – there were differences over judgements and uncertainty over the data. When the regulators required a larger margin of conservatism, or took a less favourable view of the data, their view naturally prevailed. It was possible for the original designers to overlook some relevant points – even the ACRS agreed that it did so at times.[48] It was also possible that the regulators had access to slightly more up-to-date experience in some respects than the designers (who in general would be *better* informed on the plants) because review was spread over a long time after the initial design. Moreover, new bases for conservatism were introduced from time to time, such as the requirement that a plant should be capable of withstanding the force of a supersonic aircraft crashing into it: a kind of decision likely to be made general and hence determined centrally. There were thus many reasons why licence requirements might be more stringent than the original design basis.

It is not to be supposed that when regulators required additional safeguards they were always right. There are many well-authenticated cases where they were proved wrong subsequently. A well-known case occurred in 1973 when, under intervenor pressure, the AEC derated a number of BWRs on the grounds that they might exceed the maximum permitted core temperature in a 'loss of coolant accident' (LOCA). Six months later the AEC agreed that the firms had shown that their contention that the temperature limit would not be exceeded was completely justified.[49] The leading firms in the United States believed that the requirements for licensing often added to margins of safety without adding significantly to safety. It has been seen above that this almost certainly happened in regard to gaseous effluents. It may well have happened in respect to other plant features. There were even matters on which manufacturers felt that requirements arising out of ACRS judgement might lessen rather than increase safety. For example, several years ago now, some engineers at General Electric thought it might be better to have three identical, though completely independent, systems for providing emergency core cooling rather than three different and independent systems, which the ACRS recommended. When the ACRS advised its advice was always taken – not necessarily the best way to use an advisory body – and in this particular instance the manufacturer's case was strong: probability tests could well show that one of three alternative methods was the safest. (Manufacturers sometimes complained, and did in this instance, that some members of the ACRS were too academic, but in this case the 'academic' quality – a good quality – was on the side of the manufacturers.)

The firms believed that it was undesirable to make plants 'safer than safe enough' – an unimpeachable principle provided you know what is enough. The man who wore braces and a belt was once seen as the archetypal pessimist: now that neither is needed a man who choses to wear two of each would rightly be regarded as 'a suitable case for treatment'. A Regulator will probably have an inclination to err on the safe side – to do so involves no economic penalty for him, although he will suffer with the rest of us from an excess of caution; but this is rarely identifiable or quantifiable. And the Regulators in the inter-venor period were clearly under great political pressure. People fall readily for phrases such as 'you can never be too careful' – as Bernard Shaw put it, those who always put safety first will never cross the road.

Despite conventional prejudices it is likely that when the firms said they believed their plants were safe they were right. Prudence was a major motivating factor in the activities of all those concerned in advancing the nuclear power development, if only because all recog-nised from the outset that 'if a major accident occurred it would jeopardise further nuclear development'. 'Self interest therefore as well as public interest' – as it was put in the letter transmitting the famous Brookhaven Report of 1957 'dictates avoidance of accidents'.[50] The thesis that none of the Establishment experts were able 'to make decisions free from pressure to compromise safety . . . to meet promo-tional goals'[51] was just unrealistic. Of course self-interest and public interest were not such separate forces as they are often presented as being – all the Atomic Establishment engineers, scientists, administra-tors and managers were also citizens, members of political, academic and social groups, and as such showed every sign of being a represen-tative cross-section of such groups, correspondingly responsive to common aspirations and fears, including the fear of an inadequately controlled use of atomic energy for civilian purposes. That some scientists 'who do not happen to be in the field of nuclear engineering'[52] declared nuclear plants dangerous and called themselves 'concerned' did not establish that scientists who believed the plants *were* safe were not equally 'concerned' – although they did not wear the label.

It looks paradoxical that the intervenors' case could be so weak technically and yet have a persuasive effect on many people, as it clearly did, creating uncertainty and a sense of alarm.[53] This seems, however, easy to explain. Atomic energy remained emotive. This was partly because of 'the violence with which it was introduced', leaving a legacy of guilt (with much 'self-flagellation') and of fear augmented by fallout before bomb testing in the atmosphere was banned; partly because even if there is no bomblike explosion, as all the experts agreed there could not be, radioactive nuclides which may induce cancer and leukaemia and cause genetic mutations would, if they were released in substantial quantities, present an insidious form of danger which cannot be seen and whose effects may be cumulative and de-layed. The intervenors' case was given much favourable, uncritical and unbalanced coverage in the press and on television. The sensational risks alleged and the alleged Establishment attempt to conceal them were highlighted: the probabilities were not seriously examined and the technical status of the intervenors and the 'Atomic Establishment' were not compared. From some accounts unwary readers might assume that all AEC scientists really believed the LWRs operating presented a serious risk of disaster.

The favourable treatment of the intervenors' case by the media was due partly to the skill and determination with which the case was pre-sented to them. But the political winds were also favourable – likely to encourage an uncritical reception. An environmentalist observer wrote that 'by summer 1969 it had become evident that the media were preparing to give the ecological crisis the saturation treatment accorded to the civil rights movement in the early sixties and the anti-Vietnam war protest after that.' The nuclear risks were only one of several grounds on which environmentalists opposed sites for power stations; they were concerned also with 'thermal pollution' through the increased discharge of condenser water and the aesthetics of sta-tions, cooling towers and transmission lines. The intervenors' nuclear case was congenial, moreover, to all who shared the fashionable atti-tudes of protest groups (the 'destructive reaction') – anti-establishment, anti-technology, anti-corporation, anti-elitism, anti-agencies, – the disciples of Herbert Marcuse, Kenneth Galbraith, Ralph Nader *et al.* All government regulatory agencies were under attack by 1970 as 'captives' of those whom they regulated and lawyers were muscling in on this premise to extend their political function.[54]

## (viii) THE IMPACT OF THE INTERVENORS

Whether the campaign had so far had any effect on LWR development when the ECCS hearings closed, other than delay from the long hearings and court cases is not clear. This was part of the experience which led to a recognition that the 'administrative arrangements' for dealing with the choice of sites for power plants 'were not adequate for the job'. But in the end the decisions on safety rested generally with the various organs of the AEC, and the extent to which the AEC was influenced by the case made by the intervenors is a matter for conjecture. There were instances where utilities made concessions to save time – agreeing to intervenors' demands though believing they did not add to safety. In total this probably did not add up to much, although there was always a possibility of a ratchet effect: if one utility gave way others might be forced to follow.

At the end of the long hearings on the ECCS the criteria were made more stringent in 1973 than the interim criteria of 1972. The changes were much less than the intervenors called for. Where, for example, the intervenors called for a reduction in the maximum temperature for fuel rods in a LOCA from 2300°F by 'several hundred degrees' – they aimed at below 1800°F – the new figure adopted was 2200°F. The intervenors had called for a derating of all plants by 40 per cent, but the new criteria adopted were expected to affect only a proportion of the plants operating, and the derating for these, or some of them, was put at 10 per cent or less – and only temporarily.[55] It can be said that the intervenors, by the presentation of their evidence, helped to persuade the AEC to make these and associated changes. But since all the evidence presented was already available, and came from within the AEC, from its research bodies and the regulatory staff itself, there is no reason to suppose it would not have had an impact but for the intervenors.

Alternatively, it can be argued that because of the intervenors it had too much effect – that the Regulators were led by the intervenors, and by the support given to them by the media and fears whipped up by misleading (even if sincere) advocacy, to demonstrate their care by being 'safer than safe enough'. This was certainly the view of some manufacturers.

Does 'this adversary process we have by way of public hearings in which non-technical people take part' (and over which, he might have added, a lawyer presides), Congressmen Craig Hosmer, of the JCAE, asked Hendrie, Chairman of the ACRS, in summer 1971, 'serve any real purpose in determining basic health and safety questions of a technical nature that may [arise] in licensing any reactor process?'[56] The Chairman of the ACRS and his predecessor in the same office – both treading delicately – said it was not the best way to solve engineering problems nor the best educational tool: but occasionally, he said, 'a thing will come up in the process of arguing it out in the hearing which sends us back to think about a problem again . . . ' They would not volunteer any instances. What came up in this way was not a matter of major importance in their view. 'There is some incremental gain. It may be small. However, the fact that some of these issues may be raised by the public is useful. There are items that can be discovered. They may be like trying to find one small grain in a very large bucket, but I think the possibility exists . . . ' Pushed by Hosmer, Hendrie finally said 'I am not convinced the adversary process is the best way to solve engineering problems or that it is the best educational tool . . . but in a broad discussion such as comes out there is some merit'.

As experience of such proceedings lengthened, scepticism about their value became wider and deeper. In the early stages some thought that, although public hearings were of dubious value and inefficient, in respect of individual sites they might be good for 'rule-making'. The distinction could hardly survive the ECCS rule-making proceedings.

Professor Harry P. Green, of the National Law Centre in the George Washington University, put his finger on the basic weakness in saying quite simply, in 1971, that 'the public has little if anything to contribute' on these technical matters. There was no clear-cut conflict of interests such as arose over the aesthetics of power plants and transmission lines, in regard to which a public hearing might serve a useful purpose and all parties certainly understood their own case. In regard to health and safety, all parties wanted them – the differences which existed were entirely over interpretation of extremely complex unfamiliar data, and in this area public *participation* was meaningless. Hearings on matters to which the public *can* contribute 'should proceed on the assumption that AEC will not permit operation of a plant unless [specified] standards [of safety *et cetera*] are clearly met'. As things were, intervenors themselves lacked the necessary knowledge, they could not afford good experts, 'they are forced to take very extreme positions, [and] forced to present witnesses far outside the Establishment in terms of their competency and experience in the area.'[57]

Green's worries were confirmed by lawyers who were involved in this kind of technical case. As one put it, he had for several years 'been struggling to have testimony reduced to a form where it is comprehensible to me, because if it is comprehensible to me it will be comprehensible to the average educated person.' But he had failed and 'despaired' of these problems ever being put in such a form that the average person could understand them or was 'going to want to take the time; because the amount of time involved is fantastic.'[58] There were lay intervenors who believed they could understand the technology and engineering sufficiently to make decisions – David Comey conspicuously made large claims – but the complete misjudgement of the significance of the 'semi-scale tests' was indicative and showed the claims were baseless.

In a study of the use of scientific and technological information by citizen groups in nuclear power cases made by two other members of the George Washington University and published in 1973, the part played by lawyers in the hearings was attacked from a different standpoint. The authors, who according to *Nucleonics Week* were populists, took the view that the 'consigning of citizens to helplessness before the steam roller of big government is more the rule than the exception'. They were thus not supporters of the Establishment. They described the *ASLB Hearings* not without reason as a 'charade'; the procedures 'designed by lawyers for lawyers, and ill-designed for problems of scientific or technical truth. . . . Arrogant and contentious behaviour by lawyers representing all parties unnecessarily polarised the hearings . . . and obfuscated the real issues.'[59] This is an understandable outcome of an adversary type procedure where opposing lawyers trying to do their best for clients, are also more than usually ignorant of the subject they are 'litigating'.

It gave point to the comment of T. J. Dignan, a prominent utility lawyer, also in autumn 1973, that 'radiological issues are too important to be left to the lawyers.' To which Green added later the advice: 'Close the public hearings and let the technical experts make technical decisions in the back room – like they do anyway.'[60]

Not all those who criticised the hearings would have wished to have all technical details settled in the back room. The fact remained that, however weak technically the case advanced by intervenors, there were many people – not a majority clearly, but a vocal group and not all irresponsible – who did not accept the assurances of the AEC that LWRs as licensed were *safe*. There was a credibility gap to bridge – Green thought this could be done by a 'full disclosure of benefits and risks . . . giving equal weight to optimism and pessimism'. But this formula was neither precise nor practicable: the only way to weigh risks was by probability calculations – to give equal 'weights' to the most and least hopeful guesses would leave one at 'square one'. As Congressman Hosmer put it wittily: 'Give them a copy of the latest annual AEC report and a copy of *Perils of the Peaceful Atom* and you have them confused enough to make a decision.'[61] So while it was not inherent in the AEC structure nor demonstrable in its practice that it would sacrifice safety to development, some change in institutional arrangements seemed increasingly needed to give reassurance. Such a change was also counselled by the deepening of the energy crisis. More reassurance was needed, moreover, by what may be termed the third post-1968 phase of the safety problem dominated by worries over the handling of long-lived radioactive waste, the fear of sabotage, and renewed worries over proliferation. This and the administrative changes are dealt with briefly as part of the next section.

## NOTES AND REFERENCES

1. *McKinney Report, op. cit.,* p. 41.
2. *Reactor Safety Study, An Assessment of Accident Risks in US Commercial Power Plants,* Report Summary and Appendices. under the independent direction of Professor Norman C. Rasmussen of the Massachusetts Institute of Technology (Washington: US AEC, 1974) Summary, p. 1. Hereafter cited as the *Rasmussen Report.*
3. To take one example, a great volume of research work was undertaken on the influence of radioactivity on steel used in LWRs in pressure vessels. The behaviour of the steel used in PWRs in submarines was closely analysed and significant changes in specifications were made. Among other things, it was discovered that small quantities of trace elements, such as copper, might lead to slow embrittlement. A precise check on the effect of radioactivity on the steel actually used in each vessel is kept by placing bundles of test pieces of the metal at vulnerable points and by taking out at regular intervals, one from each bundle, for examination. Inspection of welds is exceptionally rigorous both when they are made and during service. Ultrasonic examination methods have been perfected for precise periodic inspections when the reactor is out of service, and significant advances have been made in the development of in-service monitoring by acoustic methods which would record any propagation of cracks should they occur during the operation of the plant.
4. A. E. Schubert, Vice-President and General Manager of the Nuclear Energy Division of GE, in JCAE, *Environmental,* 1970, pt 2, vol. I, p. 1690. He was speaking of the plants built both by GE and 'our competitors'.
5. JCAE, *Licensing, 1971,* pt 1, p. 128. The chairman in 1971 was Dr Spencer H. Bush, Senior Staff Consultant, Batelle Memorial Institute, Pacific Northwest Laboratory, a leading metallurgist.
6. For example, JCAE, *Authorizing,* 1972, pt 2, p. 599 *et seq.*
7. The initial advance in methodology appears to have come from F. R. Farmer of the AEA in the United Kingdom.
8. *The Safety of Nuclear Power Reactors (Light Water Cooled) and Related Facilities,* Wash. 1250 (Washington: AEC, 1973) 6.44.
9. *Public Health Risks of Thermal Power Stations,* Report prepared for the Resources Agency of California by the UCLA School of Engineering and Applied Science, principal investigators: Professor Chauncey Starr and Professor M. A. Greenfield, (Los Angeles: School of Engineering and Applied Science, UCLA, 1972) p. 3.
10. It stated that the pollutants in quantities permitted by law in the gaseous effluents from oil-fired power stations (sulphur dioxide, nitrogen dioxide and particulates being the most conspicuous) would cause sixty times more respiratory deaths than the number of cancer deaths which nuclear plant effluents at the maximum permitted level would cause. The nuclear figure

(maximum one in a year) would be negligible. Risks from accidents would be considerably less for both types of plant; oil-fired plants would have many more accidents, nuclear plants might on a one in 100m chance have a worse accident. *Ibid.*, pp. 1–3 and 21–33.

11. *Rasmussen Report* (Report), pp. 16–20, 27–8.

12. *Ibid.* (Summary), pp. 20–1. One figure is from the revision made in the October 1975 edition of the *Executive Summary*, p. 10.

13. David E. Lilienthal, *Change, Hope and the Bomb* (Princeton: Princeton University Press, 1963) p. 37.

14. JCAE, *Environment, 1969*, pt 1, pp. 92–5. An AEC note said 'Doctor Teller has not insisted it is necessary to locate reactors deep underground ... He has consistently maintained that reactors built to today's standards are well designed, safe within present locations, and that reactors should continue to be built to prevent the spread of air pollution by fossil-fuelled power stations.' He had made the position clear at a New York/New Jersey Air Pollution Conference in 1967: 'The prospect of building these particularly safe reactors should not interfere with "plans" to build more reactors according to the present good practices.'

15. *Nucleonics Week*, New York, 13 March 1975.

16. *Hearings on HR 13731 and HR 13732 to Amend Atomic Energy Act of 1954 Regarding the Licencing of Nuclear facilities* (Washington: US Government Printing Office, for the JCAE, 1972) p. 386. Hereafter cited as JCAE, *To Amend Atomic Energy Act, 1972.*

17. *Rasmussen Report* (Study), pp. 185–7.

18. AEC, *Civilian Nuclear Power*, 1962, *op. cit.*

19. Lilienthal, *op. cit.*, p. 98.

20. Glenn Seaborg, Robert Oppenheimer and Philip Sporn were among those who in 1945–46 expected progress to be slow and difficult. In 1967 Seaborg thought he had been too pessimistic. Glenn T. Seaborg, *Nuclear Milestones* (Washington: AEC, 1971) vol. I, p. 103.

21. Lilienthal, *op. cit.*, pp. 21–2. 'The upsetting of experts frequently warms our hearts ... it confirms our innate sense of our own worth, of our power to create and shape events, of our existence as diverse sentient human beings whose future cannot be measured by the scientist and the engineer or laid down in graphs by the sociologist or economist.'

22. The US National Council on Radiation Protection and Measurement (NCRP) and the International Commission on Radiological Protection (ICRP). JCAE, *Environmental*, 1969, pt 1, pp. 158–63.

23. JCAE, *Environmental*, 1970, pt 2, vol. I, pp. 1223–57, contains descriptions of the structure, composition and activities of the NRCP and the ICRP. The Federal Radiation Council is discussed in *ibid.*, pt 1, p. 145 *et seq.* There was also an Advisory Committee on Biological Effects of Ionising Radiation within the National Academy of Science. *Nucleonics Week*, New York, 30 November 1972.

24. JCAE, *Environmental*, 1969, pt1, p. 646 *et seq.*, and pt 2, p. 1395. *Nucleonics Week*, New York, 4 March 1971.

25. JCAE, *Environmental*, 1969, pt 1, p. 687. H. J. Dunster, now Deputy Director of the UK Health and Safety Executive, described Gofman's linear theory as saying that if a hurricane kills 1000 people, a breeze will kill ten people. *Nucleonics Week*, New York, 27 August 1970.

26. There are several expert criticisms in *ibid.*, both in Part 1 and Part 2, for example, Lawriston Taylor in Part 2, p. 1220 *et seq.*, and an interesting panel discussion between Gofman and four leading experts, p. 1403 *et seq.*

27. *Ibid.*, pt 2, p. 1696, and *passim.*

28. See discussion by Raymond Moore, Assistant Commissioner of the Environment Control Administration, *idem*, pt 1, p. 316, and *passim.*

29. JCAE, *Environmental*, 1970, pt 2, vol. I, pp. 1394–5.

30. There was also a problem that if effluents were cut beyond a certain point the radioactive material kept within the plant could cause problems.

31. *US AEC Annual Report to Congress (1971)* (Washington: AEC, 1972) p. 49.

32. *Nucleonics Week* (18 October 1973) reported that Thomas Cochrane of the National Resources Defence Council, of which Arthur Tamplin was a leader, contacted AEC employees on the LMFBR programme, asking that every report, document and memorandum that contained information re-lꞏᵗing to the programme should become part of the published record. They asked architect-engineers' staff to supply as much information as possible.

33. All three gave evidence to the JCAE and other Congressional Committees.

34. 'Significant event' statistics were given for 1970, 1971, 1972 in the AEC's *The Safety of Nuclear Power Reactors*, 1973, *op. cit.*, ch. 2, p. 15, and further statistics on the sources and types of incident in ch. 8, pp. 22–3.

35. *Ibid.*, ch. 2, p. 14.

36. *The Times*, 30 May, 1974, carried a report of the 1973 statistics which were of the same pattern – the number of incidents was larger but so was the number of plants – and called the report 'disturbing'. With an interesting lack of objectivity the reporter failed to state that the US AEC thought such statistics reassuring.

37. A full recital is included in a paper by Anthony Z. Roisman in the House of Representatives Sub-Committee on Fisheries and Wildlife Conservation, *Interim Nuclear Licensing* (Washington: US Government Printing Office, for the Committee on Merchant Marine and Fisheries, 1972) pp. 232–6.

38. JCAE, *Licensing, 1971*, p. 112 (witness: Bush), p. 115 (witness: Hendrie). J. M. Hendrie was Head of the Engineering Division at the Brookhaven National Laboratory.

39. I am quoting from correspondence: I also discussed the problem with Milton Shaw, Merrill Whitman and others of his staff. Milton Shaw made his confidence clear regularly to the JCAE. For example, *ibid.*, pt 1, pp. 13–20, and JCAE, *Authorizing*, 1973, pt 2, pp. 1216–17.

40. The environmentalist argument on the ECCS was set out fully by Myron Cherry in House of Representatives Sub-committee, *Interim Nuclear Licence*, 1972, *op. cit.*, pp. 206–9, and the contribution about the semi-scale experiment in JCAE, *Licensing, 1971*, pt 1, pp. 406–12.

41. Milton Shaw explained to the JCAE the irrelevance of the experiment in June 1971 and referred specifically to its size and inappropriate water flow layout (JCAE, *Licensing, 1971*, pt 1, pp. 13–20). He was then in charge of the Safety Research. The report of the repeat of the experiment which confirmed its irrelevance was issued by his successor H. J. C. Kouts, very much later (*Nucleonics Week*, New York, 18 October 1973). It was possible with much diligence to find the size of the vessel in very small print in Cherry's document quoted above; he quoted Shaw in an interview stating that the nine inch vessel experiment was irrelevant. Cherry treated this and other

similar comments as just dismissing unwelcome news. The subsequent re-working of the experiment showed him to have been himself the one who dismissed unwelcome evidence that was true (JCAE, *Licensing, 1971*, pt 1, pp. 408–9). Other early reactions (including Westinghouse's) to the inter-venor claims were given in *Nucleonics Week*, New York, 6 and 13 May, 1971.

42. *Rasmussen Report*, (Report), pp. 152–8. The definitive report became avail-able in October 1975; the slightly revised figures in this have been used in the text. *Executive Summary*, p. 10; *Report*, pp. 84–5.

43. *Neucleonics Week*, New York, 9 May 1974. Edward Kennedy was reported as proposing public financing of intervenors. Antony Roisman, it was said, 'has his ear'. 'Tony's one of the guys we've been talking to.' Myron Cherry advanced the idea in summer 1971 that intervenors should be provided with access to scientific expertise and financial resources, the agency organising an inquiry thereby gaining in public confidence. JCAE, *Licensing, 1971*, pt 1, p. 393. The idea had some interesting support, cf. *ibid.*, pt 3, p. 1403.

44. The publicity methods were set out in a fascinating paper on 'Multi-media Confrontation' included in *ibid.*, pt 3, pp. 1402–11.

45. *Ibid., loc. cit.*

46. *Ibid.*, pt 2. 555. The whole document is illuminating, giving possibly the best survey of the administrative organisation.

47. *Ibid.*, pt 1, pp. 12–20, pt 2, pp. 532–3.

48. *Ibid.*, pt 1, p. 122.

49. *Nucleonics Week*, New York, 17 January 1974.

50. JCAE, *To Amend the Atomic Energy Act*, 1972, pt 1, p. 3871.

51. Jonathan Bingham, of the House of Representatives, in JCAE, *Licensing, 1971*, pt 1, p. 324.

52. See the evidence of Mrs Carl for the Lloyd Harbour Study Group which opposed the licence for the Shoreham (Long Island) plant: 'The public is no longer an amorphous mass of obstructionists for obstruction's sake. [It] includes many eminent scientists who do not happen to be in the field of nuclear engineering, but who nevertheless have expertise and philosophies which are germane to the wise deployment of nuclear facilities in areas heavily populated...' JCAE, *Licensing, 1971*, pt 1, 298. She argued that the Idaho semi-scale experiment (above, p. 67) strongly supported these eminent scientists: an interesting submission in the light of events.

53. Two academic investigators, populist in sympathy, wrote in 1973 that while it 'would be incorrect to characterise active intervenor groups as representa-tive of the general population, or even the popular will, it would be equally incorrect to dismiss their contentions as merely self-serving... They are mainly intelligent and upstanding members of communities...' But the authors also wrote: 'We have found strong evidence among citizen groups of "know-nothing-ism", blind anti-technology and anti-government senti-ments, pessimism and doom forecasting.' *Nucleonics Week*. New York, 27 September 1973.

54. See, for example, Leo Wolman, 'The Environment Past, Present and Future', in JCAE. *Licensing, 1971*, pt 4, especially pp. 1638–42, and Professor L. Jaffé, *ibid.*, pp. 1634–9.

55. The official view was that it might initially result in a 5 per cent average derating.

56. JCAE, *Licensing*, *1971*, pt 1, p. 121. The text as printed reads 'err' where I have put in brackets 'arise'. I take it 'err' was an error in transcription.

57. *Ibid.*, pp. 341, 343, 356.

58. *Ibid.*, pp. 354–1. Evidence of Arvin E. Upton.

59. The authors were Stephen Ebben and Raphael Kasper; the quotations are taken from *Nucleonics Week*, New York, 27 September 1973.

60. *Ibid.*, 4 October 1973 (at a joint meeting of the American Law Institute and the American Bar Association).

61. JCAE, *Licensing*, *1971*, pt 1, p. 356.

# CHAPTER 6

# Peak and Collapse 1974–76

## (i) THE PEAK

Orders for LWRs in the United States reached a new peak in the first three-quarters of 1974. This reflected mainly the first impact of the Arab oil embargo and its effect on fossil-fuel prices. Other factors helped. The LWR was almost universally accepted as safe. Rasmussen, as it were, clinched this. Muntzing's promise of a six-year or less lead time within three years[1] was widely derided, but a reversal of the trend to longer lead times seemed possible. Plant makers had responded to the invitation to submit designs of standard plants. These would be subjected to a more than usually rigorous scrutiny, but when the regulators were satisfied with them they could be regarded by the utilities ordering them as reviewed for safety purposes – this seemed to offer significant savings. The scheme emphasised that plant development had reached greater maturity, and the introduction in summer 1972 of the BWR/6 by GE was symptomatic. Based on evolutionary modifications of components and systems, the central feature being a new core which gave a higher average with lower peak temperature and allowed a 20 per cent increase of output from a given size of pressure vessel with an additional safety margin, this was to provide the basis of a standardised plant which should take six months less to build, and offered a potential reduction in real costs of up to 20 per cent.[2] The other manufacturers followed with parallel changes – all modified cores similarly and made additional changes.[3] It latched on to the regulators' standardising drive. The savings through 'learning', which the AEC foresaw in 1968, seemed in sight.

## (ii) THE COLLAPSE

### (a) 'Marking Time'

The scene changed abruptly at the end of 1974. Instead of a spate of orders there was a trickle (see Figure 3.1), and many orders already placed were deferred and a few cancelled. Orders for fossil-fuel plants were deferred, too, but nuclear plants took the hardest knock – of 95,000 MW of orders deferred in 1974–75, 66,700 MW were nuclear.

The turnaround stemmed from the fall in electricity consumption in 1974 and its slow rise in 1975, and from the acute financial difficulties of the utilities, who would not need new plants so quickly and could not pay for them. But, as the Atomic Industrial Forum put it, the severe setback which they described tactfully as 'marking time' had other aspects; it was not only a period of 'consolidation, a corporate euphemism for rotten business', but was characterised by 'dislocation, confrontation and escalation'.[4]

Dislocation was caused when the AEC was replaced by two bodies, the Nuclear Regulatory Commission (NRC) and the Energy Research and Development Administration (ERDA). This resulted in discontinuity and delay as new people took over top policy decisions. The change brought the split between regulation and promotion which seemed desirable in principle and was to be a source of greater confidence in regulators' decisions. 'Confrontation' was an intensification of anti-nuclear agitation sufficient to increase uncertainty about the future – the antis might win. 'Escalation' meant that the rise in construction costs was assumed to continue unabated, the lead time was taken as nine years: a plant ordered in 1975 would cost $615 per KW plus $253 escalation. A year later the figures were $695 plus $352 (above, Table 4.1).

The setback posed four questions. Were LWR plants still competitive? Was it still reasonable to expect cost reductions through 'learning' and changes in licensing procedures? How did the balance of evidence for the safety of LWR plants change? What were the 'omens' for the future of the anti-nuclear movement?

*(b) Competitiveness*
Though a few utilities attributed the cancelling or deferring of orders for LWRs to high costs and the delays, this was not prominent. Most reached the same conclusions as the architect-engineers and manufacturers, quoted earlier, that LWR plants would in most locations, despite the great rise in their capital and fuel costs, produce power more cheaply in the 1980s than the best fossil-fuel plants. An Edison Electric Institute survey was quoted in March 1976 to the effect that utility cost forecasts envisaged an average advantage of 30 per cent for LWR over coal-fuelled plants 'through 1990'.[5] Not all utilities put the margin so high. Nor would it be uniform for all sites. But almost all made it positive. There were uncertainties, especially over fuel costs. Construction costs of coal-fuelled plants had risen as steeply as LWR plant costs – perhaps more, on account of the need to have equipment to remove sulphur. Nuclear fuel costs were rocketing and in important

respects uncertain. But while several cost components (for example $U_3O_8$ and enrichment) might rise by the early 1980s to four or five times the level assumed in estimates made in the late 1960s, the coal price index had quadrupled between 1967 and 1975, and was likely to go on rising in line with wages. Reprocessing, organised by competitive groups, remained the most speculative area. No plant to reprocess LWR fuel had been operating since 1972. The first plant had been closed while new AEC safety requirements were being met, the second was based on a process which would not work, and completion of the third plant was held up because the NRC was formulating and changing its policy on plutonium and waste treatment. Spent fuel was being stored. But the cost forecasts provided generously for reprocessing and waste handling costs – and still the sums showed the LWR competitive.

### (c) *Prospects for Lower Real Costs*

Munzing had said in 1973 that licensing times could be dramatically reduced in the following two or three years, but it did not happen. The arrival of the NRC was followed by a fall in the issue of operating licences from fifteen in 1974 to three in 1975, and of construction licences from twenty-three to nine. (It was not all necessarily due to the NRC.) Congress did not legislate in favour of licensing preselected sites or major changes in public hearing arrangements. It was satisfactory that at length – over a year later than was hoped – preliminary design approvals (PDA) were given to standard nuclear designs by GE in December 1975, and by Combustion Engineering and Westinghouse in January 1976. The approval extended to the Nuclear Steam Supply System (NSSS) only. GE estimated that it would save utilities forty to fifty-five man-years and \$2.5m for preparing the first safety analysis; and more could be saved when it extended to the balance of plant. But it was not all plain sailing yet: while some NRC officials had granted PDA to Westinghouse and Combustion Engineering for standard designs of PWRs, others seemingly with no intercommunication with their colleagues drew up a new standard for one component (to deal with 'Anticipated Transients without Trip' [ATWT] in PWRs) which was in conflict with the PDA design.[6] This implied there could still be delays and back fitting.

The most solid promise of early reduction in real costs through 'learning' came from the efforts of manufacturers to increase the availability of the plants. Generating cost estimates for nuclear and fossil-fuel plants had been based on load factors which had not been achieved. Average availability of nuclear plants was better than the load factor – an indication that utilities did not succeed in running their nuclear plants exclusively as base load plants. (The average load factor in 1975 was 66.5 per cent, the availability 72.5 per cent.) But 'forced outage' statistics, which fell from 16.9 per cent in 1974 to 15.1

in 1975 showed there was plenty of scope for improved reliability.[7] Standardisation of designs plus the large flow of orders – from January 1972 to September 1974 GE had thirty-one and Westinghouse thirty domestic orders[8] – provided a strong basis for greater reliability and the firms set out to achieve it. Westinghouse's 'emphasis on plant standardisation and reliability and on productivity in manufacturing' was given wide publicity;[9] GE between autumn 1974 and summer 1975 made a vast *Nuclear Reactor Study* (known as the Reed Report) for internal use to 'achieve a superior availability factor', whose existence became known because three GE engineers who joined the anti-nuclear campaign and resigned referred to and quoted from it.[10] It could be assumed these efforts would in fact improve availability and reduce generating costs. Utilities welcomed the prospect, but as an effort to fulfil promises rather than a road to cost reduction.

*(d) Safety and the Anti-nuclear Campaign*
The period summer 1974 to summer 1976 opened with the publication of the Rasmussen Report and ended confusingly with the defeat on the one hand of the referendum on 'Proposition 15' in California, which would have stopped nuclear expansion and the operation of nuclear plants in the State, and on the other hand a decision by the Court of Appeal for the District of Columbia against the NRC whose effect in holding up licensing could be on a par with that of the Calvert Cliffs judgment. The Californian vote gave an indication that the majority of people favoured the nuclear programme: the court judgment showed how irrelevant a majority might be. In between these events, the dramatic highlights were a disastrous fire in TVA's Brown's Ferry plant in 1975, the resignation in February 1976 of three GE senior nuclear engineers and of an NRC official engaged in the supervision of a Westinghouse plant near New York (Indian Point II) and results of the first two experiments in the Loss of Fluid Test Facility (LOFT) programme.

*LWR safety*
In the wake of the Rasmussen Report, 'reactor safety as a nuclear opposition issue' appeared 'dead or dying'. Muntzing, on retiring as the NRC took over regulation, expressed a fear that the public was 'oversold' on safety: 'There will be an accident one day – its consequences will be limited, but the public is not prepared for it.' The national debate on reactor safety, he knew, would be succeeded by other nuclear controversies – this was, *Nucleonics Week* pointed out, already happening. A 'strategic switch' was being made by the opposition to the plutonium fuel cycle. The general public got 'baffled' and 'bored' by the ECCS, fuel densification, risk benefit analyses, maximum

credible accidents and probability mathematics, but was fascinated by disaster stories, lurid accounts of terrorists stealing plutonium and making bombs.[12]

Many people remained 'oversold' in Muntzing's sense throughout these two years – they treated Rasmussen as gospel, complained one expert. There were critics of the method and of specific results. The EPA said Rasmussen underestimated the health hazards, other critics said he gave too little weight to risks of 'common mode accidents' and risks of human error which he denied. The report had been made for the AEC, whose successors had to establish themselves as less indulgent to nuclear developers; the ERDA witness said they were 'cautiously optimistic about the potential usefulness of the method', which they were applying to the Liquid Metal Fast Breeder Reactor (LMFBR) and NRC's chairman in quoting the report showed he accepted it.[12] An experienced observer[13] quoted the assessment that the report was a 'useful but not definitive input' as representative. The report could not be 'the last word', if that is what 'definitive' meant, because too much was changing quickly – for the most part increasing safety.

### Brown's Ferry Fire

Some popular alarm was revived during the period by a bizarre accident: the fire at TVA's Brown's Ferry plant in Alabama in March 1975. It originated in part of the plant designed by TVA which acted as its own architect-engineer and constructor, not by GE, in the cable-spreading room below the control room, at the point where cables pass through the wall into the reactor building. The penetration through which they pass must be air-tight; the material used to stop up holes, polyurethane, was readily combustible, and the test used to prove the penetrations were airtight was to put a lighted candle by them and observe whether the flame was drawn to the hole. This started the fire. It had started small fires before. This fire burned much longer in the reactor building than was necessary because, through misunderstanding, water was not used to put it out; cable insulations burned, the cables shorted and, among other results, the fire burned out cable controls of the ECCS. This gave the alarmists a field day. Reflecting their reactions, *The Times* in London headed its account 'One Candle Threatened Nuclear Devastation', and claimed that nuclear power experts in Britain were appalled that 'one man with a candle had brought the world's biggest atomic power station to the edge of devastation.'[14] The experts were not named; the assertion was a fantasy, duly repeated in the Flowers Report.[15] Naturally the anti-nuclear lobby in the United States put it this way.[16] It emerged with certainty that alternative methods of shutting down the reactors and maintaining water levels were available and operated: Rasmussen remarked the redundancy proved greater than expected. No sudden loss of coolant required the ECCS, and no exceptional release of radioactivity occurred. The NRC special review group concluded that if a similar fire occurred in another existing nuclear plant, with a repetition of human error, the most probable outcome would not involve even small adverse effects for public health and safety.[17]

The fire was destructive, and costly for TVA. Brown's Ferry was providing TVA with its cheapest thermal power and had a promising operating record. The fire would have been more disturbing nationally had demand for electricity been increasing normally. It showed up odd practices, the Special Review pointed out: the use of combustible material 'not included in the design' in the penetrations, the use of the candle ('an unnecessary ignition source'), possibly inadequate separation of cables. Staff handling of the fire was criticised.[18] The NRC took steps immediately to deal with these. But the fire gave no grounds for changing the former assessment, that standards of design, manufacture, construction and regulation provided adequate conservatism and redundancy to ensure safety, even in face of human error, of which, it was emphasised, the probability figures and regulatory procedures took account.[19]

### Resignations at GE and NRC

The resignation of three middle-management engineers from the GE nuclear centre at San José and simultaneously of an NRC officer at Indian Point II, raised different issues. Were the resignations a serious breach in the near unanimity of experienced engineers and scientists close to the design, manufacture, operation and regulation of nuclear plants on which the layman has to rely as the guide to safety? There had been resignations before – one AEC regulatory official had resigned in the ECCS period, and Dr Sternglass, prominent exponent of the dangers of nuclear radiation in the Gofman–Tamplin time and tradition, had been in Westinghouse. These had clearly not constituted serious breaches in our sense.

All four who resigned said the plants were dangerous – Indian Point II, according to the NRC officer, was 'an accident waiting to happen', and the GE engineers said the cumulative effect of design, constructional and operational deficiences, with inevitable human error, 'make a nuclear power plant accident a certain event.' This was what Rasmussen and all the experts said, but Rasmussen said how often accidents of varying severity were likely; the figures were reassuring. The GE engineers dismissed the probability exercise, and implied that it overlooked human error, which was not true.[20]

All who resigned, particularly the GE three, presented a case to the JCAE in considerable technical detail whose significance the layman could not judge, although it sounded impressive. Their document

relied almost entirely on the Reed Report,[21] which, as the three engineers described it, was even more monumental than in GE's description. 'A task force of seventy or eighty of the most knowledgable
people that could be put together in the nuclear business' worked for
a whole year and their 'final report was overwhelming – a five foot
shelf full . . . an excellent in-depth study' which should be available,
the engineers said, to the NRC.[22] The three selected data from it which
appeared to support the case they presented, much as the intervenors
in the ECCS hearings based their case on excerpts from the vast mass
of AEC documents, which, when isolated, appeared to support their
case. The replies from GE and NRC showed that the engineers did not
present *all* the relevant data on which experts' judgements had to be
based. An English reader could detect an error when it was stated that
the United Kingdom Atomic Energy Authority had 'recommended
against the use of light water reactors' because 'the incredibility of a
pressure vessel failure cannot be proven'. This was fictional.[23] Almost
two years before the three engineers resigned the Deputy Chairman of
the AEA had reported the provisional judgement of a group of British
experts that the vessels as constructed in America were safe.[24]

No new source of risk was uncovered by any of those who resigned.
All the problems were being dealt with by manufacturers, utilities and
the NRC; none in the judgement of the NRC and ACRS (as well as of
GE and Westinghouse) was not adequately covered by existing wide
margins of safety; all were, in the view of the GE three, correctible
(though their means of correction were not necessarily adopted). 'Only
in broader areas outside their specialities, where they admitted they
lack experience', William C. Anders, Chairman of NRC, stated, 'did
these engineers express pessimism about society's ability to cope with
the issues.'[25]

The Reed Report was made available to the NRC, who 'did not
identify any new areas of safety concern' in it. 'It was evident',
Bernard Rusche, Director of the Office of Nuclear Regulation within
the NRC, said, that GE 'have honored their obligations to inform
NRC of all safety related information thus developed'; and 'all of
the matters mentioned are being considered in our current safety
review.'[26] The Chairman of the ACRS, Dade Moeller, told the JCAE
that neither he nor any of his colleagues thought anything the GE three
said warranted a moratorium on ordering or licensing plants.[27]

Thus the position, as it emerged in the JCAE examination of those
involved in February-March 1976, was that a very small number of
engineers who were qualified to have a view based on substantial
experience, disagreed with the judgement of the overwhelming
majority of engineers and scientists similarly qualified that the degrees
of conservatism in design, manufacture, construction and operation of
nuclear power plants were adequate to provide safety margins appro-

priate to the degrees of uncertainty. The majority included almost all the people involved in the manufacturing firms and utilities and architect-engineers, and the regulatory staff, all the members of the ACRS, and all the independent experts called in as consultants. There was no revelation of secrets which GE wanted to suppress or cover up. As in the ECCS proceedings, the data referred to by those who resigned was known to the regulators and being taken into account; and this was familiar to those who resigned and was referred to at length in their document.

The resignations were an attack on the NRC, for the same reasons as the AEC regulators had been attacked – an interesting twist, because the AEC regulators had been held to be compromised by the dual responsibilities of the AEC as promoter and regulator. The NRC was an independent executive agency whose function was solely regulatory: it was attacked, nevertheless, for allegedly allowing commercial to prevail over safety considerations, and for suppressing the true facts.

The conflict was brought out with particular clarity in statements concerning one of the main new safety issues, the integrity of the BWR Mark 1 containment. In developing a Mark III containment for the BWR/6 a model test had shown there were possible loads on earlier containments not known of hitherto. An elaborate programme was organised by utilities, GE and several independent consultants, with NRC staff participating, to determine what, if any, action this required in existing plants. According to the three engineers who resigned: 'The primary focus of the programme has been to prove the plants are safe enough for continued operation, not to openly assess their true safety.'[28] There could be no objection to running plants if they were 'safe enough' to run. The implication was that the aim was to take unreasonable risks for commercial reasons. If plants were found 'safe enough' this was a measure of 'true safety'. This was what GE and the NRC said in evidence was established by the first phase of the programme: the plants still had an adequate margin of safety to meet newly identified potential loads in the unlikely event of their being experienced. It was safe to run the plants subject to a change in operating procedures while steps were devised to restore the original margin.[29] The engineers stated: 'In no way would a competent structural consultant testify that the containments clearly met the requirements for integrity of the system.'[30]

William Anders described the charge of economic bias as 'simplistic'. The NRC was mandated to 'regulate in the national interest': its basic mission was to ensure public safety, but 'it is fully consistent with this ... to be mindful that *unnecessary* costs and *unwarranted* delays are not in the national interest. ... We reject the thesis that recognition of the impact of regulatory action on costs, schedules and

power needs makes impossible the unbiased decisions necessary to protect the public safety.' There were several familiar instances of derat-
ings, shut-downs and special inspection programmes – in the AEC
tradition – which underlined Anders' point that they were 'not overly
impressed by economic considerations'. The Commission was just
setting out on another review of existing plants to see whether further
'backfitting' was necessary.[31]

The near unanimity of the competent experts – designers, manufacturers, constructors, architect-engineers, utilities, consultants, regulators and their advisers – remained unimpaired. The four were not
joined by other qualified professionals. Their seventy-two page testimony and the responses it drew from GE, NRC and the ACRS showed
again impressively the immense effort given by all interested groups to
discover and measure risks and ensure safety. The question arises:
why should the three GE engineers, and the NRC official[32] have resigned? The GE engineers were greatly concerned with international
risks. They were worried by efforts to sell plants to Israel, Egypt and
South Africa, by India's atomic bomb, by the toxicity of plutonium as
well as by the bomb risk. This may have influenced their judgements
on plant safety. They were members of the Californian-based Creative
Initiative Foundation (CIF) which seemed from a distance a blend of
humanism and revivalism. One of the three agreed they all 'might have
become sensitised to hazards of nuclear plants during CIF classes and
group discussions'.[33] Sensitised perhaps to the point of exaggeration.
Errors, omissions and exaggerations certainly came into their testimony,[34] which subjective factors would help to explain.

### First LOFT results

Appropriately, soon after the engineers' testimony and the rebuttals, a
new phase opened in the ECCS saga. The LOFT facility at Idaho[35]
had been completed and results of the first two, in the long series of
experiments planned, became available. They came hot on the heels of
a statement by an NRC consultant[36] (whose colleagues disagreed with
him) that reliance on computer simulation of complex phenomena was
the weakest link in reactor safety, and he suggested that PWRs should
not be licensed because of the 'steam binding' possibility. The first
results of the LOFT programme suggested that the calculations –
resulting from a computer programme – of the speed at which emergency cooling water would reach the plenum of the core had a high
margin of conservatism. They had predicted six seconds; the experimental result was three seconds. This suggested 'that the hot down-
comer wall does not significantly impede the delivery of emergency
coolant to the core'.[37] There was a long programme of further experiments ahead – but this was an encouraging start for the designers of
the ECC systems.

(e) *Anti-nuclear Campaign: Ebb and Flow*

The four resignations had been concerted as part of the anti-nuclear campaign, and the San José resignations were a dramatic part of the campaign for Proposition 15, to be voted on in June. They were agreed on in December 1975, but timed for February, allowing intense preparation with cooperative media. As the focus of the campaign was on plutonium, high level wastes, risks of sabotage and atomic war, the 'boring' problems of reactor safety receded, but reactor closure and a moratorium on new building remained a chief object of agitation, with Nader still to the fore. By August 1975, Dixey Lee Ray forceful last Chairman of the AEC, had fears that unless the nuclear industry was organised effectively it might disappear; Nader groups were pressurising their House and Senate representatives so effectively that even Senator Pastore, Chairman of the JCAE, despite his earlier record, did nothing Dixey Lee Ray said, to push pro-nuclear bills because of letters from the Senator's Rhode Island constituents.[38] By August there were anti-nuclear bills in twenty-eight States and, in September, New York banned the movement of nuclear fuel through the streets.[39]

Nevertheless, according to a Harris opinion poll, 63 per cent of the people favoured building more nuclear plants, and 18 per cent were not sure. Harris found that politicians, polled at the same time, thought that only 42 per cent of the people would be in favour, and 38 per cent would be against. They might therefore act on the wrong hunch. Nuclear supporters began to organise. A nationwide group 150 strong – *Scientists and Engineers for Secure Energy* (SESE) – composed almost wholly of university professors, including five Nobel prize winners, was announced early in 1976, its purpose to set out the nuclear facts and clear the debate 'permeated with misconceptions and misunderstandings'.[40] Earlier a 'Labour Committee against Proposition 15' was formed in California, with the Executive Officer of the AFL/CIO Californian Labour Federation as its chairman. The anti-nuclear initiative was seen by the leaders of organised labour as a 'grave threat to our national security, to our precious energy supplies, and to this state's expanding employment.'[41] The protest front had been divided.

The result in California, 'a 2:1 majority in favour of nuclear energy', was in line with national polls, but it had not been taken for granted. It now seemed likely that confrontation could be contained, and that as the economy and consumption of electricity picked-up, utilities would order LWR plants again. Outstanding orders were of course still substantial, although spread over a longer time (forty-six plants [47,600 MW] were scheduled to come into operation in the five years 1976–1980); beyond that, a further ninety-four plants (110,100 MW) were on order, originally most for the five years 1981–85, but this had been changed by deferrals. In the November Elections other anti-nuclear initiatives were defeated.

Within two months of the Californian vote the United States Court of Appeals for the District of Columbia Circuit put another road block in the path of the licensing process.[42] It ruled that in giving a construction licence to the Midland plant the NRC 'failed to examine energy conservation as an alternative to a plant of this size', and in giving an operating licence to the Vermont Yankee plant 'had not given sufficient consideration to the issue of waste management and reprocessing to satisfy the requirements of the NEPA.' The implication of the first seemed that an agency appointed by the executive could impose forms of energy conservation of its own choosing on a locality, and that a court in the case of an appeal could in turn decide on the same issue, as though both were qualified and had a duty and right under the NEPA to do so. The implication of the second was that because, in the court's view, an AEC rule-making hearing in 1973 had not adequately probed the evidence of the AEC's Director of Waste Management, but had concluded nevertheless that the environmental impact of the fuel cycle in cost/benefit terms was 'insignificant', waste disposal should be a major topic for hearings on individual licence applications and subject to all the adversary type proceedings whose value had been effectively criticised in the JCAE.[43]

The court described the NRC actions as 'capricious and arbitrary'. Critics, worried at the growing American dependence on imported oil, might have applied the same epithets to the subjective judgments of the court. According to summaries, the judgment did not refer to the large volume of work on waste disposal by the AEC, which had been discussed at length by the JCAE and had convinced engineers and scientists familiar with it that the environmental impact could be insignificant, a view shared by engineers and scientists working on the problem in other countries.[44] The immediate practical impact of the court's judgment was that the NRC decided to have a new rule-making hearing on waste disposal and in the meantime suspended granting licences. Despite the court's ruling, the Ohio Edison Company restarted the buying of LWRs by ordering a twin 1200 MW plant from Babcock & Wilcox in August 1976.[45]

## NOTES AND REFERENCES

1. Above, p. 51.
2. The BWR/6 design of Nuclear Steam Supply System (NSSS) was accompanied by a new Mark III design of containment and a new design for the balance of plant (STRIDE – Standardised reactor island design) which was to eliminate all the variety which every architect-engineer chose to introduce. This was of particular importance for the BWR.
3. Westinghouse, for instance, introduced a new system for changing fuel faster, which would make it economically possible to change more often and use less highly enriched uranium.

4. Quoted from the American Industrial Forum's *Info*, Washington, in *Atomic Energy Clearing House*, Washington, 19 January 1976, hereafter cited as AECH.
5. *Nucleonics Week*, 15 April 1976. See also *ibid.*, 1 April and 3 June 1976.
6. *Ibid.*, 3 June 1976. The NRC had been represented on an American Nuclear Society (ANS) task force which had developed a standard for ATWT: the NRC (a different section presumably) without reference to the ANS' work, wanted a standard which would deal with situations in which several (ten or twelve) other failures might coincide with the trip failure. It was a characteristic escalation of caution.
7. NRC figures on availability, load factors, outage *et cetera* were given in *Nuclear News*, (Buyers' Guide), Hinsdale (Illinois), mid-February 1976, p. 37.
8. Based on lists in AEC, *The Nuclear Industry*, 1974, pp. 10–12. Combustion Engineering had nineteen orders, Babcock & Wilcox fourteen.
9. An article entitled 'How Westinghouse is Consolidating its International Lead', *Nuclear Engineering International*, London, December 1975, sets out their 'emphasis on plant standardisation and reliability and on productivity in manufacturing'.
10. *AECH*, Washington, 1 March 1976, p. 37 *et seq*. The GE 'highly technical' study (a twenty-one page executive summary, 205-page main report, and 713 pages of sub-reports) cast an interesting light on part of the firm's methods of maintaining efficiency. 'While each of our businesses is managed with a great deal of decentralised authority, we use a process of study and review through which the top management can obtain objective appraisals of our major business ventures by persons who are not involved in the day-to-day management of the individual business. Periodic studies conducted at corporate level exemplify the process. They are essential ... so that realistic plans and resource allocations can be made.' The nuclear study was made by 'nine of our most experienced scientists and engineers', of whom only two were from the nuclear division. They met eleven times for spells of two to three days at a time. This task force had sub-task forces, 'making in-depth studies of specific areas such as fuels, materials, mechanical systems, processes and chemistry'. The fact that such a study was made for the Chairman would not normally become public knowledge.
11. *Nucleonics Week*, 23 January 1975.
12. *Ibid.*, 17 June 1976.
13. *Nucleonics Week*'s representative at the sessions, 17 June 1976.
14. *The Times*, 31 March 1975.
15. Royal Commission on Environmental Pollution, *Nuclear Power and the Environment* (London: HMSO, 1976) p. 75. 'Very serious consequences were very narrowly avoided.'
16. For example, David Comey said in August 1975 that Brown's Ferry was in two hours of a melt-down. *Nucleonics Week*, 21 August 1975. They were still apparently putting it this way in April 1976. *Newsweek*, New York, 12 April 1976, p. 45.
17. *Report, NRC Special Review Group* on Brown's Ferry fire, 28 February 1976, Summary and Recommendations, p. 5.
18. For excessive preoccupation with the doctrine 'don't use water on electrical fires'. *Ibid.*, p. 5.

94      *Nuclear Power and the Energy Crisis*

19. The point was made vigorously both by Norman Rasmussen himself (*Nucleonics Week*, 8 May 1975) and by William Anders, at the time Chairman of NRC, in evidence to the JCAE. *Reactor Safety: A Discussion by Officials of the Nuclear Regulatory Commission* (Washington, US NRC, March 197 ), hereafter cited as NRC, *Reactor Safety, 1976*, pp. 10–11. See also p. 12: 'The Reactor Safety Study [Rasmussen] indicates that nuclear risk is very small. His study group found "no other electrical generating source has been studied so closely."' It implies complete acceptance.
20. *AECH*, Washington, 1 March 1976, p. 50. The text of the documents presented to the JCAE by those who resigned was published in the *AECH* of this date.
21. It went outside on a few points to embrace the PWR in its generalisations.
22. *AECH*, Washington, 1 March 1976, p. 64.
23. I checked this with the Authority.
24. *AECH*, Washington, 1 March 1976, p. 54 and above.
25. NRC, *Reactor Safety 1976*, Statement by W. Anders, p. 12.
26. *Ibid.*, Statement by Bernard Rusche, p. 2.
27. *Nucleonics Week*, 18 March 1976.
28. *AECH*, Washington, 1 March 1976, p. 63.
29. The safety problem only arose at all if the largest pipe in the primary coolant system was completely ruptured in a fraction of a second. The NRC stated in evidence that 'it is hard to visualise how a large pipe could break in that manner.' GE experts told the Joint Committee on Atomic Energy that 'if piping of this type ruptures it does so gradually, first leaking through a crack which, if it progresses, does so slowly'. Rasmussen put the probability of this accident as between one in 10,000 and one in 100,000 reactor-years. The NRC said 'our technical judgement is that existing Mark I containments would not fail to perform their intended function if they were exposed to the loads that would be associated with the postulated design-based accident. Being mindful of the extremely low probability that these loads would ever occur we have concluded that continued operation is acceptable.' The change in operating procedure introduced was a change in operating pressures in different parts of the plant. The NRC account is in NRC, *Reactor Safety 1976*, Statement by Bernard Rusche, pp. 4–7. GE's account fills in further details, see *AECH*, Washington, 1 March 1976, pp. 40–1.
30. *AECH*, Washington, 1 March 1976, p. 63.
31. *Nucleonics Week*, 6 May 1976.
32. I have concentrated on the GE engineers both to simplify (the issue of the balance of judgement is common to all the resignations) and because the GE resignations linked up with the Californian primary vote campaign.
33. An article in *Time Magazine*, Amsterdam, 16 February 1976, on the 'San José three' said that the CIF 'seeks to strengthen human relationships and instil in its members a reverence for life'; one who had dropped out said the 'members had a fervour about changing the world for the better'. The three planned their move in 1975 with James Burch, another CIF member who was president of Project Survival, which organised the Californian anti-nuclear vote.
34. For example, the statement about the AEA was wrong: in setting out the containment problem the document did not show the remoteness of the

risk nor did it state the categorical assurance of NRC and GE that there remained an adequate though reduced margin of safety. To give one other example, the testimony stated that vibration in fuel channels led to 'failures' in local power range monitors and a degrading of their signals. The vibration was a problem which needed to be dealt with, but GE said categorically that it did not affect either the monitors or the quality of their signals. *AECH*, Washington, 1 March 1976, pp. 43–4, 51–2.

35. Above, p. 66, and below pp. 238 and 259, n.81.
36. Keith Miller was at Berkeley and a consultant on the NRC advanced code review group. *Nucleonics Week*, 20 May 1976.
37. *Ibid.*, 20 May 1976.
38. *Ibid.*, 7 August 1975. Dixey Lee Ray had returned to a chair at Seattle University and joined the Americans for Energy Independence group.
39. *Ibid.*, 4 September 1975.
40. *Ibid.*, 3 June 1976. The Nobel Prize winners came from Columbia, Cornell, Princeton, Stanford and University of California, Los Angeles.
41. *Ibid.*, 22 April 1976. The Chairman was John R. Hennings.
42. *Nuclear News*, Hinsdale (Illinois), September 1976, pp. 29–31.
43. Above, pp. 74–7.
44. This is one of the topics to be discussed in a separate study.
45. *Nuclear News*, September 1976, p. 31.

# CHAPTER 7

# The LWR Outside the United States

## (i) THE FLOW OF ORDERS

By the early 1960s the LWR was the only reactor system with an export and licensing business. Magnox had not survived in the export market. The Canadian Candu had yet to secure its first orders in India (1967), Argentina (1973) and Korea (1975). In the latter part of Phase I, orders were placed in several countries (Belgium, France, Germany, India, Italy, Japan, Spain, Sweden, Switzerland) for LWR plants of from 240 to 400 MW. The order for Oyster Creek had been a turning point.[1] The years 1967–70 showed there was complete confidence in LWR performance. Orders were placed for much larger units, upwards of 600 MW in 1967 and 1150 MW in 1969, in the belief that they would provide cheaper power than any fossil-fuel plants. The standard of comparison was always oil, and the price of oil was still relatively low:[2] the decision on the competitiveness of the LWR was made, before the rise of 1971, and long before the price explosion of 1973.

The progressive rise in the cumulative volume of orders from 1969 to 1975 has been illustrated in Figure 3.2.[3] This only covers orders actually placed; forward plans announced by several countries showed more clearly the dependence on nuclear power envisaged. Germany in 1970 expected total additions to generating capacity by 1980 of 48,000 MW, of which LWRs would contribute 21,000 MW. France in 1970 envisaged LWRs providing 40 per cent of new generating capacity added by 1980; Sweden, 54 per cent; Japan, 30 per cent. By 1975 forward programmes placed still further reliance on LWRs; both Germany and Japan by 1973–74 expected to raise their LWR capacity to around 50,000 MW by 1985.

## (ii) COST COMPARISONS

The analysis of comparative costs of LWR and oil-fired plants in 1969–70 was set out elaborately in a report by a Consultative Committee appointed to advise the French Government on the programme for nuclear energy in the sixth plan. It excluded the AGR, without dis-

cussion, from its list of reactors which might be commercially successful in the short or middle term. The report underlined that, because
French experience with LWR design and construction was so slight,
the French nuclear industry would be behind those in the principal
centres of LWR development. This called for special encouragement
which would enable French manufacturers to supply equipment for
export as well as for the home market.[4] Initially the French had
adopted the $CO_2$/graphite (Magnox) type of reactor, a result largely of
the requirement, under President de Gaulle, that civil power reactors
should use natural uranium. The electricity supply utility (*Electricité
de France* [EDF]) had participated in a PWR plant on the Franco–
Belgian frontier, and clearly wished to move to the LWR in the sixties.
It hoped to gain experience of the BWR also by a joint venture, but
this had not materialised. The Committee's report regarded both LWR
types as equally attractive – the PWR was slightly cheaper to build
than the BWR, but with higher fuel costs. It was decided to start with
PWRs because this was the type with which they had practical experience.

By 1966 the Commissariat à l'Energie Atomique (CEA), the French
AEA, was known to think that if enriched uranium was available the
$CO_2$/graphite type of reactor was not attractive.[5] The French decision
to go the LWR route was announced in April 1968 – before the last
two British AGR plants were ordered.[6]

The German Government did not provide a systematic assessment
in a public report: it was not involved in central planning and decisions to choose reactors were made by individual utilities.[7] But officials
followed costs carefully because they were subsidising LWR reactor
development and they expressed a belief in 1970 that the largest PWR
plant being constructed in Germany, Biblis I, would have generating
costs 30 per cent lower than those in a new oil-fired plant using oil at
1970 prices. Cost prospects were less favourable for hard-coal plants
than for oil-fired plants (there were substantial subsidies in Germany
for hard-coal plants). In some areas plants using brown coal (lignite)
as a fuel could compete with the LWR – where the generating plant
was at the open-cast site, and provided that no provision was made in
the price for opening up other brown-coal deposits. This obviously
could not be the basis of expansion.

All the cost comparisons which favoured the LWR against oil in
France and Germany at this time would have had even greater force in
Japan.[8]

(iii) ORGANISATIONS COMPARED

French and German organisation and policy presented a remarkable
contrast – and provided very different routes to LWR success.

## (a) France

France was the only Western country on the Continent which made nuclear weapons. The CEA, a government body, was set up in 1945. It had no wartime experience, and had to build up the basic science and technology, initially to produce plutonium, largely for military purposes. Interest in producing power became lively after 1951 as American and British activity intensified. An EDF representative joined the CEA in 1951. Reactors designed to make plutonium were converted to make electricity also. Expenditure quickly became substantial. Much of the development remained for weapons. A PWR was developed for use in submarines; and enrichment and reprocessing had a weapons orientation. Firms were drawn in early to develop the technology industrially and build up an industrial infrastructure: but they were not initiating forces. As in Britain, by the 1960s the industrial achievement on the civil side – in developing competitive reactors – was lagging behind that in Germany, Japan and possibly Sweden, as well as the United States.

The French Government decided, therefore, to import American technology. Several PWRs were ordered from Framatome, a Westinghouse licensee in which Westinghouse had a 48 per cent holding. Later two BWRs were ordered, but cancelled because the market grew more slowly. CEA now sought to buy a large part of the Westinghouse holding: presumably the French Government thought enough Westinghouse know-how had been acquired and CEA wanted to push its own ideas of PWR design. Possibly the Government thought dependence on Westinghouse without GE competition a danger. It therefore adopted a policy of establishing national monopolies – in the fast breeder and turbo generators as well as the LWR. It remains to be seen how the industries will fare, and whether the shot in the arm was enough. In 1976 they were riding high and getting export orders.

## (b) West Germany

The West German organisation presented a stark contrast to the French. Germany was not permitted to have any nuclear development until the restoration of sovereignty in 1955 and a military nuclear development was excluded. Most European countries now, recognising the forthcoming importance of nuclear power, had set up some form of state institution 'to avoid' (as an Organisation for European Economic Co-operation [OEEC] spokesman put it) 'being left behind'.[9] Most had achieved very little. The importance which the Germans, in the Government and outside, attached to it was symbolised by the appointment of a Minister for Atomic Energy, the first in the world, a post filled, for only a brief spell, by the redoubtable Franz-Josef Strauss. The

Minister had only a small staff, and no executive agency like the AEA, AEC or CEA was set up to carry out the detail of government strategy and tactics. Instead there was a vast network of committees under the German Atomic Energy Advisory Committee, which had five sub-committees and innumerable sub-sub-committees, gathering together experts on all conceivable problems posed by atomic energy and representing the academic world, manufacturing industry, the electrical supply industry, and federal and state (*Länder*) governments.[10]

The object was to evolve a programme of federal support for what was the first task – catching up with the advances in both the basic science and the technology which had been made in other countries, above all in the United States and to a lesser extent Britain, to whom the Germans looked for help.[11]

Many groups in Germany had projects for which they sought support – for research institutes, for specific research studies, for particular reactor experimental development. The federal Government had no technology to pass on and no attachment to a technology. It was accustomed to subsidising private competitive enterprise industries, where there were special transitional circumstances – coal, steel, shipping and shipbuilding had benefited – and the nuclear industry was seen as a special-circumstance industry, where development costs were beyond the resources of manufacturers, utilities, universities and *Länder* governments. But the Federal Government was averse from central plans of the French and – in atomic energy – British kind. The State Government of North Rhine Westphalia, then Social Democrat (SPD), did favour a programme like the British Magnox and urged a big federal expenditure on it, but utilities and manufacturers thought such a choice premature – it could only yield a 'host of white elephants', and the Atomic Energy Advisory Committee presumably agreed.

The new Ministry with its Advisory Committee was quickly involved in supporting much scattered activity. Several research institutes soon emerged and several projects to explore reactor types got under way. Research was decentralised – there was to be no Harwell. This reflected the Federal Republic's constitution, since financing universities was a function of the *Länder* government, which became initial sponsors of many nuclear research projects. Decentralisation was also symptomatic of the diverse initiatives from which research developments grew. The first major initiative came from Karl Winnaker, Chairman of the Hoechst chemical concern; it resulted in the Karlsruhe Atomic Research Institute. The second came from the North Rhine Westphalian Government which planned a centre for the nuclear research of all universities in the *Land* at Jülich; it did not work out quite according to plan.[12] While the decentralised research, with some competition in the early days, was symptomatic of diverse initiative, the organisation of the institutes emphasised the integration of sectors in German nuclear development. There were industrialists on the managing boards of the institutions; industrial groups organised and managed some of the projects in the institutes and it was broadly assumed that prototype – and large scale experimental – developments would be carried out by firms in the works. There were none of the class divisions which could be observed in the British set up.

Reactor development projects also reflected the integration of sectors. A project, under way in 1958 to develop what became the German type of HTR, the pebble-bed reactor, was sponsored by a group of small Düsseldorf utilities, and drew its technical inspiration from Professor Schulten at Jülich; its design and construction was by a company formed jointly by Brown Boveri and Krupp (BBC–Krupp), and the finance came from the Düsseldorf group, the company and the federal and *Land* governments.[13]

At the outset, as in the United States, several different reactor systems were tried by different groups; and there were projects which failed or were not followed up: Siemens worked on a pressure-tube heavy-water moderated system (a 100 MW unit was built), but they dropped the work. Similarly the Allgemeine Elektrizitäls Gesellschaft (AEG) worked on LWR super heat, and Interatom on a sodium cooled thermal system. Some of this work had useful side effects for the FBR later. In this early work, the Federal Government contributed financial help, directly and probably by tax remissions.[14] In the mid-1960s the building of large prototypes was subsidised by the Federal Government directly, by the European Atomic Energy Commission (Euratom), from funds of the ERP and by tax and tariff remissions.[15]

The LWR in Germany was an outcome of the initiatives of utilities, the leading German electrical engineering groups –AEG and Siemens – and the American firms who pioneered the two LWR types, Westinghouse and GE, and who had close links in Germany, Westinghouse with Siemens, GE with AEG. In 1955 Heinrich Mandel, then a young engineer rapidly making headway in RWE (Rheinisch-Westfälisches Elektrizitätswerk AG) the principal electricity supply utility, of which he was to become manager, strongly advocated the LWR system as the most promising. In 1956 this became the basis of RWE policy.[16] Several utilities, including RWE, joined in promoting studies by German engineering firms – financed partly by the Federal Government – of the probable cost of various foreign systems. In summer 1958 RWE and the Bayernwerk[17] ordered a 15 MW experimental BWR from AEG and GE. Successful experience with this, which began operating in June 1961, led to an order by the same partners for a 240 MW BWR at Gundremmingen in 1962. In 1964 a second slightly larger BWR with oil superheat was ordered, and a PWR from Siemens of similar capacity.[18] These were the plants which had subsidies from many sources. In spite of much subsidisation, the nuclear plant builders made substantial losses, which became clear by the early seventies.

In introducing the LWRs, with American technology at the base, Siemens and AEG introduced variations in design: Siemens and Westinghouse parted company. Siemens and AEG claimed that some of their changes made their plants safer. When the German Government decided these must be adopted it provided a form of protection.[19]

The German arrangements were much closer to the American than to the British. There were several utilities operating like ordinary commercial companies (although there were large local government holdings in most of them). They had autonomy in choosing generating plant, subject only to safety and environmental regulations. A utility might well be prepared – as in the United States – to act jointly with one or more other utilities for a specific object, as when RWE and Bayernwerk joined to buy the BWR, but this was a matter for its own discretion. The largest utility, RWE, was in 1962–63 about one-third the size of the CEGB in Britain in terms of electricity generated and sold. It regarded it as economic and practical to buy both types of LWR – as indeed did other and smaller utilities. The utilities probably had a larger role in promoting the LWR in Germany than in the United States but that is essentially because the American manufacturers had established the systems, and were selling them to foreign utilities after succeeding in selling to their own.

The German manufacturers also played a powerful, creative role in assessing, adopting and adapting the American systems. They, like the utilities, were free to choose; and while receiving Federal and international financial support they had autonomy in their development work, were expected to carry through all stages of development from experimental-scale reactors onward, and participated in management of the major research institutes. The Federal Government placed no barriers to the entry of any German manufacturing firm into the reactor business (it would not automatically subsidise any firm that did enter); it did not create designated groups; it did not put any restriction on foreign companies who would bid for contracts with German utilities; and it relied primarily on private industry for initiatives both in regard to systems and fuel development. Competitive initiatives were welcome: they occurred not only with the LWR, but with nuclear fuel. Here Siemens initially assembled fuel from purchased components; AEG and GE formed a company, Kernreaktorteile (KRT), to manufacture LWR fuel, more particularly BWR fuel – it was hoped for all Europe – and a specialist fuel company NUKEM was established by four partners; the chemical firm Degussa, the non-ferrous metal firm Metallgesellschaft, the British mining firm Rio Tinto-Zinc Corporation Ltd. (RTZ), and RWE. This became the dominant firm; it absorbed KRT, and GE was drawn in.[20]

Thus under these German arrangements the choice of the system to provide the first generation of reactors rested with the groups best able to assess the technical and economic merits of what was on offer and who stood to lose, disastrously, if their judgement proved wrong. Ministers and civil servants had no part in the selection except to ensure safety: they did not, as in Britain, get involved in making final decisions in matters on which they were not qualified to form a judgement.

In the 1970s the German arrangements became less decentralised, with more industrial consolidation and ostensibly more positive government action. This betokened more a change of circumstances than of purpose and philosophy, though one minister on the left of the SPD, impressed by environmentalists, seemed inclined, like some British MPs, to make personal judgements on safety.

Industrial consolidation came most conspicuously when AEG and Siemens merged first (in February 1969) their turbo-generator manufacturing businesses (together with the sale but not the manufacture of nuclear supply systems) and later their nuclear manufacturing businesses as well, into a joint company, Kraftwerk Union (KWU). The motivation was simple. Both firms knew that to produce the higher capacity units now being ordered and reach the standard of efficiency in the United States they needed larger, more automated, precision plant whose economic operation required a scale of activity larger than they could secure individually. The situation was common to the whole of Western Europe. Operations were not satisfactorily remunerative. Both firms initially would probably have preferred to merge internationally, not with each other, which would have established a more rational structure, but national policies in France and Britain ruled this out.[21] Both firms, particularly AEG, had lost heavily on nuclear contracts. An early outcome of the merger was an increase in prices; a notable illustration of this was that the price for the Biblis B PWR plant when ordered by RWE late in 1970 was 60 per cent higher than the price of Biblis A ordered only eighteen months earlier.[22] Utilities naturally regretted the lessening of competition and were inclined to place more orders with the remaining turbo-generator makers, Brown Boveri and MAN.[23] An order was also given in 1970 to a newcomer to PWR manufacture in Germany, Babcock–Brown Boveri Reaktor GmbH (BBR) using the American Babcock PWR technology: a second order was given to them in 1975. But this was not as valuable as competition between the BWR and PWR, with one of the American majors having a direct impact.[24] When the one joint company offered both types of reactor, the competitive edge was likely to be less, and shortly after KWU took over the nuclear activity it ceased to offer

BWRs. It would offer them later, it said in 1976, after the Gundrem-mingen plant (ordered but not yet licensed) had worked:[25] this looked like the Greek calends.

The Federal Government, in its fourth nuclear programme (1973–1976), stated that the subsidy for the LWR would be DM 96m during this period and then end. Its purpose had been to help the industry reach the standard of American technology, which was difficult in the smaller market. By 1975, KWU claimed this had been achieved: component supply was up to the high level of precision required; the training of component makers was over.

When the subsidy ended, the Federal Government's direct concern with the LWR was limited, as in the United States, to regulation for health and safety. It remained, however, much involved in the HTR and FBR development programmes, and in the expansion and development of the ancillary fuel cycle industries. Since the commercial prospects of the HTR were uncertain and those of the FBR remote there was nothing to attract large investment from utilities or manufacturers, and the part played by the Federal Government looked bigger than it had been with the LWR. In all these areas there was much internal discussion and some collaboration, and the Federal Government was naturally involved. But the basic character of the German organisation was not changed: utilities still retained unimpaired their right to decide what plants to buy and when. When the Federal Government sponsored development on an industrial scale the work was left to privately-owned industries or firms – and not just to one firm: for the HTR it was BBC; for the FBR, Interatom. By the end of 1976, a reprocessing plant was being designed for a group of utilities by a group of manu-facturing firms, Bayer, Hoechst, Gelsenberg and NUKEM.[26] The German participants in Urenco Centec, the Dutch-British-German concern to produce enriched uranium by the centrifuge process, were the last three of these plus MAN and Interatom,[27] and there were still several separately organised centres of research. The monolithic structure characteristic of the United Kingdom, to which the French seemed to be returning, had been avoided.

## (c) Sweden

Swedish organisation and policy has a special interest because a monolithic system was deliberately discarded. In 1945 the Swedish Government established an Atomic Energy Committee which became a research council, and an Atomic Energy Company was formed in 1947, four-sevenths of the capital being owned by the state, three-sevenths by private industry. It was regarded as a nationalised industry. Several major industrial companies became active in nuclear development by the early 1950s, among them the important electrical engineering firm Allmänna Svenska Elektriska AB (ASEA).[28] The Atomic

Energy Company decided to develop a heavy water system for use in Sweden, initially to provide district heat, but ultimately power. By spring 1956, two years after the Company had started operating its first small experimental unit of this type, a conflict could be detected between it and ASEA. 'Interest in Sweden,' *The Times* recorded at this date, 'seems to concentrate on American work, on the principle that Britain is concerned with quick results, America with a cheap supply of power'. ASEA, 'interested in advanced types of reactors, had plans for cooperating with an American firm, although the form of State control ... seemed to make this difficult'.[29] The famous lecture by Sir Christopher Hinton did not divert Swedish interest from America, although a close study of Magnox was made. The conflict developed and persisted. The ASEA annual report for 1962 stated that it had on request put in a tender for a pressurised heavy water reactor of the Atomic Energy Company type, but put forward as an alternative a boiling water concept with nuclear superheat.[30] The Atomic Energy Company's work was criticised as costly, and technically unsatisfactory.

The manner in which the conflict was resolved was enlightening. Sweden had a mixed electricity supply system – a public sector, efficient and well managed, and a private sector, also well managed. A private sector company, the Oskarshamn Power Group, formed for this purpose by eight Swedish private utilities, ordered a BWR from ASEA in January 1966.[31] This proved a turning point. Subsequent orders for nuclear plants in Phase II, including orders from the State Power Board, were all for BWRs from ASEA or PWRs from Westinghouse; the heavy water line of development was dropped, the Marviken heavy water reactor being abandoned at a late stage of construction, and the Atomic Energy Company was refused funds to research on the Fast Breeder. The successful development was entirely the outcome of private enterprise – on both sides, buyer and seller.

This produced no shock because although ruled (for decades) by Social Democrat governments, most Swedes were ready to recognise that state enterprises were often inefficient. It was, however, generally believed that the success of ASEA was only achieved because there was a private utility sector. It required this important element of decentralisation. The Swedish Government accepted the judgement of the utilities and the manufacturer, and allowed the use of the American PWR and the local BWR; and the State Power Board, like the private utility, chose a LWR.

No other organisation in Europe holds the same interest as the French, German and Swedish, though the readiness to choose both the BWR and the PWR in most of the other countries – Italy, Spain Switzerland, Finland for example – was instructive. The utilities were free to choose.

## (d) Japan

In Japan, although there was a formidable array of government agencies and committees promoting or controlling nuclear development, the determining forces in the choice of the LWR were the nine utilities, and the five main electrical engineering groups. The first initiative in introducing nuclear power was taken by the Government, who set up a research institute in 1956 and in 1957 established a special company to construct a Magnox plant at Tokai Mura, whose purchase was a government decision, greatly influenced by Hinton. Construction started in 1960: it was not a success. No further nuclear plant was ordered until the mid-1960s: the first BWR was ordered in 1966. Subsequently orders were placed by utilities directly with GE and Westinghouse, with whom they had had close links. Of the five electrical engineering companies, three similarly had close associations with the two American firms; they largely constructed the plants ordered in the United States and acquired the know-how, so that by 1971 they took orders directly from the utilities, operating as licensees, and 90 per cent of the components were made locally. The Japanese companies, Mitsubishi (W), Toshiba and Hitachi (GE) had taken licences also for the FBR work of the American companies. By 1971, Mitsubishi was tendering for PWR export orders. Sites had now been chosen for upwards of fifty LWR plants. For all the early LWR development there were government subsidies.

British observers still argued in 1971 that a government agency might buy an AGR, and a government agency was building a prototype Advanced Thermal Reactor (ATR) analogous to the SGHWR. A government research organisation, with manufacturing firms collaborating, was working on the HTR and FBR concepts. Four years later, at a joint meeting in London between the Japanese Atomic Industries Forum (JAIF) and the British Nuclear Forum (BNF), the Japanese programme was even more dominated by the LWR than in 1971. Research on other reactor systems, still proceeding in 1975, was overshadowed by it: the AGR was not mentioned, the ATR looked rather like a sideshow, the FBR and HTR remote. When research projects were mentioned, fusion and fuel recycling seemed to rank above all these, but 'the research and development of LWR have the highest priority'. It was said that R and D expenditures by the industrial firms concerned reached over 100 billion yen (£170m) in the year 1974 – about one-sixth of turnover.

Construction and manufacture of LWRs in Japan was by *groups* of firms. There were two groups making BWRs, one making PWRs. Manufacture of the principal components was part of an integrated process: there was no question of component supply being open to competitive tender. Contracts for nuclear plants were not based on competitive tender: utilities placed orders with firms with whom they had worked in the past.[32]

In 1971 the British Nuclear Forum thought there was scope for British firms to supply components for Japanese nuclear power plants. British Nuclear Fuel's (BNFL) sale of Magnox fuel for the Tokai Mura Magnox plant and its contracts to reprocess spent Magnox fuel from Japan were presented as a model of export selling. But the AEA was the only firm making Magnox fuel (in France an analogous fuel was made, but the precise design was made by AEA for GEC) and Britain was the only country with spar᠆ capacity for reprocessing Magnox fuel. The bid made in 1970 to get orders for reprocessing BWR fuel also was jumping the gun: no one, anywhere, had the plant for doing this. Whatever the scope for selling components to Japan in 1970, by 1975 it had lessened dramatically. Japanese component makers were looking for export markets, and Japanese main contractors were hoping to export LWRs.

## (iv) DELAYS AND PROSPECTS

Delays occurred in the construction of LWR plants outside the United States, but not on the American scale, and a number of the early big units, in Japan and Germany for example, were completed on time within schedules which confirmed that the four-and-a-half to five-and-a-half years' construction time set out by the AEC as a practical standard (Figure 4.5) could be achieved.

The large flow of orders made regulation not a new but a growing preoccupation for governments. It was conspicuously so in Germany in 1970–71.[33] Here they were conscious of special local problems – plants could not so easily be remote from populated areas as in America; many had to be on the Rhine, and cooling problems were acute. As Siemens and AEG departed from American designs, plants could no longer be regarded as made according to American regulatory requirements.[34] Vessels were made locally; hence testing their integrity was a particularly heavy responsibility. It appeared as though a tug of war developed between two groups, among officials and ministers and, as in the United States, it was said that excessive concessions not really increasing safety were made by the regulators: a leading British safety expert said this added needlessly to costs.[35] The tug of war in Germany induced delays in licensing; already noticeable in 1971 they were much more disturbing by 1976.

Environmentalist movements sprang up wherever nuclear plants were proposed, and were indeed a United States export (in terms of persons and propaganda) and they contributed to delays. The constitutional arrangements were less favourable, however, to the prolongation of delay by their methods outside the United States. They needed central government support.

Towards the end of Phase II steps were taken towards establishing international safety standards. In March 1976, Japan and Germany agreed to become partners with the United States in the LOFT experiments at Idaho, all three participating in planning, conducting and financing the experiments.[36] Two years earlier, GE and ASEA agreed to collaborate on safety research for the BWR, which would presumably result in a wider interchange than the LOFT agreement, but this in turn might be broadened.

The governments of Japan and the Western countries on the continent were satisfied that the LWR and the associated fuel cycle industries were safe and experience of LWR costs led them to recognise in 1973 that a more rapid expansion of LWR capacity if it could be achieved, coupled with conservation, would be their most effective strategy in response to OPEC oil policies. There was an apparent departure from this when the Swedish Social Democrats, who had launched a large nuclear programme, were defeated by a coalition, the leader of whose largest party was vigorously anti-nuclear. The other parties, however, were not with him on this, and no immediate change of course occurred.

Officials of the European Community produced early in 1974 a fanciful target of 220,000 MW capacity to be reached by the Nine, including the United Kingdom in 1985. This was soon revised to 200,000 MW, and in December to 160,000 MW. The last figure was simply an addition of national programmes, and included 15,000 MW for Britain. British officials' attitudes to these successive estimates were scornful. The British quota was already looking improbable, and it was likely there would be smaller lags on the Continent. Falling demand was already contributing to this. A Select Committee of the House of Lords on the European Communities thought the difficulties were such that a more modest programme might easily be delayed by a decade. Dr N. L. Franklin. Managing Director of Britain's Nuclear Power Company, and Mr C. Allday, Chief Executive of British Nuclear Fuels, both thought 125,000 to 130,000 MW 'just about attainable'.[37] *The Times* in a scathing comment on the Report correctly saw its background: 'The United Kingdom has had too many unhappy experiences in its attempt to leap forward to the nuclear age for it to be tempted again by an excessive promise of atomic power.'[38] To these unhappy experiences we now turn.

## NOTES AND REFERENCES

1. See, for example, *State Science and Economy as Partners*, a symposium of papers collected by the West German Federal Ministry of Scientific Research (Berlin: A. F. Koska, 1967) p. 1142.
2. The average value of crude oil imports in to the United Kingdom was £7.01 per ton c.i.f. in 1970. Fuel oil imports were slightly more expensive; they had been considerably lower. (In 1967–69 the average was £6.21 per ton c.i.f.).
3. Above, p. 27.
4. *Rapport Présenté par la Commission Consultative pour la Production d'Electricité d'Origine Nucléaire sur le Choix du Programme de Centrales Nucléo-Electriques pour le VI^e Plan* (Paris: Ministère du Développement Industriel et Scientifique, 1970) pp. 11–14. Hereafter cited as *Choix du Programme, 1970*.
5. See Burn, *op. cit.*, p. 125. A transitional view was described by A. D. Little Inc. in *Growth of Foreign Nuclear Power*, study for the AEC (Washington: AEC, 1966) pp. 57–9. At that point the EDF had ordered a Magnox type with a more advanced fuel element analogous to a type advocated by TNPG in 1964 (below, p. 164). This they thought might compete with 1000 MW oil-fired plants. But if it did not the AGR route in the EDF view offered 'no advantage in their economic frame of reference'. It was agreed by the EDF that LWRs could 'probably produce power at lower costs' than gas-graphite plants, but their adoption would involve big capital outlays on enrichment facilities and equipment manufacture.
6. *Choix du Programme, 1970, op. cit.*, p. 7.
7. The utilities themselves, however, did form plans which were looked at as a whole, and governments were necessarily involved for safety and environmental reasons, and concerned with the implications of plans for energy supply and security.
8. Coal supplies in Japan were negligible.
9. *The Industrial Challenge of Nuclear Energy* (Paris: OEEC, 1957) especially the useful paper on 'Programmes of the Other OEEC Countries' (that is, other than the United Kingdom, United States, Germany and France), by Professor Leander Nicolaides (Greece), who had been appointed chairman of a group of three to study these developments in 1954.
10. *The Times*, 1 August 1956. *State Science and Economy as Partners, op. cit.*, p. 157.
11. *The Times*, 7 January 1956.
12. *State Science and Economy as Partners, op. cit.*, p. 140.
13. *Ibid.*, p. 141.
14. Overall expenditures by the Federal Government on nuclear programmes were as follows.

The German Federal Government's Expenditure on Nuclear Programmes
from 1956 to 1972 (DMm)

|  | *1956–62* | *1963–67* | *1968–72* | *1973–76*<br>*(estimates*<br>*in 1973)* |
|---|---|---|---|---|
| Basic research | 706 | 1540 | 2530 | 2035 |
| Nuclear technology development | 440 | 1470 | 2770 | 3367 |
| Safety and radiation protection | 30 | 60 | 260 | 360 |
| Other (including Euratom,<br>IAEA, Eurochemic) | 276 | 731 | 594 | (365)[a] |
| Total | 1452 | 3801 | 6154 | 6127 |

[a] The subdivisions in this column do not correspond exactly to those in the other three columns, so this is calculated by difference. It includes fusion research, but many international items seem to be spread in the other items.

15. *State Science and Economy as Partners, op. cit.*, p. 143.
16. Information from Professor Mandel.
17. *State Science and Economy as Partners, op. cit.*, p. 141.
18. *Ibid.*, p. 142.
19. Professor Mandel was arguing in 1973 that it had become important to have international standards for safety in order that competition between suppliers of plants could be maintained. What would protect the German industry at home could militate against it in export sales; Mandel saw the problem in this light. He referred to this topic in an address as President of the *German Atomic Forum*, Bonn, 1973, p. 14.
20. I found that close and warm relations had developed by 1973 between GE and ALKEM, a daughter company from NUKEM, on an adjacent site, which manufactured fast breeder fuel and plutonium recycle fuel.
21. I discussed the problem with the main firms in all the countries concerned in the middle sixties.
22. The figures were given to me by RWE. Nominally Biblis A, ordered in May 1969, was a KWU sale; but KWU only started operating in April and all the selling, including the pricing, had been done by Siemens.
23. Of these Brown Boveri were much the more important.
24. AEG appeared to continue to have access to all the technical developments of the BWR after entering KWU. They had full data on BWR 6 and the Mark III containment.
25. *Nucleonics Week*, 6 May 1976.
26. The plant would be built by a private enterprise consortium, Uhde, Lurgi, NUKEM and St Gobain.
27. The functions of the firms were shown in *URENCO/CENTEC. The Organisation and its Services* issued in June 1976.
28. There is a brief account of early Swedish development in *The Industrial Challenge of Nuclear Energy, op. cit.*, pp. 225–6.
29. *The Times*, 16 May 1956, in an article 'Bases of Sweden's Industry'.
30. *ASEA Annual Report 1962*, p. 28.

31. There was a brief account in *Nuclear Engineering International*, January 1966, p. 41. This describes some features of the plant without drawing attention to the interesting politics of the situation.

32. The British Nuclear Forum produced a valuable survey: G. H. Greenhalgh, *Nuclear Power in Japan* (London: British Nuclear Forum, 1971), being the outcome of a visit by a group from the Forum. A paper by Shiro Mochizuki to a London Conference of the BNF and JAIF in September 1975 gives an equally valuable review of the position four years later.

33. The first conspicuous instance was when the West German Government decided to defer a decision to allow a PWR plant to be built adjacent to the BASF chemical works at Ludwigshafen to supply process steam, as well as power for the works and for the public supply. It was turned down as being too close to a large centre of population and also as being both a risk to the chemical plant and at risk from the chemical plant.

34. *Fourth Nuclear Programme 1973–1976 of the Federal Republic of Germany* (Bonn: Federal Ministry for Research and Technology, 1974) p. 72. 'The independent development of the LWR line ... has increasingly called for independent safety research ... safety is becoming internationally important as an economic argument. Hence the result of safety studies on advanced systems ... are no longer exchanged as liberally as they used to be.' I think this was more true of the PWR at this time, and the initiative in checking interchange did not I think rest with Westinghouse.

35. Hence, he concluded, Britain should obtain LWR technology directly from American sources.

36. The agreement with Japan, which came after that with Germany, was announced by NRC in Washington on 25 February 1976. OECD had formed an International Energy Agency, who fostered cooperation in the LOFT programme.

37. Report of the Select Committee of the House of Lords on *The European Communities EEC Energy Policy Strategy* (London: HMSO, 14 May 1975) p. vii, pp. ix–x, Qs 384–406, 993–8. Hereafter cited as *SC, House of Lords, EEC Energy Policy, 1975*.

38. *The Times*, 16 September 1975.

# PART II

# From AGR to SGHWR

# CHAPTER 8

# Preamble

The history of the AGR in Phase II was a source of progressive disillusion. In 1967, though there were warning lights, the AGR was still expected to provide competitive power within a few years and widely, if less generally, to secure export orders. Its development potential was still said on high authority to promise sensational reductions in the cost of power within a decade. But some design difficulties were causing delays: others followed each other in constant succession, repeatedly putting back the date at which one of the plants would work. Two finally began to work at low power early in 1976. Domestic orders had come slowly. None came after early 1970. There were no export orders. In 1973 the CEGB announced it would buy no more AGRs. When those on order worked they would, initially at any rate, produce less than their designed power, and the designs would not be satisfactory to repeat. The CEGB decided to buy instead PWRs of American design. They would, they concluded, produce cheaper power than the AGR. They were, moreover, proven reactors.

The Board reckoned without its masters. Its abandonment of the AGR was not challenged. But successive British Governments still insisted on deciding which reactors should be used and who should design and construct them. The Labour Government in 1974 decided that the CEGB must have another reactor developed by the AEA and, as the HTR was not ready for commercial use, it must be the SGHWR, of which a 100 MW prototype had run for four years. It was a type the CEGB had dismissed as obsolete and which the new National Nuclear Corporation (NNC) did not wish to make.

Parallel to the disillusioning experience with the AGR, there was a continuous effort by the Government and Parliament to find the right form of organisation for the British nuclear industry. This reflected the disappointment over the industry's failure to achieve the promised success – especially the failure to win export orders. The reorganisation which took place showed a failure to understand fully what was wrong. There were two major changes of organisation. First, in 1968–69 one consortium was eliminated and two were reconstituted.[1] The AEA became a participant in each, with a representative on their boards and parts of the AEA staff were absorbed in the new companies. The arrangement was an uneasy compromise. The wasteful discontinuity in development when the AEA handed over to consortia was eliminated, but this had little immediate importance; the AEA retained a dominant position in development, competition between autonomous companies was not created, there was no organic link between British and foreign companies and design and manufacture of fuel remained with the AEA.

Persistent failure in AGR development with cumulative losses and no successes in other operations led to a second change in 1973, when the Conservative Government decided to set up a single design and construction (D and C) company. There was now almost universal, if resigned, support for a national monopoly. Even the CEGB agreed. The lack of orders gave it an air of inevitability. There were different views of the form it should take. The government chose to set up a company, the National Nuclear Corporation (NNC), in which half the capital was provided by the General Electric Company (GEC), which would provide strong management and had considerable financial strength. Again it was a compromise. It could appear to be a move toward the successful type of company organisation in the United States and Germany, but it was still remote from this in form and even more in spirit. The state still retained a participation. Britain had indeed scored another first: she was the first major industrial country to give a monopoly to one company to design and construct nuclear plants, although the argument that the potential size of the domestic market made it imperative would ostensibly have had as much force in Germany, France, Italy, Sweden – even possibly in Japan. France ultimately went the same way and in Germany KWU emerged as the dominant design and construction company, but without a monopoly. When the Labour Government also refused to allow the principal utility, the CEGB, to choose its own plant and the design company to make the reactor of its choice, but insisted on the manufacture and purchase of the SGHWR, Britain was scoring further 'firsts'. In response, GEC proposed to reduce its participation in the NNC from 50 to 30 per cent, as the risk was less acceptable and the prospect of successful business had been reduced. The Government took over the 20 per cent which GEC relinquished and became the largest shareholder.

Four questions arise out of this series of events.

First, why did the long-drawn-out 'disaster' of the AGR – as many former supporters of the reactor now called it – occur? Second, why did it take so long for the probability of failure to be recognised? Nuclear power was not neglected, indeed it was the subject of intense discussion and policy making; nevertheless, the most significant data were not observed quickly, or not understood quickly, by those who scrutinised developments and those who made decisions. The public's watchdogs were consistently wrong. Third, why did the United Kingdom, alone of the countries (apart from Canada) who decided to use nuclear power, refuse to use the LWR? Were all the others wrong in judging it to be safe as well as economic? Was it the high-water mark of insularity and national state socialism? Finally, how will government policies, in choosing both the organisation of the nuclear industry and the reactor policies to be followed, influence British energy costs and supplies? But first it is necessary to chronicle, within the limits of the published records, the course of AGR development.

## NOTE

1. The Composition of the Consortia and their successors.

*Formed in 1954*
> AEI–John Thompson Nuclear Energy Company
> Nuclear Power Plant Company
>> C. A. Parsons, A. Reyrolle, Sir Robert McAlpine & Son, Clarke Chapman, Head Wrightson, Strachan & Henshaw, Whessoe.
>> English Electric, Babcock & Wilcox and Taylor Woodrow Atomic Construction Company [became the Nuclear Design and Construction Company].
> GEC–Simon-Carves Atomic Energy Group

*Formed in 1956*
> Atomic Power Construction Company (APC)
>> Richardsons Westgarth, Fairey, International Combustion Co.

*Amalgamations in 1960*
> The Nuclear Power Group (TNPG) was formed by AEI–John Thompson and the Nuclear Power Plant Company.
> [All member firms of both consortia became partners.]
> United Power Company (UPC) was formed by the GEC–Simon-Carves Group and APC.
> All member firms became partners.

*Withdrawals 1963*
> Richardsons Westgarth withdrew from APC

*1964 (end)*
> GEC–Simon-Carves withdrew from UPC, leaving APC (which had retained its identity and name within the UPC).

*New Companies formed in 1968–69*
> The Nuclear Power Group (TNPG)
>> All former partners became shareholders, except AEI, which had been absorbed by GEC. The AEA had 20 per cent of the shares, and the Industrial Reorganisation Corporation (IRC) became a shareholder until its dissolution.

British Nuclear Design and Construction (BNDC)
The same partners as the former English Electric Group, but as English Electric was absorbed into GEC-English Electric, GEC came into the operation. The AEA had a 20 per cent holding and the IRC took a share until its dissolution.
[The APC was in effect eliminated at this stage.]

*Reorganisation in 1973*
National Nuclear Corporation was announced as the single D and C company, its shareholders being GEC (50 per cent), British Nuclear Associates (35 per cent) and the Government – represented by the AEA (15 per cent). British Nuclear Associates had seven shareholders drawn from the two former companies, TNPG and BNDC, which were brought out by the new company; the shareholders were:
Babcock & Wilcox
Clarke Chapman
Sir Robert McAlpine & Son
Taylor Woodrow
Head Wrightson
Strachan & Henshaw
Whessoe
After the decision to adopt the SGHWR GEC reduced its participation to 30 per cent, the Government increased its to 35 per cent.
An operating company, wholly owned by the NNC, was formed, the Nuclear Power Company, sited at Whetstone and Risley where BNDC and TNPG had been operating.

# CHAPTER 9

# The Course of AGR Development

## (i) *Prelude and Prototype*

The decision to construct a prototype AGR in Britain was taken in 1957.[1] The AGR has been presented as the natural successor to the Magnox system: they belong to the same family, since both use graphite as moderator and $CO_2$ as coolant. Magnox, however, used natural uranium as fuel – the AGR enriched uranium. The choice of Magnox was imposed in 1953 by the 'properties and economics of available pile materials'[2] – heavy water and helium were too expensive and no enriched uranium was immediately available. Most Harwell scientists doubted whether a natural uranium, gas-cooled reactor was worth building since it was 'unlikely that more advanced reactors would use gas cooling'.[3] Even Hinton was dubious: and the go-ahead for Calder Hall was given, in March 1953, because more plutonium was urgently needed for military uses, and some should also be stockpiled for future fast reactors[4] – on whose development Hinton seemed ready to concentrate almost exclusively.[5] Plutonium was then valued at £3100 per ounce, several hundred times the price of gold, and on this basis the by-product cost of electric power from Calder Hall would be extremely low.

At this stage Harwell engineers 'never transposed this special case into an unduly rosy view of the economics of a primarily power producing'[6] Magnox reactor. But the project passed from Harwell to Risley when construction was decided upon. And, quite rapidly, rosy views were taken by some Risley engineers – never by all – who concluded that the $CO_2$/graphite system offered great scope for rapid development. The official history does not yet cover this transition. By the end of 1953, the first account published of *Britain's Atomic Factories* spoke of Calder Hall as 'this first unit in what may well become the power generating project of the UK',[7] and, in proposing the second reading of the Bill to establish the AEA in March 1954, Sir David Eccles said it was 'in the Calder Hall family of reactors that we see the earliest prospects of producing economic results'.[8] In both cases the Calder Hall family was to lead up to the fast breeder reactor. The power-generating project duly came as The Nuclear Power Programme of February 1955 and by 1956–57 competitive power was forecast for 1962–63.[9] The extremely optimistic view of the development potential of the $CO_2$/graphite reactor system – not universal throughout the AEA – accounted for the decision to go ahead with the AGR in 1957. 'The most satisfactory choice for continuous development in the UK would clearly be to follow the main line; that is, uranium-fuelled graphite-moderated gas-cooled reactors, as Dr Rotherham, Head of the Research Development Branch at Risley, put it in 1957, without defining the gas. The conclusion was invalid if the Harwell scientists were right who believed that when enriched uranium was available other systems were better.[10] The French came to the same conclusion as the Harwell scientists when they were free to choose – they dropped the $CO_2$/graphite system and chose the LWR.[11]

When the AGR prototype design was started in 1957 no Magnox plant more sophisticated than Calder Hall had been fully designed, and Calder Hall itself had only nine months' operating experience. It had been completed in three and a half years – 'exactly within the time limit' which Hinton had promised, as Margaret Gowing records.[12] The scale decided upon for the AGR prototype was 33 MW(e) gross (100 MW(t)), stretched in operation quickly to 41 MW(e). There were, I have been told by one closely concerned, advocates of a larger scale; what determined the choice has not, as far as I know, been published.

To complete the context the American position is crucial. In 1957 an 80 MW(e) PWR started operating at Shippingport (Pennsylvania). It was completed within three years of the first work on site, which was started before design was complete.[13] A 100 MW(t) experimental BWR had operated since 1956 at the Argonne Laboratory; GE started operating in 1957 at their Vallecitos Atomic Laboratory a 50 MW(t) experimental BWR, linked to a 5 MW generating plant supplying power to the Pacific Gas and Electric Company, who helped to finance the plant. It was built in about eighteen months. A 160 MW(e) PWR plant and a 180 MW(e) BWR plant were under construction for normal utility operation. The BWR – Dresden I – reached full power in 1959, three and a half years after site preparation started and five months *before* the scheduled date. This was a more ambitious plant than Calder Hall.[14] In 1958 two more BWR plants were ordered which, apart from providing power, were to test further advances in design: Humboldt Bay was the first direct cycle plant and the first with pressure suppression containment; Big Rock Point was to test high performance fuel as a step towards higher power densities. Both reached 75 MW(e), about twice the Windscale AGR capacity, and began to supply power early in 1963.

The specification of the AGR prototype was defined by March 1958; construction started in November 1958, power was first produced in

December 1962, full power by February 1963. Thus the designers of the AGRs for the Dungeness B tenders (required by February 1965) had less than two years' operating experience. The plant had a steel pressure vessel, and its heat exchangers were outside the pressure vessel, the usual practice for Magnox plants in 1957–58.

Dr Gordon Brown, chief project engineer on the Windscale prototype, emphasised in 1969, because there was a misunderstanding on this, that the AEA's work as designer at Windscale was confined to the reactor.[15] This was the pressure vessel and its contents (the core and its components) and the refuelling facilities and some parts affected by it. The rest, including heat exchangers, circulator pumps, main gas ducting and the electric equipment, was designed by private industry. The AEA set the parameters to which the designs must conform, the pressures, temperatures, gas volumes, corrosion conditions and so on.

In spring 1959 Hinton thought the 'promise of the AGR was such' that the CEGB 'might be wise to run the risk of placing orders for AGR reactors before they have got operating experience on their prototype'. It was a 'risk we are considering'.[16] Was it an aberration? He wanted lower costs quickly, and feared the FBR would not be in industrial use till 1970.[17] Three years later he told the new Select Committee on Science and Technology, of the British House of Commons (hereafter referred to simply as the Select Committee), that the 'experience which has come to light... has not been experience which would induce me to take that risk' – although he did not think it so bad as to rule out orders in the future. But there were alternatives, American light water reactors and the Canadian Candu, and he could not now order without a year's operation of the AGR on full load.[18]

From 1960 the AEA Reactor Group began making design studies for a larger – twin 500 MW – AGR plant 'to obtain firm assessments'. The first, in 1960, followed the Windscale model: a second at a later date (a description was published in April 1963) was based on the use of a prestressed concrete pressure vessel, with a leakproof steel lining to prevent gas escaping, of the type adopted in the designs for the last two Magnox plants, Oldbury and Wylfa. Oldbury was expected to be on full load in summer 1966. (It came on full load in December 1967.) In these plants the boilers were brought within the pressure vessel, being arranged behind shields round the core and the plan was adopted for the 1963 AGR design study. This allowed the elimination of the outer containment hitherto needed, and lower capital costs. The coolant gas, after flowing through the core to absorb the heat, flowed – in the design adopted by the AEA – into a 'hot box' above the core and then had to make two sharp 90° turns to flow down into the boilers. The AEA experts emphasised that other configurations were possible – the 'best criteria for economical design and construction' of concrete pressure vessels 'are not well understood or at all rationalised'.[19]

In the evolution of the AEA design studies there were regular com‐
mittee discussions with the consortia, who could suggest improvements.
One made by the Nuclear Power Group (TNPG) in August 1961,
'based on its own studies', was that larger clusters of fuel rods in the
fuel elements than those chosen by the AEA would give better heat
transfer performance. They proposed thirty-six-rod, three-ring clusters.
The AEA said in November 1961 that they 'were not convinced that
the value of large clusters warranted the cost of the development work,
estimated to be £100,000. Discussions on the 36-rod clusters continued
at various collaboration committees, but heat transfer experiments
were not put in hand.' As late as 24 July 1963 the AEA at the Nuclear
Power Collaboration Committee 'hoped that no one would want to
press tests on 3-ring clusters.'[20]

(ii) TENDERS AND CONTRACT

The CEGB took the first step towards ordering a commercial-scale
AGR plant in April 1964, asking for tenders for one for a second
nuclear plant at Dungeness – Dungeness B. The Government had just
announced the Second Nuclear Power Programme. Shortly afterwards,
with government approval, tenders were also invited for LWRs for
the same site. The CEGB had shown great interest in light water
reactors after the Oyster Creek and Nine Mile Point orders; the pros‐
pect that one might be bought led the AEA to send one of its principal
experts to explore with GE and Westinghouse the terms on which the
Authority might make these companies' fuels in Britain, if the need
arose.[21]

The AEA, as International Combustion Limited (ICL), one of the
constituent firms in the Atomic Power Construction (APC) consor‐
tium, put it, 'were disturbed at this prospect of competition between
the AGR and the by then well developed light water reactor systems,
because... there was a risk that the development potential of the
AGR would not be recognised.' They now became extremely interested
in the promise of a thirty-six-rod cluster and other more advanced
parameters, and decided to make yet another fullscale cost study.[22]
They decided in July to give a contract for this to another of the con‐
sortia, the United Power Companies (UPC), which had announced it
would not tender for Dungeness B.[23] ICL was given a contract to de‐
sign the boiler in June, when the AEA 'had stressed that there was a
possibility that this design might, if it proved sufficiently attractive,
form the basis of an order.'[24] The AEA informed TNPG, NDC and
the CEGB in mid-July that it was giving this contract to UPC; all
showed great concern, and the AEA, having again stressed that UPC
would not be tendering for Dungeness B, decided late in September
'not to proceed with the exercise' with UPC.[25] The possibility of an

order was, in ICL's version, 'politically unacceptable', so 'work on the boiler design was stopped'.[26] UPC decided in October, however, that while not making a tender it would 'submit a priced design study to the CEGB based on the work terminated in September, and centred round the 36-pin fuel element developed by the AEA'. In January 1965 APC decided that the design study should be submitted as a 'formal tender'.[27] When this decision was taken the GEC-Simon-Carves group had withdrawn from UPC (they had disagreed with the decision to tender), leaving APC on its own again.

In all this the AEA and APC, as the late Sir Harry Legge-Bourke, a member of the Select Committee, put it, were 'working very closely on the AGR'.[28] Dr Gordon Brown confirmed this: 'the AEA at all times ... were closely associated with that work'.[29] By an odd coincidence, in November 1964 the CEGB allowed tenderers to depart from certain parameters in the specifications for which the tenders were asked, and set out the conditions to be complied with if they did so. It looked to TNPG in retrospect as though this had been to ease the way for APC, but Sir Stanley Brown, who had succeeded Hinton, said that the CEGB had only heard 'through the grapevine' that APC would submit a tender three weeks before it was due.[30] They had not been working with APC and the AEA. The contract was awarded to APC, who also departed from the specified parameters by raising the pressure to 450 psi and the capacity to 660 MW. The AEA was credited with the novel idea, the 36-rod cluster, which they had for so long resisted, around which the enterprising APC built their design. The initiating role of TNPG was passed over in silence by the AEA, APC, the Select Committee and the Institution of Professional Civil Servants (IPCS).[31] It was an illuminating instance of the false values adopted in AEA/consortia comparisons.

TNPG and British Nuclear Design and Construction Ltd (BNDC) did not adopt the new parameters: in view of the type of tender the CEGB asked for it was hardly practical to do so. Gordon Brown explained to the Select Committee that, as director of gas-cooled reactors in the AEA, he had seen that all the information which the Authority 'had developed at that time', and which the APC used, was made available to all the consortia.[32] But since some of the information was developed by the APC they had a time advantage and greater intimacy with the work. The episode caused bitterness (the APC, it was said, had with the connivance of the AEA 'pulled too many fast ones') and some observers believed this created stresses which lessened cooperation in the industry when APC ran into desperate difficulties.

This was probably a minor matter. But it was of great significance that because of a shortage of time the APC tender 'was not a tender in depth to the degree that some of the other tenders were'.[33] Gordon Brown put it this way in 1969, referring now to his experience as a member of the panel that assessed the Dungeness B tenders. He moved from the AEA in September 1965, it is said with Government encouragement, to become Managing Director of APC, to give the additional strength to the company which the CEGB required. Sir Edwin McAlpine put it more roughly: the tender, he said, was 'put in perhaps in rather a hurry ... and obviously must have missed out many things in the pricing.'[34] Gordon Brown rejected this view although he emphasised that 'the AEA, and I have the highest regard for their team, ... were directly associated with all that work'.[35] As I read it, he confirmed that the Windscale team were associated with the drafting of the APC tender. Mr H. W. Jackson, who was Chairman of APC in 1969, conceded that it 'would have been better if they had been able to put in several million pounds of contingency',[36] to cover unforeseen engineering difficulties.

Colonel George Raby, Chairman of APC when it secured the Dungeness B order, had emphasised before the Select Committee in 1967 some of the other circumstances in which the tender was drawn up. 'We think our state of perfection is pretty good [Legge-Bourke had gained the impression that Raby thought the technology so perfected that it needed no more competition to perfect it] but we have not yet built a big AGR and it will be two years before we do so. We think we have done our sums and all the work so that we do not get awful shocks in two or three years' time, but we have only got the 40 MW job at Windscale. We are now dealing with 600 MW reactors. We have extrapolated them, but not from the 40 MW. The biggest we have at the moment operating in this country, and it is a Magnox job, is something like 250 MW or 275 MW.'[37] He did not say that these relatively big 'Magnox jobs' did not include one with a concrete pressure vessel incorporating the boilers in the vessel.

Gordon Brown hinted in 1969 at what we may call the differential aspect of the jump in scale from the Windscale 'prototype' – for the fuel and refuelling technology the requirements of the 600 MW unit were comparable to those of the 40 MW unit; but for the circulators and the boilers 'you are going from a 100 MW to a 1500 MW boiler and from 300lbs psi pressure to 448lbs at Dungeness B and 600lbs at Hinkley Point.'[38] On this analysis extrapolations in the nuclear area were simple, those in the area covered by the consortia were formidable.

International Combustion, who designed and made the boilers, added further clarification. 'Due to the short period of time between the APC Board's decision in October 1964 to submit proposals and the tender submission date of February 1 1965 very little development of the preliminary design was possible other than to produce sufficient information on which to price.' At the time of quoting, it was believed that this lack of development 'did not constitute an undue risk as the reactor island comprised a pressure vessel ... developed on the later Magnox stations [which were not as yet completed or even nearly in operation and in whose design APC had no part] with a fuel element concept developed and tested [not however tested in its final form] in the Windscale prototype reactor.'[39]

The Dungeness B tender was thus not only not in the same depth as the others, but it was drawn up rapidly, without the advantage of a prototype. The AGRs ordered later were not ordered from a sketch plan nor was construction started without detailed designs, but they too had no basic prototype – and gained nothing from Dungeness B experience from which TNPG expected to learn.

### (iii) TENDER ASSESSMENT PROCEDURE

It was reasonable to assume when the Dungeness B contract was awarded to APC that, however unreliable the comparison of AGR and BWR costs might be, nevertheless an AGR which could do what was promised had been designed. The *Appraisal* published by the CEGB was explicit on this. 'The tenders were assessed by experienced teams of engineers and scientists from the Generating Board and the UKAEA.'[40] (Their names were not published, but I was told that the Board's main representatives had originally been in the AEA.[41]) 'Each section of each offer was considered by the appropriate specialists without knowledge of the tender prices.... The offers were examined to ensure that there was adequate technical information to support the design, that the stated performance would be obtained, that the engineering was of the required quality' and so on. 'When the design was below the minimum acceptable standards ... a sum of money was added to the tendered price to remedy the deficiency.'[42] The principal changes considered necessary (and obviously regarded as practicable) were listed. In 1967 the procedure was set out in still greater detail for the Select Committee on Science and Technology. Each tender was divided into seventy-one sections, and teams of AEA and CEGB experts produced reports on each section, and there were 250 reports on each of the full tenders. Section heads of the CEGB and the AEA then produced a single report on each of the seventy-one sections of each offer. An Assessment Panel used these reports to prepare full reports on each tender, and a Steering Committee summarised these into a full report. 'Such effort is essential if the operation, maintenance and safety of the station is to be fully satisfactory ... and the cheapest cost of power achieved.'[43] In the *Appraisal* it was stated that APC's 'bold advances' beyond the Windscale technology were 'well substantiated',[44] that the 'technology on which the offers were based' was 'found to be fundamentally sound in every case',[45] that 'neither the gas-cooled nor the water reactors presented technical problems which could not be solved on the Dungeness B time scale' – that is, operation of the first reactor by 1970.[46] All this gave a sense of complete assurance based on an immensely careful examination of the technical soundness of the AGR design, as though there was no reasonable ground for uncertainty. Indeed, quite specifically the assessors accepted recent and hurried design innovations for the AGR, although they would accept none made in the United States since 1963. One of the AEA participants said to me later: the decisions were honest – but wrong. In 1974 Sir John Hill, who succeeded Penney as chairman of the AEA, told the Select Committee that the 'appraisal of the various designs was carried out on an absolutely meticulously fair and proper analysis'[47] – which is consistent with the view that the conclusions were honest but wrong, but gives a different impression.

(iv) FIRST DESIGN DEFICIENCIES

Although the Dungeness B decision was widely accepted uncritically, and with relief too, the *Appraisal* published by the CEGB was subject to two immediate criticisms: one that it was biased in its treatment of LWRs, the other that 'the extrapolation from 30 MW to a full-scale power unit was too rapid.'[48] To the first we return. The contracting firm quickly found that the second was, as International Combustion put it, 'to some degree valid'.[49] 'To some degree' was presumably a face-saving form of words. By the end of 1966 complex difficulties began to emerge in the boiler design. 'The full effects of increase in gas pressure and temperature [which were the bases of the cost reductions promised] and of methane inhibition of graphite corrosion [and the crucial factor in the durability of the reactor since the graphite would be sealed permanently into the concrete pressure vessel for upwards of thirty years and no repairs of significance would be possible] were not known at the time.' The gas flows were found to present conflicting requirements of rigidity in the fixing of boiler tubes to prevent excessive vibration and of scope for expansion of the tubes, while the dampness of the $CO_2$ caused much higher corrosive rates than the AEA – who provided the 'fundamental data required for detailed design' in this respect – had initially indicated. Efforts 'to overcome the problems associated with corrosion growth, static adhesion, fretting and vibration' by changes of materials and minor changes of design proved unavailing. Early in 1968 it was seen to be necessary 'to adopt a wholly new design philosophy', within the constraints of the dimensions set by the unworkable design and the wish to use as many as possible of the components already fabricated. During the redesign, model tests showed that the temperature variations between different parts of the boiler were different from those initially specified by APC, and difficulties arose 'in the area of the superheater and the reheater'. Drastic redesign brought site work on boilers to a halt in 1968–69.[50]

Parallel to these troubles there was a more widely publicised source of trouble, but not of *additional* delay. One of the leakproof steel liners to the concrete pressure vessels was made too small and both suffered severe distortions when penetrations were welded in. This left the penetrations too small to allow pieces of equipment, including boiler components, to be inserted and involved a long and costly reconstruction.[51] Unlike the boiler design problems, which ICL said were 'unforeseen and largely unforeseeable', this was faulty work by subcontractors, contributed to by an undesirable subdivision of the job and readily avoidable.

### (v) THE SELECT COMMITTEE INVESTIGATES

The boiler design problem was becoming acute before and during the first Select Committee study of nuclear power problems in 1967, and having its major delaying effect during the second in 1969. The Committee did not refer to it in 1967, although since some members visited Dungeness B they presumably knew of it. During the second they seemed to accept assurances that the troubles were of limited and local significance.

They may have been lulled in 1967 by Penney's reassuring approach. The AGR, he said, was currently the AEA's 'main priority. It always happens in this kind of work that some emergency seems to arise, there is a great commotion about something, it looks of course as if it was not right, and then of course everything converges on it and it always melts away.'[52] At the same session Penney told the Committee that he thought it would be possible during the next six years to reduce AGR generating costs by 30 per cent in terms of 1965 money.[53] They received glowing accounts of the promise of the AGR from the CEGB, the consortia, ministers, civil servants, inter-departmental working parties and study groups.[54]

But there were warning notes. Mr F. H. S. (later Sir Stanley) Brown and Mr E. S. Booth, of the CEGB, very frankly revealed in valuable evidence, which took some of the covers off the Dungeness B affair, that the CEGB had been at first concerned with the ability of APC to 'tackle and programme' (Dungeness B) because of the shrinkage of their resources since their last contract. One of the CEGB's conditions had been that APC should strengthen their resources and recruit 'very senior personnel'. 'They had so much design work on their hands to implement a sketched out design into working detail that they were fully stretched.' Several members of the Select Committee berated the CEGB for not giving their second contract – for Hinkley Point B – to APC. 'To give them a second station immediately [Stanley Brown said] was asking the impossible.'[55] Moreover, their last Magnox plant, Trawsfynydd, had been out of service so long that they had not the best reputation for reliability. This plant had been costing the CEGB £1m a month for replacement electricity alone.[56] These important remarks by the CEGB seemed to be received with incredulity: they did not appear to fit in with majority preconceptions.

TNPG stated categorically in 1967 that it was doubtful whether in its present form the AGR was exportable – the plain implication was that it was too costly. They and APC had tried to export it and failed for two years. Mr Eric Lubbock, now Lord Avebury, who was on the Select Committee, thought this was merely defeatist.[57] But even Raby pointed out that while LWR capital costs were rising their fuel costs had come down – 'that is where it is hurting us'.[58] TNPG also pointed out that data published on the heat transfer characteristics of fuel developed for the AGR by the AEA showed results still 15 per cent below design specification.[59] The Scientific Member of the National Coal Board (NCB), Mr L. Grainger, a nuclear scientist who had held important posts in the AEA,[60] warned that technical difficulties facing AGR development were underestimated. So far additional costs at Dungeness B had come from so-called engineering problems, but there were significant nuclear problems (which were also engineering problems) concerning fuel, which Grainger set out impressively, whose solution was difficult and could not be fully assessed before Dungeness B had been operating for a few years.[61] It might well be that costs could be raised and output lowered. The warning seems to have been disregarded by the Select Committee – was it thought merely partisan? – and was dismissed by the Ministry of Power working party on nuclear power costs and a group set up to examine its findings, who both claimed (one in making, the second in approving, an optimistic forecast of AGR costs), that they had made a fair balance between optimistic and pessimistic technical possibilities. How could they know? It 'could be against the national interest' the examining group urged 'if decisions were taken in a spirit of fearing the worst from all new techniques.'[62]

The warnings that the AGR might be too expensive to export, the growing evidence that the Dungeness B design was not adequate without major changes, and the suggestions that problems could still arise over the fuel, should have had an impact on the Committee's analysis. Most members seemed to assume that the AGR had become a good export prospect when the CEGB preferred it to the LWR – but the assumption could, in the light of experience, prove shaky. A majority of members thought the cost of building AGRs could be reduced if there were only one design and construction body, no competing consortia, and no break between AEA and commercial plant development. There were structural weaknesses. But many members of the

Committee were too easily persuaded that big reductions in the cost of AGR plants had been possible had APC been given a second order immediately.

The contract for the second AGR plant was given to TNPG by the CEGB who did not ask for competitive tenders for a price roughly 10 per cent below the Dungeness B price. It was argued by several members, with the help of figures provided by the Chairman of APC, that had APC received the second order and the promise of a third which would have allowed the firm to enjoy all the cost reductions offered by 'replication' (a not very precisely defined term picked up from the Americans at the Geneva Conference on Peaceful Uses of Atomic Energy in 1964, and dignified in Britain as a new philosophy[63]) the public would have been saved £14.4m.[64] This was the price paid, two members of the Select Committee, Norman Atkinson and Tam Dalyell, said, for the system whereby the CEGB gave chosen contractors turns – 'Buggins's turn next'.[65] The CEGB's reply, referred to earlier (was incisive and impressive. APC's last job for the CEGB (Trawsfynydd) had turned out badly – not a good recommendation for being granted a monopoly. When they started on Dungeness B they had neither detailed plans nor adequate staff. They had to recruit, but even so were hard pressed to get detailed plans out for the job they had got, and could not possibly have taken on the second job when the CEGB wanted to place the order (1966). The Board queried the low cost figures which Raby told the Committee APC could have offered because of replication – 'conversational estimates for works of this magnitude can be, and usually are, grossly misleading'. They had good reason for scepticism.[66] Advocates of a single authority seemed to dismiss all this without consideration. Yet it was already doubtful whether APC had a workable design to repeat, (it is now certain they had not) or sufficient design strength to bear the load of another order, or the readiness in fact to offer concretely the low tender assumed,[67] or the ability to deliver on time a second plant – without which great losses would be incurred by APC and the CEGB and the mythical £14.4m would disappear. While these uncertainties were not frankly faced the discussions of the Select Committee were in the realm of fantasy.

In 1969 the Select Committee examined the 1968–69 reorganisation of the nuclear industry, under Anthony Wedgwood Benn, as Minister of Technology. It had by now a lot of information on what had gone wrong at Dungeness B. There was no mention of anything wrong at Hinkley Point B yet, and the rather misleading statement that nothing was wrong with the nuclear part of Dungeness B was given in evidence, reinforcing similar statements by ministers in the House of Commons.[68] Nevertheless, Dungeness B apart, the AGR was clearly, from the evidence, off its pinnacle. The grotesque claims of 1965–67 were no longer repeated. The AEA concluded from its initial operation that the SGHWR had a better performance than the AGR;[69] a joint AEA–CEGB study was said by the IPCS to show SGHWR generating costs to be at least 7 per cent below those of an AGR.[70] TNPG rated the HTR and the SGHWR as 'significantly more economical than the AGR'.[71]

Nevertheless, the Select Committee remained 'concerned to find that there appears to be no real spearhead to the drive to sell British reactors abroad': this was 'a field in which British effort was seriously uncoordinated, with serious results'; there was 'no specific organisation for overseeing the national effort in this particular section of the export market', though the Ministry of Technology was 'well aware of the importance to the industry of overseas orders, especially at this time of shortage of domestic orders'.[72]

## (vi) EXPORT PROSPECTS

The premise was still that the British industry had reactors which it could have been expected, and could still be expected, to sell. Was this premise justified by the evidence before the Select Committee?

There were a number of well-publicised instances in which British consortia had made bids to sell AGRs in competition with LWRs – in Belgium, Germany, Greece, Spain, Italy, – and they had failed. TNPG had said in 1967 that they and APC had failed because the AGR was too expensive – that a gas-cooled reactor using $CO_2$ was inherently too costly because of the bulk of materials required, including much stainless steel.[73] TNPG regarded a switch to helium and coated particle fuel as relatively simple.[74] The AGR's high capital cost was not significantly offset by lower fuel costs or by the higher availability which the CEGB expected from on-load fuel changing, but which Continental utilities rated as much less valuable,[75] possibly valueless. Mr R. W. F. Guard had argued in 1965, in his enlightening comments on the *Appraisal*, that it was misconceived, and that on-load fuel changing was valuable for Magnox where fuel rods with a hole in the can must be removed at once, but not for the AGR or LWR, where fuel rods with holes can be left undisturbed until a refuelling break. Plants could have very good load factors (80 to 90 per cent) with an annual shut-down for refuelling. The charging machine itself was only 'as reliable as its smallest micro switch and was likely to cause "outrage"'.

APC said that they and the German branch of Brown Boveri (BBC) were able by redesign greatly to reduce the capital cost of the AGR in Germany, and in 1969 BBC were again testing the German market – with prices which, BNDC later told me, frightened them, they were so low. But the offer to the chemical firm Badische Anilin (BASF), who wanted process steam as well as electric power, was still 20 per cent above the price asked by Siemens for a PWR.[76] It is, indeed, hard to conclude from the evidence to the Select Committee in 1969 that AGR prices competitive with LWR prices could have been offered and justified. The Dungeness B price was acknowledged to have been too low. By now, as the IPCS said, 'the well publicised difficulties on the Dungeness B station can hardly have improved the export prospect of the AGR'.[77] In retrospect one may wonder what would have happened had an export order been obtained – could the domestic troubles with the AGR have been avoided? Success would presumably have been an additional disaster.

The Select Committee concluded that the gas reactors had not been 'attractive to buyers overseas', and would have had to be 'significantly cheaper' than an LWR to attract buyers. 'There might arise of course situations in overseas countries where the AGR would prove to be an economic reactor', but this was not presented as very promising. Among the gas-cooled reactors, the HTR 'must be given priority as the leading large reactor for export. Otherwise foreign companies and utilities will continue to order the reactor types to which they are accustomed.'[78]

So in 1969 the Select Committee finally did not envisage an export drive succeeding with the AGR. Everything hinged on the HTR and SGHWR. The SGHWR, they agreed, would not sell for export easily, unless a British utility was operating one,[79] so there was no case for arguing that the exporting effort had been inadequate and there were technical problems to solve before a commercial HTR could be offered.

The premise implicit in the Select Committee's argument for a single company to secure exports of British reactors, that there were reactors to sell which the existing organisation did not sell, was without foundation. Was it reasonable to assume that the SGHWR or HTR would soon become saleable? This question is returned to later, but Lubbock, in some probing questions to BNDC, showed welcome alertness to the danger of counting technical chickens before they were hatched. BNDC had a new boiler disposition for Hartlepool. They 'did not anticipate any major trouble'. 'The trouble is [said Lubbock] we hear this with every successive advance. ... We heard it with Dungeness B, we heard it with the prototype fast reactor.'[80]

(vii) ALL AGRS IN TROUBLE: AND THE CEGB CHANGES COURSE

The Ministry of Technology announced in October 1969 that Britain would instal a further 4000 MW of AGRs, 'a most timely declaration of confidence', the Science Editor of the *Financial Times* wrote.[81] One plant, Heysham, was ordered under this programme. The second proposed, Sizewell, was finally not ordered, initially, because demand for power fell off. The AGR continued to fall in esteem.

By 1970 all British gas-cooled reactors, save the earliest and simplest, were in substantial technical trouble. All Magnox reactors, except the Calder Hall and Chapelcross units and the Berkeley station, were found in 1969 to be suffering from corrosion where the pipes carrying the coolant gas entered and left the pressure vessels.[82] The problem was discovered at Bradwell in autumn 1968 and affected primarily bolts and fixings.[83] 'Further deterioration' was being 'prevented' by reducing the temperature at which the reactors were operated, which reduced their output by 10 to 20 per cent. The Minister of Power, then Mr Roy Mason, had been assured, he told the House of Commons, that AGRs would not be affected: the manufacturers were 'at pains to point out that design differences mean that dioxide corrosion just cannot take place'.[84]

Meantime the last two Magnox plants, which were the first with concrete pressure vessels, both ran into difficulties in the insulation of the vessels. The Wylfa design took account of difficulties at Oldbury, but – possibly because it ran at higher temperatures and pressures – experienced new troubles in 1970. Oldbury still manifested insulation weaknesses in 1970–71.[85]

By the end of 1970, the two AGR plants ordered since Dungeness B were having problems and delays which were serious if not on the Dungeness B scale. At Hinkley Point B and Hunterston they were threefold.[86] First, despite statements that AGRs were immune from the Magnox corrosion problems, precautionary steps were taken. Parts of the mild steel liner were protected early in 1971 with an overlay of stainless steel: this required dismantling part of the graphite core. (The same was needed at Dungeness B.) Cool air was ducted round some of the boiler support structure. By 1972 it was feared after experiments that 9 per cent chrome steel used in the boilers would suffer from accelerated corrosion due to the $CO^2$; and the only remedy for this was seen to be derating the plant, as with the Magnox units.[87] Second, insulation again presented immense difficulties. The problem was more severe than at Oldbury. A great volume of gas at 600 psi (compared with 350 at Oldbury and 450 at Dungeness B) was forced by the circulators into a relatively small 'hot box'[88] and round two right-angled bends into the boilers. It resulted in enormous turbulence and vibration with noise levels 'greater than those found at the tail pipe of a jet engine'. It was, a leading participant in AGR development remarked, an almost insoluble problem – the proper solution was a design which avoided gas flows on this pattern.[89] The third Hinkley Point problem also concerned gas flow. When a fuel rod was changed, which happens often in on-loading charging, there was a large upward flow of cold gas 'which would cause the gas outlet temperature to flutter, creating boiler control problems'. To avoid this required a costly sophisticated bit of plant – an addition to the cost of on-load charging whose supposed advantage, estimated unimportant by Continental utilities, had turned the balance in the Dungeness B assessment in favour of the AGR.

The BNDC plant at Hartlepool was delayed when the Inspector of Nuclear Installations decided that their new design for concrete vessels was not safe. It was the first time the Inspectorate had asserted itself in this way.[90] The situation at Dungeness B continued to deteriorate. In mid-1970 the new-design boilers were not delivered (the design had been offered in a tender for Hartlepool) and the gas blowers were uncompleted three years after the contract date through difficulty over gas seals.[91] The boilers suffered from the chrome steel problem and other metallurgical difficulties.[92] Management of this development had been transferred to BNDC in 1969, and financial responsibility was taken over by CEGB – but it was said BNDC could not immediately throw great technical strength into the job, while APC staff left rapidly – and the work suffered from some demoralisation.[93]

At the end of 1971, Hinkley Point B was spoken of as a year behind the planned time schedule; soon it was two years, then three. Rectification of troubles took longer than was forecast; and there were new difficulties.[94] Mr (later Sir) Arthur Hawkins, who succeeded Sir Stanley Brown as Chairman of the CEGB in 1972, broadened the picture in evidence to the Select Committee in December 1973. 'They have once-through boilers in a very difficult situation where for instance it is difficult to control the water levels – you are facing two almost impossible reconciliations. You get corrosion outside in one condition and corrosion inside in the other.'[95] He emphasised the difficulty of access: 'Have you crawled through Hinkley Point? I have. It is so difficult to maintain, that the problem is keeping it in service for the fuel design life.'[96] It was now recognised that the AGR 'is inherently a difficult system and less economically attractive than was at first supposed'. The 'CEGB would not order another AGR of the designs we are now building... there are points in the design which would make the operation of those reactors difficult.'[97] There were also unquantified and unresolved problems of graphite corrosion and carbon deposit in the coolant circuit arising from the use of $CO_2$.[98] It was assumed that one plant would run at 18 and others at 10 per cent below capacity. Hawkins emphasised the 'difficulties were still in design, it was just misleading to say it was in engineering, not nuclear; whether there would also be faults in quality control would not be seen till the plants worked.'[99]

Troubles in gas circulators caused further delays to Hinkley Point B in 1974, and early in 1975 turbulence where the gas stream turns 90° into the boilers, caused new trouble and broke part of the equipment to control the gas flow – so the commissioning of the plant and of Hunterston B was put off again.[100] The first units at Hinkley and Hunterston finally began to send a little electricity on the grid – at 1 to

5 per cent capacity levels – in February 1976. By then, it was said, 'we had learned an enormous lot more about the gas flows in the last twelve months'. The CEGB in its *Annual Report* for 1975–76 said 'progress has been made in understanding the complex gag behaviour. However, because engineering solutions had to avoid major changes to the basic design they minimised rather than eliminated the movements and the associated wear.' Tests 'indicated an acceptable minimum life for the modified gags.'[101] At Dungeness B the earliest date for start-up of the first unit was put in 1975 at 1977. By 1976 it had become 1978. 'The construction of the concrete pressure vessels is complete, the CEGB stated in its Annual Report for 1974–75: 'Insulation of the first reactor is almost complete, and that of the second is well advanced.' This was ten years after the order was placed; the foundations for the pressure vessel had been completed by early 1968. The troubles of the AGRs built by the two stronger consortia had been lighter. Erection of both Hartlepool and Heysham was expected in April 1976 to be completed by 1979.

Two strands of this unfinished story may usefully be identified at this point. First, the AGR presented design problems whose significance was not perceived by those who recommended it in 1963–65, promoted its adoption in 1965 and sponsored construction based on a sketch design. Second, there was by 1972–73 an almost general readiness to agree that the system was inherently more complex than was at first appreciated, and therefore would be more costly than LWR systems even when the design problems had been overcome.

(viii) COSTS AND PRICES

The cost of AGRs, like the cost of LWRs, rose dramatically in Phase II. This was a result of a different mix of similar factors, inflation and higher interest rates, development difficulties, changes of design, and regulatory problems. Development problems were the major factor with the AGR and were all concerned with making plants currently being constructed work, not with making plants of higher capacity and greater economy. Regulatory problems were minor.

This section is concerned with two aspects of the cost situation: first, the implications of the trends for the long-term economics of the AGR; second, the total cost of making the AGR plants ordered, which is still not fully known.

We deal first with the long-term comparative cost trend for the AGR. The estimated construction cost per KW of AGRs ordered, at the date of ordering, and with no adjustments for inflation or revision, are set out in Table 9.1.

Stanley Brown pointed out how the first AGRs fell in price against the inflationary trend. But Heysham spoiled the picture. It was identical with Hartlepool and only 14 per cent of the 54 per cent increase could be explained by inflation. The increase was mainly to get, it was hoped, a reasonable rate of profit.

TABLE 9.1

Original Cost Estimates for AGR Plants[a] (excluding initial fuel and IDC)

| Plant | Date | £ per KW |
|---|---|---|
| Dungeness B | 31 March 1965 | 75 |
| Hinkley Point B } Hunterston B } | 31 March 1966 | 72 |
| Hartlepool | 31 October 1968 | 71 |
| Heysham | 30 November 1970 | 108 |

[a] Figures provided to the Select Committee by the CEGB on 8 January 1976, which are discussed below, p. 137. They show an increase of over 54 per cent for Heysham. At the time the order was placed the increase was said to be 40 per cent. *Financial Times*, London, 5 December 1970.

There were no more recent figures for the AGR based on contracts, because there were no more orders. Further increases would have occurred, if only because inflation became more acute. The cost of oil-fired plants ordered rose by almost 60 per cent between the end of 1971 and the end of 1974,[102] primarily due to inflation (including the effect of dearer imported materials). For the AGR additional costs arising out of redesign and modification also continued to increase after 1970–71.

The consortia and utilities were interested in determining the cost of a redesigned AGR if one were ordered in 1972–73, and some of their assessments became available in the form of comparisons with probable costs of other types of nuclear plant. The most widely publicised were given to the Select Committee by the CEGB in December 1973. Estimated costs for the LWR, HTR, SGHWR, Magnox and AGR were compared: the figures are set out in Table 9.2 (These are present-worth costs, not comparable, therefore, with the figures in Table 9.1.)

The PWR and Magnox figures were the most solid: these were the only two 'proven reactors'. The claims for the AGR in the Dungeness B *Appraisal* (circulated in six languages to drum up business), on which the Minister of Power based his assertion that Britain had made the 'greatest breakthrough of all time', were abandoned completely.

The comparable present-worth costs of the AGR and BWR in the 1965 *Appraisal*[103] were £127.68 per KW and £136.49 with generating costs of 0.46d (0.19p) and 0.49d (0.20p) per KWh. These assumed lower interest rates and higher availability than the 1973 figures, but a shorter life. The 1973 figures indicate a large increase for all reactor types due to common causes by the end of 1973 (there would have been a further rise of 50 per cent by the end of 1975), but provide no basis for a precise comparison of rates of change in real costs. Clearly, however, the British engineers likely to be best informed concluded that the LWR had substantially better cost prospects in the near future than the AGR or other reactors which were not yet being made on the scale of a commercial prototype.[104]

TABLE 9.2

Estimates of Present-worth Cost of Nuclear Plants[a] and Generating Costs in Britain in 1973 (1973 prices)

| Reactor type | PW cost (£/KW)[a] | | Generating cost (p/KWh) |
|---|---|---|---|
| | Whole plant | Nuclear steam supply system | |
| | £ | £ | p |
| Magnox | 366 | 116 | 0.72 |
| AGR | 293 | 89 | 0.57 |
| HTGR | 260 | 60 | 0.51 |
| SGHWR | 262 | 67 | 0.51 |
| PWR | 233 | 50 | 0.46 |

[a] The estimates were all for 2500 MW stations with two units, except for Magnox which was for two 500 or 600 MW units. They were for a fourth plant of a kind to be constructed in each case in which 'benefits accruing from technical and commercial development would have been realised and costs settled down'. The assumptions were 10 per cent discount rate, 25-year plant life, and lifetime load factor of 64 per cent. *SC Science and Technology, 1973–4*, Q. 349. Slightly fuller statements were given, p. 192, and in *Report of NPAB*, p. 26. Unlike cost figures in the United States these figures make no allowance for escalation: comparisons must therefore be made with American cost figures *less* escalation. They do cover interest during construction. Various aspects of these figures are discussed below, p. 229.

Fuller details included in a paper submitted to the Select Committee[105] showed that the fuel cost for the AGR, forecast in 1965 as below that for the BWR, was in 1973 estimated above that for the PWR. The cost of the initial charge for the AGR was put at £15.44 per KW for Dungeness B, compared with £17.26 for the BWR. For Hinkley Point B, the figure was put at £11.21.[106] In 1973, the AGR figure was £19; the PWR, £14. What Guard and Grainger had foreshadowed appeared to have been confirmed. It was of the utmost significance because it had always been agreed that the capital cost per KW of the AGR would exceed that of the LWR, but it had been stated that fuel costs would be lower.[107]

The other significant relativity was between the costs of generating in nuclear and fossil-fuel plants. The present-worth cost per KW of the Hinkley Point B plant had been put at much below that of the Drax coal-fired plant by the CEGB in spring 1967: £140m compared with £186m.[108] The major cost items in both rose faster than the general rate of inflation, but the estimated balance in favour of the AGR was almost certainly lessened and may have been eliminated by 1973, before the oil and coal price explosion. The change in ground rules adopted by the CEGB, higher discount rates and lower average load factor for nuclear plants, militated against the AGR.[109] But the AGR case was exceptional, and if the comparison were made between the LWR and fossil-fuel plants, before the oil price explosion, the nuclear costs remained lower in Britain as in other countries.

The CEGB, whether to bolster the case for more investment in nuclear power or to defend past investment decisions, claimed that during 1971–72 the best Magnox plants, Dungeness A and Sizewell A, already produced power at lower costs than the best coal or oil-fired plants.[110] The current prime costs were much lower for the nuclear plants, because it is the essence of the economics of nuclear plants that initial capital outlays are higher per unit of output and ultimate fuel costs much lower than for fossil-fuel plants. The figures given to establish the CEGB claim, however, described as 'actual production costs', included capital costs. These were based on the cost of the investments in the prices current when the plants were built, between 1960 and 1966.[111] In 1971–72 prices, the cost would have been about 60 per cent higher. For some accounting purposes, the 'historic' basis for estimating capital costs is found to be useful, but to add prime costs in 1972 prices to capital cost in 1960–66 prices is useless as a means of measuring the comparative use of resources. Resources used in the past (on which the capital costs were based) do not become less as the value of money subsequently falls.[112] The CEGB's figures were useless either for judging past investment or making new decisions. Nevertheless, it has repeated the exercise with a bare cautionary note.

The Department of Trade and Industry labelled these 'Accounting Cost Comparisons' and pointed out that they took no account of inflation. It provided the Select Committee with a comparison of costs

TABLE 9.3

Generating Costs at 1 January 1972 Money Values (p/KWh) in Magnox, Coal and Oil-fired Plants in Britain

|  | 8%<br>*Interest rate* | 10%<br>*Interest rate* |
|---|---|---|
| Magnox | 0.56–0.94 | 0.64–1.07 |
| Coal | 0.37–0.62 | 0.39–0.65 |
| Oil | 0.40–0.43 | 0.42–0.46 |

based on January 1972 prices – replacing the CEGB's historic cost valuation. This showed Magnox costs much above those of coal-fuelled and oil-fuelled plants of the same vintage. The figures are shown in Table 9.3.

The Department also published figures to show that, as currently assessed, the last of the AGR plants being built could have costs just below those of oil and coal-fired plants. See Table 9.4.

TABLE 9.4

Estimated Generating Costs at 1 January 1972 Money Values (p/KWh) in Nuclear, Coal and Oil-fired Plants under Construction in Britain

|  | *8%* *Interest rate* | *10%* *Interest rate* |
|---|---|---|
| Drax (coal) | 0.42 | 0.45 |
| Grain (oil) | 0.37 | 0.40 |
| Heysham (AGR) | 0.36 | 0.42 |

This made it a knife-edge decision between oil and nuclear. But the Department of Trade and Industry pointed out that LWR, HTR and SGHWR plants were all expected to have lower generating costs than the AGR.[113]

We turn now to the second problem: the probable cost of the AGR plants ordered in the years 1965–70. Our concern is primarily with the increase in real costs, not in higher costs due to inflation (although the total cost in today's prices is a guide to the volume of resources used), nor in the increase as shown in utility accounts. The CEGB provided some figures to the Select Committee early in 1976 which made the sum look almost simple. These are given in Table 9.5.

The CEGB said Heysham was the only plant where substantial replication was projected. This was true within the CEGB confines; but Hunterston B, ordered by the South of Scotland Electricity Board (SSEB) in autumn 1967, was based largely on replication of Hinkley Point B. Table 9.5 should, for this context, be extended by inclusion of a column for Hunterston B, broadly identical with that for Hinkley Point B.[114] If this is done the total 'current estimated actual costs' would be £1020m.

These figures show varied, but striking, increases in direct construction costs discounted for inflation, useful for conventional accounting. They do not give the total direct construction costs measured in prices for one base year, for example, 1965, or for 1975, and they do not include all the costs involved. They exclude interest during construction, which was enormously increased by the lengthening of the construction periods. They also exclude the cost of providing alternative supplies of power in the years during which the AGRs were

TABLE 9.5

CEGB Estimate of AGR Capital Costs (November 1975)

| | £ million | | | |
|---|---|---|---|---|
| | Dungeness B (1200 MW net) | Hinkley Point B (1250 MW net) | Hartlepool (1250 MW net) | Heysham (1250 MW net) |
| Date of original estimate | 31.3.65 | 31.3.66 | 31.10.68 | 30.11.70 |
| Original estimate at prices at dates given above | 89 | 95[a] | 92 | 142 |
| Current estimated actual costs (i.e. at prices actually paid or anticipated to be paid in future) | 280 | 140 | 220 | 240 |
| Current estimated costs at constant prices ruling at at date(s) of original estimates | 186 | 126 | 158 | 174 |
| Current estimated actual costs based on March 1975 prices[b] | 415 | 270 | 310 | 290 |
| Percentage escalation at constant prices | 109% | 33% | 72% | 23% |

[a] The figure was revised downwards, it is understood, to £93.6m in 1967, reflecting a reduction in the TNPG tender price. The estimates from Hinkley carried the cost of gas turbines beyond the need for emergency power; the MW figures refer to nuclear capacity only. See *SC Science and Technology, 1967*, p. 392.

[b] These figures were not in the table as originally published, but have been provided by the CEGB subsequently.

programmed to supply power at low fuel costs but did not. They allow nothing for the extra costs to the utilities of financing the supply of nuclear fuel, or for the cost to the AEA fuel-fabricating plants of providing capacity before it was needed and for a volume of output which would not be needed. They allow nothing for the R and D costs of the AEA, responsible for all the initial development and for a substantial continuing programme after 1965, or for the R and D work of the CEGB.

There are difficulties in estimating the total cost, and only approximate answers can be given, but sufficient to give orders of magnitude.

Table 9.6 (p. 139) gives an estimate of total construction costs in 1975 prices including interest during construction. Both the CEGB and the SSEB annual accounts give expenditures on nuclear plants, but the CEGB figures do not give separately expenditure on AGR and Magnox plants from 1965–66 to 1971–72. I am indebted to the Board for providing the figures, including their expected expenditure from 1976–77 to 1979–80 and later.

Interest during construction (IDC) is capitalised, annual interest payments being treated as part of annual capital outlay: they represent the values which the expenditures would have had in other uses. The CEGB does not capitalise IDC in its accounts; it covers annual payments out of revenue, so that they do not become a capital liability in the accounts on which a return should be earned interest. But the CEGB included IDC in its estimate of generating costs at the Dungeness B plant and has done so continuously since, using the discounted cash flow method.[115] This is right, and the logic requires that interest payments be capitalised.

The choice of the interest rate presents difficulties. I believe the figure should be an appropriate rate of return for a long-term DCF calculation, and have used the test rate of discount selected by the British Treasury first in1967 for the appraisal of projects in the public sector. It was then fixed at 8 per cent (which I have used also for earlier years) and raised to 10 per cent in autumn 1969.[116] The Treasury appeared to assume that an addition should be made for inflation. The interest rates may be taken to reflect inflation, but only partially – and certainly not at 1974–76 rates. Nevertheless I have made no addition for inflation, partly in order to err in these estimates on the low side.

The Table gives expenditure on AGR construction to 1975–76 – necessarily approximate since many assumptions are involved – and a forecast of the total which will be spent provided the rest of the programme is completed within current forecasts and assuming total expenditure as in Table 9.5.

In 1965 prices the construction cost excluding IDC of the five AGR plants, whose price per KW was expected in 1967 to fall until 1980, and did fall to 1969, would have been about £440m. The estimated cost of £1550m in Table 9.6 is approximately £610m in 1965 prices, a rise of 40 per cent in real cost. IDC rose much more, from about £100m to £220m in 1965 prices, an increase of 120 per cent in real cost.

Had electricity consumption not fallen far below forecasts, the delays of from four to eight years in completing the AGRs would presumably have led to power shortages. As things were, enough power could be obtained from existing fossil-fuel plants, but the fuel cost was much higher than it would have been in the AGR plants. The utilities had to meet an additional cost in providing the supply, and the whole of this can be regarded as a cost of the delays. The sums involved are large. The CEGB said it cost them £1m a month to provide replacement fuel for Trawsfynnyd (500 MW) in 1966–67:[117] by March 1973 they put the cost at £3.5m a month per 1300 MW reactor.[118] The figure rose as fossil-fuel prices rose. The total cost for the CEGB to the end of 1975 in current prices was probably in the order of £800m, with a prospective further £950m; in 1975 prices, the figures were £975m and £790m.

TABLE 9.6

Utilities' Expenditure on Construction of AGR Plants
(including IDC capitalised)

| | In prices at time of payment (£m) | In 1975 prices (£m) | IDC[a] (£m) |
|---|---|---|---|
| 1964–5 | | 1 | |
| 1965–6 | 7 | 19 | 2 |
| 1966–7 | 17 | 42 | |
| 1967–8 | 28 | 67 | 3 |
| 1968–9 | 33 | 77 | 6 |
| 1969–70 | 70 | 163 | 12 |
| 1970–1 | 97 | 207 | 22 |
| 1971–2 | 94 | 183 | 35 |
| 1972–3 | 124 | 223 | 50 |
| 1973–4 | 96 | 146 | 66 |
| 1974–5 | 85 | 109 | 81 |
| 1975–6 | 75 | 76 | 98 |
| Total to 1975–6 | 726 | 1313 | 375 |
| Estimated for 1976–7 | 90 | 80 | 75 |
| 1977–8 | 88 | 73 | 62 |
| 1978–9 | 72 | 55 | 33 |
| 1979–80 | 44 | 29 | 10 |
| | 1020 | 1550 | 555 |

[a] I have assumed in calculating IDC that the interest on expenditure during a year may be counted as due on the whole expenditure for one half of the year.

It may be argued that while there is an undoubted commercial cost it cannot be attributed wholly to the choice of the AGR unless by another choice the whole of this extra cost would have been avoided. The BWR programme was a possible alternative. From experience overseas it is likely that a BWR programme would have had delays (though relatively small), and the load factor would have been below 75 per cent. As a rough guess I am assuming it would have provided, in the period 1970–76, 60 per cent of the power expected from the AGRs, and on this basis the cost due specifically to the choice of the AGR may be put at 60 per cent of the total cost of replacement fuel as set out above. This reduced figure, set out together with the total replacement cost in Table 9.7, is used in estimating the total cost of the AGR programme – again partly because it is on the low side.

TABLE 9.7

Cost of Electricity from Alternative Sources

|  | In 1965 prices | | In 1975 prices | |
| --- | --- | --- | --- | --- |
|  | *Expenditure to end of 1975 (£m)* | *Final total expenditure (£m)* | *Expenditure to end of 1975 (£m)* | *Estimated total expenditure (£m)* |
| Total cost | 390 | 700 | 975 | 1745 |
| Additional cost of power which could have been available from an alternative nuclear investment (rounded) | 235 | 420 | 585 | 1000 |

The initial fuel for the AGR programme will cost substantially more than the 1965–67 estimated prices (around £70m plus interest after delivery and before commercial operation).[119] How much more published records do not show. The fuel is bought from BNFL on terms which are virtually cost plus, so the effect of inflation on refining and fabrication costs and enrichment will be passed on. Delay in the programme adds to cost cumulatively – because much fuel has been made and some delivered several years before operation. This fuel will therefore have to be financed for longer before it is used than was planned – some of it, the CEGB accounts imply, for four or five years before its use, and this cost, analogous to IDC, must be capitalised. The rate of interest also exceeds the rate current when the plans were made. A delay of five years would add nearly 60 per cent to the fuel investment. The figure of £200m in 1975 prices, used in Table 9.8 for initial fuel cost, seems a reasonable indication of scale according to available figures, but is probably low.[120]

There remains the cost of R and D. The AEA direct expenditure on the AGR to March 1975 was £137m, of which nearly half was spent before March 1965. In 1975 prices the expenditure was about £275m. One should add interest compounded and something for general reactor research contributed to the AGR.[121] The grand total, in 1975 prices would be around £400m. The CEGB also had a substantial R and D expenditure supporting nuclear activity. The expenditure (including capital expenditure) on all its R and D rose from £6.4m in 1961–62 to £29.5m in 1974–75; the total in 1975 prices was £334m. Of this, it is assumed, £100m was for nuclear work.

Table 9.8 sets out the probable total cost of the AGR programme in 1975 prices.

TABLE 9.8

Total Cost of AGR Programme (£m, 1975 prices)

| | |
|---|---|
| Construction cost including IDC[a]) | 2100 |
| Initial fuel cost | 200 |
| R and D – AEA | 400 |
| CEGB | 100 |
| Sub-total | 2800 |
| Cost of electricity from alternative sources | 1000 |
| Total | 3800 |

[a] The IDC included in construction cost is not in 1975 prices.

This is substantially higher than the cost of the Magnox programme (in 1975 prices) for a slightly higher output.

Hill, in evidence to the Select Committee in 1974, challenged the use of the term 'failure' of the AGR programme, simply because there had been so large an expenditure as yet unrequited. When the plants were operating they would save £200m on oil imports a year: how could such a benefit to the balance of payments be called a failure? When the AGRs worked their generating costs would be lower than those of oil-fired stations built since the AGRs were ordered.[122]

If the total capital cost of these plants including all the costs of all the delays and redesigns and remedial work at its up-to-date prices were included, the second claim was wrong. Hill neglected the fact that the resources used could have been used in other ways which would have been remunerative for several years. Presumably a BWR ordered in 1965 would (on the analogy of Oyster Creek and Dresden II) have operated from 1971–72. Saving on oil imports could have started before the crisis prices of 1973. The AGR programme protracted part of the balance-of-payments drain which could have been eliminated,[123] and the financing of the programme added to the government borrowing which added to balance-of-payments difficulties. No one in the Select Committee referred to the misconceptions in Hill's challenge; the proceedings read as though no one spotted the weakness, and some members appeared to find it comforting to be told the AGR had not been a mistake.

## NOTES AND REFERENCES

1. R. V. Moore and J. D. Thorn, 'Advanced Gas Cooled Reactors: an Assessment', *Journal of the British Nuclear Energy Society*, London, April 1963, p. 97.
2. Gowing, *op. cit.*, vol. ii, p. 263 *et seq.*
3. *Ibid.*, p. 289.
4. *Ibid.*, p. 291.
5. *Ibid.*, p. 289.
6. *Ibid.*, p. 291.
7. K. E. B. Jay, *Britain's Atomic Factories* (London: HMSO, 1954) p. 83. The foreword by Duncan Sandys is dated December 1953.
8. *Hansard*, 1 May 1954, col. 852.
9. Above, pp. 10–11, and Sir Christopher Hinton, *The Future for Nuclear Power*, Axel Ax:son Johnson Lecture) (Stockholm: Royal Swedish Academy of Engineering Sciences, 1957) p. 20: 'In the next batch of stations, completed say by 1962–3, the advantage will probably have passed by a small margin to the nuclear station.'
10. I quoted this passage from Dr Rotherham in Burn, *op. cit.*, p. 104, commenting that this was a rational explanation of what happened but that the procedure was not self-evidently the logical way of attaining lowest costs. The CEGB were asked by the Select Committee to comment on this, which they did (*SC Science and Technology, 1967*, Appendix 16, p. 403). Whether Rotherham wrote the comment or agreed with it I cannot say, but the Board did not appreciate what the problem was. They claimed that the favourable relative costs of the AGR had been *established* by the Dungeness B Assessment, which confirmed conclusions of the AEA economist Mr Jukes in 1957, and that the AGR had '*achieved*' the extensive development which was promised for the $CO_2$/graphite system in 1957. It is interesting that both these positions still appeared tenable to the Board at this date. Nothing had been established about the performance of the large AGR, nothing had been achieved.
11. Above, pp. 97–8.
12. Gowing, *op. cit.*, vol ii, p. 291. She does not bring out fully the extent to which the design, construction and erection was by contractors from private industry, and completion on time depended on all these firms being on time.
13. JCAE, *Development, 1955*, pp. 90–101, 138–90.
14. Dresden I was also on a new site, whereas Calder Hall was adjacent to the Windscale reactors and fuel reprocessing plants, so that many services including communications were available. Commonwealth Edison asked GE to design Dresden I towards the end of 1955; at that time 'the technology did not exist for Dresden. Dresden had to come from the development of a whole package of inventions.' Experimental work at Argonne provided the starting point. (Papers by Murray Joslin of Commonwealth Edison and George White of GE at the Japan/US Industrial Forum Meeting, Tokyo, 5–8 December 1961.) The Vallecitos BWR was built to 'develop the necessary fuel technology and check out many design features' for Dresden I. The nearest one can get, probably, in comparison for timing is

to say that the end-1955 position on Dresden was comparable with the June 1952 decision to make a feasibility study for a full-scale Pippa (at Calder Hall). But the Pippa team had already been quite a long time on the job. (Gowing, *op. cit.*, vol. II, p. 289.

15. Report of the House of Commons Select Committee on Science and Technology on *UK Nuclear Power Industry*, Report, Minutes of Evidence and Appendices (London: HMSO, 1969) Q. 631 and pp. 218–19. Hereafter cited as *SC, Science and Technology, 1969*. It had been the same at Calder Hall, where, for example, the coolant circuit was designed by Parsons (as well as the turbo-alternators), heat and radiation-resistant cable by BICC, leak detection equipment by Metro-Vick, heat exchangers by Babcock & Wilcox, charge machines by Strachan & Henshaw, *et cetera*. See Rolt Hammond, *British Nuclear Power Stations* (London: Macdonald, 1961) ch. 4, *passim*.

16. Report of the House of Commons Select Committee on Estimates on *UK Atomic Energy Authority*, Report, Minutes of Evidence and Appendices (London: HMSO, 1959), Q. 2375 (22 April). Hereafter cited as *SC Estimates, 1959*.

17. *Ibid.*, Q. 2374.

18. Report of the House of Commons Select Committee on Nationalised Industries on *The Electricity Supply Industry*, Reports, Evidence and Appendices (London: HMSO, May 1963) Qs 1090, 1106 (9 May 1962). Hereafter cited as *SC Nationalised Industries, 1963*.

19. Moore and Thorne, *op. cit.*, p. 101.

20. *SC Science and Technology, 1967*, p. 428, 'The Historical Record of the 36 Rod Cluster Fuel Element', note by TNPG Ltd. This is an illuminating document on the history of the AGR, referring to and quoting from Collaboration Committee minutes. I referred to the events, of which I had been informed by TNPG, in 1967, but there was no documentation at that time (Burn, *op. cit.*, p. 61, 88). The complete minutes of this committee should be enlightening. TNPG also claimed to have recommended in 1958 against the costly work on beryllium cans which cost £10m and failed.

21. I visited the firms in the United States at approximately the same time and we met shortly after returning to London and compared notes.

22. *SC Science and Technology, 1969*, p. 219.

23. *SC Science and Technology, 1967*, p. 428.

24. *SC Science and Technology, 1969*, p. 219. I was told on a visit to Windscale in August 1964 that the Authority was making a design study and asking firms to quote their price for about 80 per cent of the components. My note of the meeting proceeds: 'I pointed out that, although they may be doing this, since in fact they are not offering an order they may not get a very realistic price.' In retrospective I think my scepticism not unjustified.

25. *SC Science and Technology, 1967*, pp. 428–9.

26. *SC Science and Technology, 1969*, p. 219.

27. *Ibid., loc. cit.*

28. Sir Harry Legge-Bourke, a member of the Select Committee, gave the information in a question to Sir Stanley Brown. He stated that after visiting the Dungeness B site, he came away very firmly with the impression that, 'had it not been for the very special relation built up over the AGR between APC and AEA [APC] would not have got the contract for Dungeness B.' *SC Science and Technology, 1967*, Q. 920.

29. *SC Science and Technology, 1969*, Q. 624.
30. *SC Science and Technology, 1967*, p. 429 and Qs 919–20. The CEGB of course knew much earlier that APC were going to submit 'a proposal not in conformity with tender specifications for its own version of the British design' because this was public knowledge (for example, see the *Financial Times*, London, 15 December 1964). At that date it was said that APC still proposed to submit an LWR design based on Combustion Engineering's LWR.
31. See, for example, *SC Science and Technology, 1967*, Q. 1081. The Institution's Deputy General Secretary, Mr J. Lyons, said 'it is relevant to what I have been saying about the Authority's ability to contribute in this field' that is, in core design) that the idea that APC used so successfully came from the Authority'. It is possible that Lyons did not know the facts.
32. *SC Science and Technology, 1969*, Q. 623.
33. *Ibid.*, Q. 623.
34. *Ibid.*, Qs 224, 624.
35. *Ibid.*, Q. 624.
36. *Ibid.*, Q. 627. How many millions would have been needed it is still too early to know – the plant has not (at the time of writing) yet worked.
37. *SC Science and Technology 1967*, Q. 651.
38. *SC Science and Technology, 1969*, Q. 631.
39. *Ibid.*, p. 219.
40. *CEGB Appraisal*, p. 3.
41. I was told this by one of the AEA members of the team.
42. *CEGB Appraisal*, pp. 3, 6.
43. *SC Science and Technology, 1967*, pp. 394–5.
44. *CEGB Appraisal*, p. 2.
45. *Ibid.*, p. 3.
46. *Ibid.*, p. 23.
47. Report of the House of Commons Select Committee on Science and Technology on *The Choice of a Reactor System*, Report, Minutes of Evidence and Appendices (London: HMSO, December 1973, January 1974) Q. 796. Hereafter cited as *SC Science and Technology, 1973–4*.
48. See, for example, Guard, *op. cit.*, pp. 440–1.
49. *SC Science and Technology, 1969*, p. 219.
50. *SC Science and Technology, 1969*, p. 220.
51. *Ibid.*, pp. 213–4, 220–1.
52. *SC Science and Technology, 1967*, Q. 2.
53. *Ibid.*, Q. 52.
54. *Ibid.*, p. 44 (CEGB), pp. 72–3 (BNDC), p. 123 (APC), p. 412 (Benn) pp. 470–4 (Ministry of Power Working Party on Nuclear Power Cost), Q. 263 (Stanley Brown), Qs 990–1 and 999 (Benn).
55. *Ibid.*, Q. 898.
56. *Ibid.*, Q. 864 (Stanley Brown).
57. *Ibid.*, Q. 570. Eric Lubbock seemed to regard foreign buyers as so incompetent that if TNPG said that the AGR seemed unsaleable in export markets because it had not been bought, this would lead foreign buyers not to buy it. Similarly he appeared to suppose that my analysis, in sug-

gesting that the AGR costs in producing power would be higher than LWR costs, would hurt export prospects. Mr Benn seemed prepared to agree. (See *ibid.*, Q. 998).

58. *Ibid.*, Q. 685.

59. *Ibid.*, p. 429.

60. Grainger was responsible at Springfields from 1951 for establishing the R and D Laboratories concerned with Magnox and DFR fuel elements: he was chief metallurgist and deputy director R and D at Risley from 1955, and head of the metallurgy division at Harwell 1959–62. *Nuclear Engineering International*, London, November 1962.

61. *SC Science and Technology, 1967*, Q. 1383 *et seq.*, and pp. 431–5, 491–5.

62. *Ibid.*, p. 489 (and more generally pp. 485–93).

63. *New Scientist*, London, 3 September 1964, p. 551 (article by David Fishlock).

64. *SC Science and Technology, 1967*, Q. 470. Dalyell explained how he arrived at the £14.4m. In app. 27, p. 424 Raby's figures are set out.

65. For example, *ibid.*, Q. 923–4.

66. On this the CEGB said (see *ibid.*, Q. 865) that APC had submitted a bid for Hunterston B – the third AGR Plant – and of the price tendered for this the chairman of the South of Scotland Board authorised Stanley Brown to say that 'he does not recognise in any of the tenders received by him prices as low as that given by APC' in evidence.

67. See previous note and pp. 124–6.

68. *SC Science and Technology, 1969*, Qs 625–6.

69. *Ibid.*, Q. 172 (Kronberger), Q. 917 (Moore).

70. *Ibid.*, p. 86.

71. *Ibid.*, Q. 261.

72. *Ibid.*, pp. viii–xi.

73. I had discussions on this with TNPG in 1965, 1966 and 1967. I gathered that they had 'never thought as well of the AGR as Kronberger did'.

74. *SC Science and Technology, 1969*, Qs 792–4.

75. Professor Mandel, who controlled RWE's reactor policy and was the most influential utility leader with great influence outside Germany also as a consultant, told me in 1966 that he believed all plants would have to close for maintenance once a year, which the CEGB treated as avoidable.

76. I was informed by BASF.

77. *SC Science and Technology, 1969*, p. 86.

78. *Ibid.*, p. ix.

79. *Ibid.*, p. ix–xiii.

80. *Ibid.*, Q. 369.

81. *Financial Times*, 30 October 1969.

82. *The Times*, 24 September, 1969 (reporting an announcement by the Minister of Power in the House of Commons).

83. *Annual Report of the CEGB, 1968–69*, London, p. 3.

84. *Financial Times*, 15 November 1969.

85. *Annual Report of the CEGB, 1966–67*, p. 27, 1971, p. 41, *Financial Times*, 16 June 1970. The problem was intensively studied experimentally by the TNPG for the Oldbury plant, and they believed in 1965 they had 'optimised' the design 'with confidence in its ultimate performance'. See R. D.

Vaughan, 'Experience with Integral Gas-cooled Reactors', a paper given at the symposium on the Evolution of Proved Nuclear Power Reactor Concepts held in Rome, June 1965. At Wylfa in June 1970 when 'three or four recirculating fans were in use . . . the imbalance set up a mode of vibration that had not been anticipated which dislodged a plate securing the insulation'. It was decided 'that the area of the four blowers must be stiffened and strengthened.'

86. *The Times*, 30 March 1971; *Financial Times*, 27 September 1971, quoting Sir Edwin McAlpine.

87. *Nucleonics Week*, 3 February and 10 February 1972. *Financial Times*, 20 March 1973. The boiler problem was less serious for Hartlepool and Heysham because their boilers could be removed and for Heysham different steel could still be used.

88. Above, p. 119.

89. TNPG had developed an extremely costly method of insulation for Oldbury, based on stainless steel sheets and foil, and requiring almost 'watchmaking' refinements in application. They hoped to find a cheaper solution, using fibres, for Hinkley Point B. The ultimate solution was again immensely costly.

90. *Financial Times*, 13 November 1970.

91. *Financial Times*, 19 June 1970. For a design for Hartlepool APC included Brown-Boveri pumps.

92. *The Times*, 4 November 1972.

93. In 1969 the financial burden of the design and construction failures made it clear that APC – and its main constituent companies – could not complete the contract. The CEGB acquired the capital of APC for a nominal sum, and made a new contract with BNDC to complete the work as managing agency for the Board. The main partners in APC were to pay £10.5m of the additional costs involved in completing the station. *CEGB Annual Report, 1970–1*, pp. 24–5. Gordon Brown told the Select Committee in May 1969, 'We [APC] have had 31 per cent of our technical staff resign since July of last year. We have had 19 per cent of our senior staff resign in that period of time. We had twelve very senior staff resign last month. It is not only morale. You cannot do long-term planning and this is what your technical staff want to be associated with.' See *SC Science and Technology, 1969*, Q. 605.

94. See, for example, *Financial Times*, 20 March 1973.

95. *SC Science and Technology, 1973–4*, Q. 352. The choice of once-through boilers (Benson boilers) which were an advanced type of boiler, had been made in order to reduce the number of penetrations in the vessel. Vaughan, *op. cit.*, p. 5. The paper was concerned with the Oldbury plant. This was a clear case where experience was necessary to sort out the gains and losses from a design choice which incorporated an advanced technology and had a desirable object; every penetration weakens a vessel and requires 'engineered safeguards'.

96. *SC Science and Technology, 1973–4*, Q. 360. Vaughan, in the paper quoted in the above note, p. 9, spoke of 'the concentration of equipment within the confined space of the pressure vessel vault' as presenting *constructional* but not operational problems.

97. *Financial Times*, 3 August 1972, quoting Hawkins's evidence at an earlier meeting of the Select Committee. The full evidence is enlightening: Report of the House of Commons Select Committee on Science and Technology on *Nuclear Power Policy*, Report, Minutes of Evidence and Appendices (London: HMSO, 1973) Qs 147–57, 218–19. Hereafter cited as *SC Science and Technology, 1972–3*.

98. *Nucleonics Week*, 10 February 1972. *SC Science and Technology, 1973–4*, (18 December 1973), Q. 149. In their 1974–75 Annual Reports (pages 19 and 32 respectively) both the AEA and the CEGB stated that research and development on the carbon deposition on fuel-pin canning steels and on radiolytic oxidation of graphite had 'indicated that some coolant compositions [having lower methane and higher carbon monoxide content] inhibit the $CO_2$/graphite reaction and only produce small amounts of deposit at high can temperatures'. On this basis an 'agreed definition of gas composition for the early operation of the stations had been agreed'.

99. *SC Science and Technology, 1973–4* (18 December 1973), Q. 355. He 'hoped and believed this would not be too serious'.

100. *Financial Times*, 17 February 1975. 'A violent vibration within the reactor, fierce enough to be felt by the commissioning team separated by many feet of concrete from the site of the trouble, occurred on components used to regulate the gush of very hot high pressure cooled gas from the fuel assemblies at the heart of the reactor.' Since only a few of the 'gags' were damaged an optimistic view was that the method of assembling some of the gags was at fault.

101. *CEGB Annual Report, 1975–76*, p. 23.

102. The estimated cost per KW of oil-burning plants reported in *The Times* and *Financial Times* and based on CEGB handouts were: 1964, £48; April 1970, £61; June 1970, £60–£70; August 1971, £65; November 1971, £88; July 1972, £106; November 1973, £110; November 1974, £137.50. This sort of figure may be taken as roughly indicative but it does not provide an accurate index.

103. *CEGB Appraisal*, Table 1.

104. Hinton disparaged the figures by implication because they did not come from the AEA and may have relied largely on estimates by British plant manufacturers. Below, p. 155. But he thought that when the design parameters of the AGR had been written down to the safe levels tested at Windscale the cost advantages over the best Magnox would be marginal. But Magnox must be rejected because it is 'likely to prove expensive'. *SC Science and Technology, 1973–4*, p. 186.

105. *Ibid.*, app. 6, p. 192.

106. Replacement fuel plus operating costs were higher in the BWR in the 1965 comparisons, but lower for the PWR in the 1973–4 comparisons. (This is *not* a result of differences in BWR and PWR fuel costs.) The Dungeness B figures came from *CEGB Appraisal*, Table 1. The Hinkley Point B figures came from, *SC Science and Technology, 1967*, p. 392.

107. The comparison is more complex than the figures given show. There would be more heat in the BWR than in the PWR core; it is really the overall lifetime cost of fuel which must be compared. The CEGB figures did not give this, but they showed that fuel and operating and maintenance costs

in the PWR would have a present worth of £47m and that of the AGR would be £52m. It is probable that the difference was largely fuel cost (though partly maintenance), which meant that estimated total fuel costs were lower in the PWR than in the AGR.

108. *SC Science and Technology, 1967,* app. 2, p. 392.
109. For example, the discount rate of 10 per cent instead of 7½ per cent raised the capital cost but lowered the present worth of coal or oil consumed over the lifetime of the plant.
110. *Nucleonics Week,* New York, 20 April 1972.
111. The basis of the figures is shown in an article by Stanley Brown, 'The Nuclear Power Programme', in Lord Sherfield (ed.), *Economic and Social Consequences of Nuclear Power* (Oxford: Oxford University Press, 1972).
112. The CEGB calculation assumes a progressive fall in the real value of depreciation provisions below what was originally intended, and a diminishing real return to all who provided money to finance the investment, including a diminishing return on all internally generated funds.
113. *SC Science and Technology, 1972–3,* p. 97. It is perhaps not the clearest possible discussion of the principles involved, which are now more popularly understood than they were then. Why one coal-fired station out of eight had an exceptionally high figure just above the lowest Magnox was not explained.
114. The expenditure on Hunterston B seems likely from the latest SSEB accounts available at the time of writing to be close to the £140m in Table 9.5.
115. See, for example, *CEGB Appraisal* and *SC Science and Technology, 1973–4,* p. 392.
116. The Department of Trade and Industry in cost estimates in 1973 used both 8 per cent and 10 per cent in calculations of capital costs. *SC Science and Technology, 1972–3,* p. 97.
117. Above, p. 126.
118. Report of the Nuclear Power Advisory Board, *Choice of Thermal Reactor Systems* (London: HMSO, 1974) p. 15. Hereafter cited as Report of NPAB, 1974. The CEGB 'estimate that with the AGRs the cost of fuel to replace the electricity production lost through delays will exceed the original estimated capital cost'. Fuel prices rocketed further after this.
119. The cost of the initial fuel charge for Dungeness B was put at £15.4m in 1965, £16.01m in 1967; for Hinkley Point B the cost (in 1967) was given as £11.21m; I recall no explanation of the difference – it could be in the content of enriched uranium. The estimate of £70m is based on the average of the two: but possibly a lower figure, £60m (assuming that all the later plants would follow Hinkley Point B), should be adopted.
120. CEGB payments for initial fuel from 1970–71, to 1975–76 were £82.6m, of which £31.8m was in the first two years. £9m was planned for 1976–77. To these figures expenditure for Hunterston B must be added which would bring the expenditure, in current prices, to £100m up to 1975–76 (in 1975 prices about £155m). Interest at 10 per cent for (on average) two and three-quarter years would bring the total to £200m. There was going to be further expenditure – some already announced for 1976–77. The large CEGB expenditure in 1970–71 and 1971–72 would have been partly for

Hinkley Point B, partly for Dungeness B – so that the average period during which fuel needs to be financed may well exceed two and three-quarter years.

121. The figures are based on an estimate of reactor-years operation lost annually. At 75 per cent load factor (to be expected in early base load operation) a 600 MW reactor would have produced upwards of 4000 GWh a year. Average margins between working costs of nuclear and fossil-fuel plants are published by the CEGB; they were 0.175p in 1966–67, 0.22p in 1971–72, 0.54p in 1974–75, 0.56p in 1976–76. The excess cost of power bought from fossil-fuel stations would exceed the average margin appreciably since it would come from stations below average efficiency, relatively low in the merit lists.

122. *SC Science and Technology, 1973–4,* Q. 799.

123. If the expenditure on AGR had been used outside or partly outside the energy field, it could also have provided exports; the BWR option is taken for simplicity.

CHAPTER 10

# Sources of the AGR
# Disaster

(i) WHOSE RESPONSIBILITY?

When Arthur Hawkins told the Select Committee on Science and Technology in the British House of Commons that the AGR was inherently a difficult system and less economically attractive than at first supposed, the Chairman, Mr Arthur Palmer, asked why the CEGB had been so over-confident earlier. Hawkins replied that the question should be put to the AEA, who had advised the CEGB, who had been under 'somewhat fairly heavy pressure',[1] whether exclusively from the AEA was not clear. The question could rightly be put not only to both the Authority and the Board, but also to ministers responsible for the decisions which turned out badly and, too, to the Select Committee itself, which for long attacked the consortia for failing to export the unsaleable. But Hawkins was justified in implying that the AEA had primary, initiating, responsibility. The politicians who created and continued the organisation had a more fundamental responsibility.

The key to the AGR edifice was the Dungeness B *Appraisal*, summarising the grounds on which the reactor was ostensibly preferred. This was a CEGB document, but appeared to reflect AEA domination. Nevertheless the two jointly concluded that the AGR as currently developed would produce power more cheaply than a BWR, that the advantage would be greater later because of its development potential and that the AGR did not present technical problems which 'could not be solved on the Dungeness B time scale', a short scale on any reckoning, but all the more so since the successful tender was, as the AEA and CEGB knew but not the outside world, only a sketch design. The *Appraisal* was an unqualified commitment to these conclusions.

It is now clear to everyone that all three judgements were wrong. It did not need hindsight to discover this. It was seen by a number of critics at the time; it was a major theme in *The Political Economy of Nuclear Energy* in 1967. In its comparison of American and British reactors, the *Appraisal* seemed plainly biased; in its acceptance of

150

extrapolation from a 30 MW prototype to a 600 MW commercial-scale plant with few of the main features of the prototype, it seemed, to many engineers, rash and unwise. Hawkins in effect confirmed that the contemporary critics were right: he told the Select Committee in 1973 that the major mistake over the AGR had been to build a 600 MW reactor plant on the basis of a 30 MW unit, increasing pressures, temperatures, fuel 'and all that', and to compound the error by building not one but four plants (for the CEGB) with three different designs.[2] It was worse, for a fifth plant was built for the SSEB.

If you had wanted to try out three different variant designs, all that was needed, on the American approach, was three different reactors. Among the immediate engineering critics of the LWR/AGR comparison, some came from the consortia, but their remarks were perforce muted. The TNPG view was put in 1967 by Mr S. A. Ghalib, the Managing Director, to the Select Committee, although with a little obscurity. Broadly, he said, the *Appraisal* compared yesterday's BWR with tomorrow's AGR, implying that if you had chosen tomorrow's BWR for the comparison it would have been in front. He thought their own comparison of yesterday's BWR with yesterday's AGR, for which the CEGB had asked tenders, gave the right comparative results.[3] I gathered in 1965 that TNPG put the margin of advantage at about 10 per cent in favour of the BWR.[4]

This percentage assumed that initial development problems were solved. It did not cover the AGR costs which arose from delays and redesign due to excessive extrapolation. But the narrative shows that, with one exception, all the serious delays and extra costs in the AGR development arose primarily from the major changes in parameters, for which there was little or no experimental backing; increases in coolant pressures and temperatures, an enormous increase in the volume and rate of flow of coolant, and the use of a concrete pressure vessel enclosing both the core and the boilers (which, if costs were to be kept down, required a concentration of much equipment in a confined space and led easily to the kind of gas-flow layout referred to earlier).[5] These changes resulted in the subsequent corrosion, turbulence, vibrations, destruction of insulation and control equipment, cold spots, and so on, whose occurrences have been traced. Troubles of these kinds were inherent in the decisions taken over Dungeness B. No doubt the degree of trouble was increased by the hurry in which the design incorporating the new fuel bundles was made. In so far as the troubles lay in corrosion, these were in the area of design for which the Authority was directly responsible: the choice of materials, a problem specifically referred to in the *Appraisal* as implicit in the move to higher ratings and temperatures.[6] Design problems arising out of the mass and flow, not the corrosiveness, of the coolant were in the purview of the consortia, but the judgement that all problems could be solved within the time scale of the Dungeness B contract necessarily embraced these too. They were probably problems with which the AEA engineers were not closely familiar – the AEA never designed much of a plant, even of the nuclear island, even in their own prototypes[7] – but if their competence was limited their judgement should have been matched to this.

In the rest of this section the parts played in the AGR disaster by the main groups involved, AEA, CEGB, consortia, ministers, and civil servants, are examined *seriatim* in this order. Because of its initiating role, and its influence as principal adviser, the AEA has pride of place.

## (ii) THE ATOMIC ENERGY AUTHORITY'S PART

The AEA chose the AGR, developed it in isolation in a rather small 'prototype' inadequate for extrapolation to 600 MW, assessed its relative qualities and prospects very highly and promoted it strongly for use in the Dungeness B plant and as the standard reactor for the second nuclear plant programme. As principal adviser to the British Government and the CEGB on all nuclear matters the Authority recommended the AGR as a competitive reactor promising lower costs than the BWR, and it persuaded the Government and the CEGB to make 'commercial' AGRs of upwards of 600 MW capacity the basis of the Second Nuclear Power Programme. It was the exclusive developer and exclusive promoter of the AGR.

In doing all this the AEA made a long string of serious misjudgements. The assurance of those developing the AGR at Windscale in summer 1964 is breathtaking in retrospect. (The assurance expressed about it by Penney to the Select Committee in 1967 is hardly less so.[8]) They were asserting in 1964 that the gas-cooled system, of which the AGR was now the spearhead, was the most advanced reactor system available for nuclear power plants now and in the immediately foreseeable future; that there was more experience with the gas-cooled system than with any other system; that except for the fuel all the features of the AGR were similar to those in the Magnox stations; that in its essential engineering features the AGR closely resembled the Magnox, so that all improvements made in the design and construction of Magnox plants 'can be directly applied'; that it was the safest system; that Windscale experience had confirmed (!) that the AGR as currently developed was competitive with fossil plants; that its potential was such that 'granted success in certain fuel element development already started later versions will compete with the most advanced types of nuclear stations, such as fast breeders, now being developed'.[9]

These assertions were all without foundation. The AGR 'spearhead' of this 'most advanced system' was not in fact 'available' in 'the foreseeable future'. The experience with gas-cooled reactors was limited to lightly stressed plants – the highest pressure at which an operating plant worked was 150 psi – and only three plants larger and more advanced than the elementary Calder Hall type had operated at all, none before summer 1962. The most conspicuous change under development for Magnox plants which would operate at pressures over 300 psi was the concrete pressure vessel; the first did not start up until Christmas 1967. The design was known to present difficulties and was certainly not 'optimised'.[10] There were difficulties over the insulation of the Oldbury vessel (which operated at 350 psi) in 1966–67, and still more with the Wylfa plant completed three to four years later in 1971. By this time all save one of the CEGB's Magnox plants had had to be derated by 10–20 per cent because of unforeseen corrosion problems; and the last two, the most highly stressed Magnox plants, had much less successful load-factor records than the earlier ones. The AEA showed no awareness that this might happen. No AGR prototype above 30 MW was completed or in the process of being constructed in 1964, and the 30 MW unit operated at 300 psi, compared with the 450 psi for the first and 600 psi for the subsequent 'commercial' AGRs, and it had a steel pressure vessel. The 'gas-cooled' experience relevant to a 'competitive' AGR was trivial compared with the prototype and experimental bases for the LWRs which were impressively planned and integrated.[11] The claims for future cost reductions in later AGRs were unrealistic before the first large AGRs existed; and the claims for greater safety had no significance if the LWRs were 'safe enough', which was accepted by the Dungeness B *Appraisal*.

Nevertheless it was against the background of these extravagant, unsubstantiated, unjustified claims that the fateful decisions of 1965 and succeeding years, up to the end of 1970, were taken: the decision to order a twin 600 MW reactor plant for Dungeness B without a real prototype; to order it without any experimental verification of the AGR system at high temperatures and high pressures, and several years before any experience was obtained from concrete pressure vessels; to order the Dungeness B plant before it was designed in detail and to start construction before there was a complete design; to design the plant before there was adequate knowledge on the corrosive character of the coolant; to order four further twin unit plants of two variant designs shortly after the first, before the bases on which an improved design could rest had been substantially strengthened and before the work on the first order provided any feedback for the successors; to continue ordering after the system had been rejected by foreign buyers, after much redesign of the first unit had already proved necessary and after the AGR had been labelled as a relatively high-cost system by those within the AEA developing the SGHWR.

By 1973 – still some years before any AGR was to work, but eight years after the crucial decision to adopt it was made – it was almost universally recognised (with Hill alone perhaps half demurring) that this series of decisions had been a major blunder. The second nuclear power plant programme, which excluded the only 'proved' commer-cial system, had repeated the twin errors of the first, identified in *The Political Economy of Nuclear Energy*; premature concentration on one inadequately proved and developed reactor system, and premature building of a long programme of so-called commercial plants based on this system. Eight years after the repetition, this was admitted and presented as a discovery. The troubles were on an unprecedented scale, and it was no longer possible to close one's eyes, although it was still possible to at least half repeat the mistake.

To explain fully the source of the AGR disaster, it is necessary to explain why the AEA made the string of misjudgements and why its misjudgements were accepted for so long as the basis of major deci-sions. We turn now to the first of these questions; the second is dealt with in discussing the parts played by the other groups involved.

The misjudgements over the AGR were not isolated instances of AEA misjudgement. The forecasts of Magnox prospects in 1956–57 – included by Hinton in his Stockholm lecture – were notoriously over-optimistic both as to the trend of costs and the timing of development. The cost of power from the first Magnox plant in the first programme was above 80 per cent over the forecast of 1957. The cost forecasts were wrong in regard to the nuclear plants (though for long this was denied), the efficiency of the coal-fired plants with which they were compared, the cost of materials (coal, graphite), the value of the main by-product (plutonium).[12] The exaggerated optimism of Hinton's 1957 figures was set out clearly by Mr K. L. Stretch, Manager of Calder Hall from 1954 to 1957, in an article published in December 1958. 'Unless the facts are more carefully assembled and the analysis is carried out more critically they cannot justify the value presently attached to them.'[13] He documented this comment conclusively.

The date when nuclear power would be competitive was put later repeatedly by the AEA and by Hinton in evidence to Parliamentary Committees. In 1959 Hinton, who in his Stockholm lecture put the date at 1962–63, moved it on to 1970: by 1962 it was to be the 'early 70s'.[14] The fast breeder reactor was at one time spoken of for industrial use by 1965.[15] By 1959 this had, Hinton said, receded 'towards 1970'; but the Dounreay directors put it at 'early seventies'. By 1962 Hinton had found the development disappointingly slow, and did not expect it before the second half of the 1970s. He could not feel certainty about any date. In 1967, Penney said the prototype would be on power by 1971; that made for an order for a commercial plant by 1974, 'so the first one is on line in 1979'. 'The French are several years behind us, but they now have a tremendous effort . . . but I do not think they could have a prototype of our size [250 MW] for four or five years after ours.'[16] Mr R. V. Moore, Managing Director of the Reactor Group, saw the danger coming after the Prototype Fast Reactor (PFR) was finished (hopefully in 1971): 'If the same time delays were introduced between PFR and the [first] civil fast reactor as occurred between the Windscale AGR and the first [commercial] AGR then our lead would be lost.'[17] The time gap between the first full-power operation and the order for Dungeness B was two and a quarter years. The French however, despite Penney, won the race for the first prototype: the British were several years late. Moore had been concerned with the wrong question. There is now no pretence anywhere that commercial operation is in sight: the first 1000 MW plant can only be a high-cost prototype, and not available at the earliest until the late 1980s.[18]

Hinton said the forecasts he was giving in 1962 were 'the best forecasts by the best people', and in 1973 he regretted no reliable estimates of costs of different systems were now possible because there was no longer a strong reactor development division in the AEA. The best estimates by the best people, who had misled so consistently for so long, were no longer available. 'We have to rely', he said, 'on a single industrial consortium whose estimates may possibly reflect what they think to be in their commercial interests, rather than what is in the national interest.'[19] All the Magnox and AGR disasters fell presumably into this charmed category – the national interest.

The vigour in stating the case in favour of the AGR in 1964 was increased because it was challenged. As a result of the narrow front, it was the only competitive reactor the AEA had to offer at the time. Compared with the LWR, it was a somewhat late developer, and there was a fear, as H. W. Jackson put it, that because of this the greater virtues of the AGR might be overlooked.[20] The AEA wished to avoid this. The CEGB had appeared close to preferring the BWR in 1964, and one (probably two) of the consortia appeared to rate it above the AGR at that time.

A critical situation was created for the AEA by the Government's announcement in April 1964 of the Second Nuclear Power Programme: 5000 MW would be ordered, to come into operation between 1970 and 1975. The Government, the White Paper said, would decide later, after expert comparisons, whether the plants would be LWR or AGR.[21] There was no urgent need in the demand situation for a programme at this date. The White Paper suggested that it was needed in order to give orders to consortia and keep adequate capacity available for large future activity. There was another way of doing this: asking and financing firms to develop prototypes of different systems. The AEA was disappointed that the programme was not larger; it was said to favour 10,000 MW. It had inherited from the Ministry of Supply's

Atomic Energy Division a continuing zeal for regularly-growing nuclear power plant programmes, to supply cheap power and provide the basis for an export business in nuclear plants. The idea that Britain should keep some of the export business out of American hands goes back at least to 1951.[22] And the Authority assumed it would provide the technical base for the programme. But if the LWR were chosen, this assumption would no longer be valid.

One had therefore to choose, it was argued in the AEA, between closing down British development, thereby throwing overboard the work of a great number of British scientists, or excluding American-type reactors. If Britain, which had started all this development (a misleading assertion) and had more experience in making nuclear power (the American gloss was 'in making high cost Kilowatt hours in which we are not interested'), could not hold her own with the Americans now, the prospects would be worse later on.[23] It was in this emotive atmosphere that the hurried sponsorship of larger fuel bundles and higher temperatures and pressures (hitherto turned down)[24] occurred and all the AEA development leading to the APC bid.

These circumstances encouraged putting the case for the AGR with maximum force. Many people in the AEA also felt drawn towards putting the case against American achievement with maximum force. There was almost a sense of neurosis. The sources of this were not difficult to trace. The initial post-war phase in which the McMahon Act ruled out the giving of atomic information by the United States to the United Kingdom and continuing the wartime collaboration naturally caused resentment: it prompted a disposition to say 'well, we can do very well on our own', and soon it was not just 'very well' but 'better'. Finally, the sense that the United States had such enormous resources was oppressive; any chink in the armour was looked for with zeal, and successes were disparaged. Faced with the challenge of the LWR in the United Kingdom itself, the British delegation to the third Geneva Conference on Peaceful Uses of Atomic Energy in September 1964, led by Penney, was, according to *The Economist*, 'whipped into the party line, with daily early morning brainwashing in their hotel. Once it was only the Russians who did this sort of thing.'[25]

There were these subjective factors. But clearly advocates of the AGR believed there was a strong objective basis to the claims they made. Penney must have regarded his 1967 claims for the AGR (and the FBR) as completely realistic: so perhaps did the developers, who would have advised him. How at these various levels the AEA could make this series of misjudgements and remain convinced, even in 1967, that in the AGR it had a technical and economic lead (although perhaps not a temporal one in the short run) remains to be explained.

The explanation seems threefold. First, the AEA had less skill and experience than the American firms in this type of development, and was less aware of what was needed for success. Second, there was no competitive environment in Britain: there had been, as emphasised earlier, no time or cost pressures until now, and no bases for comparisons. Third, the division of functions between Authority and consortia, which lessened efficiency, probably reduced the Authority's caution in forecasting.

The AEA had less skill and experience than the American firms in three respects. The Authority started on nuclear engineering development work several years after the major firms in the United States (whose experience started during the war); it had no corporate tradition in the conduct of this kind of development, which is a massive responsibility for top management as well as for lower echelons (though a few of its engineers, like Hinton, had some, though not completely parallel, experience in private industry before the war), whereas the major American firms had great experience; and it did not embrace so wide a range of the activities needed in developing a reactor from the initial research work to the construction and proving of a full-scale competitive reactor as the American firms. It did not have experience of as wide a range of the research and engineering technologies which had to be integrated. This was illustrated by the extent to which it subcontracted design work on its own reactor plants to private concerns.[26]

These relative deficiencies, although not inherently surprising, were not widely recognised inside or outside the Authority, and they were at odds both with the awe with which the Authority was commonly regarded and with what even warm supporters appreciated might be regarded as arrogance in some of its postures.[27] But there were occasional insights from within. Hinton, giving evidence to the Select Committee on Estimates in 1959, and arguing in the familiar way that 'we do not have the resources to attempt the diffused approach . . . adopted by the Americans', said that although the AEA had always had a reasonable share of really good young scientists 'the problem is . . . to get the older, more experienced men who can direct, guide and coordinate the work of these younger people. It is extremely hard to recruit these people from industry. . . . Industry usually gets these top men by training them. The real difficulty has been in getting these senior people, and I see no quick solution to this problem.'[28]

Hinton's evidence formed an illuminating complement to an analysis of weaknesses in AEA development practice by Stretch in the article quoted earlier. He said there was 'a widespread belief in the Authority that all engineering progress is made in the drawing office and the laboratory' coupled with a lack of appreciation of the operational aspects of development work, and that this was a major source of over optimistic estimates. The time needed for rigorous operational proving was badly underestimated and design lacked feedback from operational experience 'which assumed greater importance as nuclear reactors became more complex'.[29] Stretch, as works manager at Calder Hall until he joined the NCB in September 1957, was an operations man, seeing problems from that point of vantage. He was not arguing the Americans were better; indeed, he was conventional in assuming that Britain had a lead, but he thought the AEA method must result in relative failure.

The two analyses make comprehensible both the excessive optimism in forecasts and the exaggerated value put on early Magnox experience, the lack of any foreshadowing in 1964–65 of either Magnox or AGR setbacks, the abandon shown in adopting the new fuel assembly and higher temperatures and pressures for the AGR without extensive testing, the under-estimate of the solidity of American LWR development and the tendency to neglect American work – until it could no longer be disregarded. In 1958–59, leading AEA figures made it clear that they were not aware that a 200 MW BWR prototype was nearing completion, with satellite reactors and other development tests supporting it,[30] and the Dungeness B report gave a fantastically misleading picture of the development of fuel for the BWR.[31]

When the Windscale developers sponsored the higher ratings in 1964, in face of the BWR challenge, they said they were adopting the American practice, which the CEGB did not favour, of promising technical achievements which you knew were within your reach but which could not be proved from tests actually made and so could not form the basis of specifications.[32] For them there was nothing more incautious in pushing the AGR from a 30 MW prototype, plus largely irrelevant Magnox experience, to a 600 MW commercial plant in 1964–65, than in pushing the BWR from the 200 MW Dresden I plus relevant 50 to 70 MW satellites and other large-scale experimental facilities to test fuel and components to 640 MW at Oyster Creek. Mr J. C. C. Stewart, then the Reactor Member of the Authority, later Deputy Chairman of the Nuclear Power Company (NPC), seemed to sustain this view in 1967 before the Select Committee: 'I do not', he said, 'regard the extension in gas-cooled technology into the AGR as being any greater or less than the extensions brought about between the early water reactors and those that are now being ordered.'[33] There was an enormous difference in the degree of proving behind the two developments – the Americans did what Stretch thought necessary, the AEA did not – but this had yet to become widely recognised. Penney's remark in 1967 on the AGR troubles – 'there is a great commotion about something, it looks as if it was not right, and then everything converges on it and it always melts away'[34] – implied that nothing could go disturbingly wrong. Hill, his successor, once conveyed the same impression to me in a discussion, saying that technical troubles can always be solved within two years: it was the organisational faults that were difficult to correct. The two types are sometimes

– indeed usually – related, and technical troubles, as with the AGR, can come in a cascade. The principle of designing plants in a rapidly growing technology where plants take several years to build, so that they will incorporate advances which will be completed as the plants near completion, is admirable so long as the promises can be kept. By and large, the American firms kept them; the British industry did not.

Readiness to rely on drawing-board development without feed-back from experience would explain the readiness of the Authority to support the APC tender, which was only a sketch design with many engineering ends not identified and tied up, and was drawn up by the weakest consortium which was not able to complete the design in detail (a very imperfect one as it turned out) unless it recruited more senior staff, which it did partly from the AEA. This bore no relation whatsoever to American practice.

The complete lack of a competitive background compounded the threefold deficiency in development, skill and experience. The AEA developers had nothing with which to compare their work: no standards save their own image. As a result of the narrow front, there was no possibility of internal comparisons within the Authority. As a result of the AEA monopoly there was no domestic comparison from outside the Authority. There was no competition from overseas (except in export markets where Britain invariably lost). There was no competition at all so long as the United States was not offering a competitive reactor (because the AEA had the development monopoly); and, as Dungeness B showed, competition could even then be heavily weighted in their favour. The AEA developers appeared very incompletely informed on the development of the LWRs, which was to some extent inevitable, but probably accentuated by unreceptiveness. They had no precise knowledge of the extent, speed and solidity of other development projects.[35]

The absence of competition deprived not only the developers but the members of the Authority, the top management, and all those who at different levels from the chairman down advised ministers, civil servants and the CEGB of any bases of comparison. They had no criteria with which to test the claims made by the developers that their product was better than their rivals'. They could not, as events proved, judge the validity of the claims of developers within their own organisation for their own reactor. This lessened the possibility of good internal management, and made the position of the AEA advisers ambiguous at all levels. There was an inherent conflict between the interests of the Authority as developer and adviser; and when the advisers lacked any basis for independent assessment, the prospects for the arrangement working well, in spite of the conflict, were greatly lessened. When the AEA used its developers also as advisers the position was almost farcical.

The original criterion of 'competitiveness' was the cost of generating power from coal, not from another type of reactor, nor from oil, which was cheaper than coal. For Britain, it was an easier target than for the United States, since British coal was much more costly. But although 'competitiveness' in this sense was something to aim at, and was used as such, it was not a substitute for commercial competition: it did not put a developer under pressure. It was soon realised that Hinton's 1957 figures, promising competitive power by 1962–63, were wrong – the nuclear too low, the coal too high. Hence fewer Magnox plants were ordered. This was harmful to the consortia, who had less business, and to the CEGB who suffered because its nuclear power was not as cheap as was promised. It did not matter to the AEA. Competitive nuclear power was put off for a decade – development would continue on the same course. The cheapness of coal-based electricity underlined the fact that there had been – and was – no urgent need in the 1950s for nuclear power. It would become necessary, possibly in the 1970s; hence continuous development was needed so that nuclear plants could be built which would work successfully when power supplies from coal (and oil) could not be increased to meet growing needs. When that day came nuclear power would not *have* to be competitive, although it would be an advantage if it were. In any case, development would in time bring down its cost.[36] So the programme would keep the AEA in business. As Penney said, rightly for the AEA, this was a costly, but not a risky, business.[37]

It was a cosy position so long as the Government, Parliament, the Select Committee in due course, and the press remained indulgent and were prepared to accept explanations of delay which absolved the Authority of responsibility as developer and adviser, and so long as foreign competition was not allowed.

There remained a third probable source of AEA confidence in misjudgement: the division of functions between it and the consortia. The AEA were, reasonably, concerned over the loss of time and money involved in the transfer of their know-how to the consortia. This was wasteful; but it could not have contributed to the inadequacy of the AEA's development of the AGR, the source of most of the subsequent trouble. It is possible, however, that the division of responsibility for development between AEA and consortia, which was paralleled by (and possibly also reflected) the reliance of the AEA on subcontractors for most of the non-nuclear engineering development for their own prototypes, led to a lack of caution on the part of the AEA in forecasting. Uncertainty about the engineering development might conceivably have had the reverse effect and induced excess caution in forecasting. This did not occur. The AEA felt a greater responsibility for 'nuclear' design than for the rest – they pointed out that the early design troubles of the AGR were 'not nuclear'. They seemed to imply that the 'non-nuclear' problems, concerned with corrosion, turbulence and vibration, insulation and the like, were not only not the Authority's responsibility but of a lower order of merit – although by now, as the challenge from other countries showed, nuclear engineering was becoming routine, and the general engineering problems became the most crucial in determining the time and cost of the final stages of nuclear plant development. It is possible that a sense of diminished responsibility may have been induced in the AEA by the consortia arrangement and may have reduced the seriousness of its assessment of the overall engineering problems of the AGR, and obscured the high cost of dealing with these and the length of time required.

(iii) THE CEGB: BOND OR FREE?

The CEGB, being nearly the monopoly buyer, had a commanding position in the choice of nuclear plants – if left with a free choice. Ostensibly it chose the AGR freely. The Board produced and published the Dungeness B *Appraisal* and was committed in public to its judgements and assessments. For some years, leading members of the Board referred to this comparison of AGR and BWR costs as though it was authoritative and final. The Board had accepted APC's departure from specifications based on the Windscale prototype in order to introduce 'well substantiated advances' which might be available before the plant went on load. The Board also stated that nothing later than the Oyster Creek design, rated as only 515 MW, could be accepted for a BWR. Thus, while APC was allowed to 'push the technical limits' of the AGR in the way attributed to the Americans, TNPG and GE were *not* allowed to 'push the limits' of the BWR, in ways adopted in the United States by leading utilities. The advances turned down for the BWR have been proved by several years operation.[88] Such decisions – they were not isolated – contributed conspicuously to the bias of the *Appraisal* against the BWR.

How free the CEGB was in all this is not certain, and the Dungeness B performance, in retrospect damaging, must be read in two contexts: what had gone before, and what Hawkins said subsequently in 1973.

The nationalised utilities had been treated by successive governments as subordinate participants in discussions of atomic energy policy – the first ten-year programme of 1955 was shown to the British Electricity Authority only for comment; its successor expressed a preference for a smaller programme than that of 1957. From 1960, under Hinton's leadership, the CEGB secured downward revisions of this 1957 plan.[39] In 1962, Hinton made it clear the Board did not regard choice of the AGR, for whose development he had been responsible, as automatic: its progress had not pleased him. He was attracted by the

Canadian heavy water reactor. By January 1964, he and others in the Board were impressed by GE's most recent development of the BWR, which it was thought now offered cheaper power than the AGR. The consortia were also impressed, and TNPG, with the cognisance of the CEGB and perhaps encouragement from some Board members, began negotiations with GE (with whom they had a link through AEI) to explore how a BWR could be built in Britain. English Electric approached Westinghouse in association with the Westinghouse licensees, Rolls Royce, in the same vein for the PWR.[40]

The Board had for some years been trying to reduce the number of firms designing and constructing heavy electrical plant. Big reductions in generating costs could be obtained by using generating plants of very high capacity, but efficient production of such units required larger, though fewer, design teams and production facilities. The Board was disturbed by the many failures in new big plants. From this background the Board hoped to avoid having three AGR designs by reducing the number of consortia to two – APC was expected to drop out. At the end of 1963 the idea was being canvassed that there should be only one AGR and one BWR design. One consortium thought the AGR might be built on a design contract (in effect, though it was not put in this way, as a large prototype), they thought it might well turn out to be a one-off job only. At this stage the CEGB was not going along with the AEA, although no doubt there remained friendly links. At some point, the CEGB actually told the three consortia, according to Sir Harry Legge-Bourke, that the AGR was not suitable for Dungeness B.[41] There had been disagreement over the AEA's choice of the Steam Generating Heavy Water Reactor (SGHWR) to develop as a reserve against the improbable failure of the AGR. Hinton thought the Board, as the major consumer, should have been consulted.[42] This was one of the topics of disagreement which became publicised when Hinton gave evidence to the Select Committee on Nationalised Industries in 1963.[43]

The transition from these contentious AEA/CEGB relations to the agreement marked by the *Appraisal* is not readily explained from published documents. Quintin Hogg, as Minister responsible for the AEA, told the House of Lords in summer 1964 that the public manifestations of divergence of opinion between the AEA and the CEGB caused him great concern. He had set up an inter-departmental committee to 'reconcile the differences'. Reconciliation was a private process. Hogg was encouraged by the disappearance of public disagreement.[44] In retrospect continuance of disagreement might have been more salutary. At the end of 1964, Hinton retired 'at his own request', he was succeeded by Stanley Brown.[45] It appeared to release some tensions and some criticisms. By now there was a new (Labour) Government.

To fill in the gap between the institution of reconciliation by Hailsham and the agreement on Dungeness B there is only the guidance given by Hawkins – and rumour. 'The United Kingdom chose the AGR under somewhat pretty heavy persuasion,' Hawkins said in 1973, 'because at that time in the internal reports ... there was a very strong case for the light water reactor compared with the AGR.' In 'going AGR the CEGB did so under the strongest pressure from the AEA'.[46] But the AEA could not compel the CEGB. Was the CEGB, without the influence of Hinton, completely persuaded by the AEA – or was there ministerial action in support of AEA pressure? Were the changes in the personalities at the top important? No doubt there was a hard core of AEA–AGR supporters in the Board – many formerly in the AEA. Arthur Palmer, chairman of the Select Committee, said in March 1973 that when Fred Lee, as Minister of Power, made the Dungeness B announcement in 1965 it was 'widely rumoured that the CEGB would have liked to have considered the selection of a light water reactor even if it meant taking an American reactor, perhaps manufactured in the country under licence, but it was made plain to them by the authorities that they were expected to take a British design.'[47] It is hard to reconcile the hardheaded, rationalising, exploring and receptive attitudes in the Board in early 1964 with the AEA–CEGB consensus of May 1965 without the aid of a *deus ex machina*.

Although the CEGB was not responsible for initiating the AGR (though Hinton personally was) and assessed it correctly at the outset in 1964 when the Board showed unwillingness to choose it, nevertheless once it had been persuaded it contributed greatly to building up and keeping alive the AGR mythology. Engineers in the consortia found a remarkable euphoria in the Board after the decision was made. The *Appraisal* was fantastically favourable to the AGR, 'more lavish in its praise of the AGR than a manufacturer's sales brochure'.[48] Yet the CEGB knew what the public was not told, that there was at this stage only a sketch plan for Dungeness B, that there was much less detail in the APC tender than in the other tenders offered, and less than was normal for a conventional station which posed less novel problems. They knew that the contractors were not adequately staffed when the contract was awarded to do the detailed design. As Palmer said in 1972: 'Perhaps the contractors should not have been chosen ... as they had inadequate knowledge, capacity and skill.'[49] Even so the Board, in its document, averred there were no technical problems to prevent completion of the plant by 1970. In retrospect, it is clear they were wrong and that the 'meticulously fair and proper analysis', of which Hill spoke, missed a whole mass of difficulties within the core and outside it, or misjudged their importance, which began to show up almost immediately the detailed work on design and construction began. Yet the theme that the *Appraisal* was a sound assessment, and had established the superiority of the AGR over the LWR, was sustained by the CEGB for several years. Whatever the motivation, this no doubt contributed importantly to what Palmer called the 'over-confidence' in the AGR.

The euphoria was presumably over by 1969 when the CEGB was persuaded that both SGHWR and HTR would give better results than the current generation of AGR.[50] But recognition of this did not lead to a reassessment of the *Appraisal;* errors and misjudgements were not yet openly acknowledged.


(iv) THE CONSORTIA IN THE WINGS

The consortia had no part in making decisions on the choice of reactor. The British organisation ruled out the possibility. But they could, in submitting tenders (and thereby indicating what they thought they could do) and in circulating their own assessments of the relative values of alternative reactor systems, influence the judgements of those who did make the decisions and encourage or discourage enthusiasm. So the question does arise: did the consortia contribute to the over-confidence in the AGR? And there is a further question. Did they so over-estimate their capacity that by failure in performance they were partly responsible for the AGR trouble? Clearly the APC, whose Chairman, Colonel Raby, had been Deputy Director (Engineering) at Harwell – Cockroft's right-hand man, the *Financial Times* said[51] – played a vital part in getting the AGR adopted. One may wonder how otherwise success could have been achieved – a consideration which no doubt weighed with the AEA in encouraging APC participation and giving it aid and with the CEGB in giving them the contract although they only tendered a sketch design. APC did not initiate over-confidence in the AGR, but drawn into the act by the principal promoter it stoked it up.

TNPG and BNDC, the consortia which CEGB wished to continue, both put the LWR above the AGR, in their bids for Dungeness B, as offering lower generating costs, and this was not a contribution to over-confidence in the AGR. TNPG had argued in 1964, and in retrospect may have been right, that a further improved Magnox, with a more sophisticated fuel of a type analogous to the one used successfully in EDF4 in France, would have had lower costs than the proposed AGR.[52] If in 1964 utilities in Britain had been as free to choose as those in America one of these consortia, possibly both, would probably have offered an LWR and secured an order.[53]

But in their evidence to the Select Committee in 1967 the two applauded the APC action, calling it an outcome of competition, a 'very successful advance', a 'sound engineering proposition'.[54] TNPG hardly succeeded in concealing that its assessment of the AGR had not changed. But once the contract had been awarded in 1965 the firms had decided not to criticise the *Appraisal* publicly. Some members of both continued to express doubts privately. When, however, the national monopoly developer and the national monopoly buyer were in agreement and enjoyed strong government approval (or acted under government compulsion), the consortia, who could only get orders for the AGR, upheld this system in public. It was another penalty of the British system. The two may have helped thereby to sustain the over-confidence.

How aware the consortia were of the risks in extrapolation involved in the AGR is uncertain. Both probably thought they were getting help for concrete vessels for the AGR from Oldbury and Wylfa – indeed, they did learn much, but not enough. APC's belief that their tender involved them in no 'undue risk' may have helped to encourage over-confidence in the others. Although the CEGB knew them as the weakest consortium, they acquired a great cachet by taking the Dungeness B order, as their influence on several Select Committee members showed. Possibly the two stronger consortia did not assess the difficulties fully.[55] TNPG hoped to gain from Dungeness B as a prototype and of course got nothing. It is unlikely that they warned the CEGB of the danger of serious delay – it was not, indeed, a propitious time to do this, since they wanted orders. But all this could not be regarded as a major contribution to over-confidence.

It remains to examine how far the consortia and the firms who owned them and firms supplying main AGR components were responsible for the AGR troubles because of bad management and bad engineering. On some occasions, in giving evidence to Parliamentary committees, Hill seemed to imply they carry most of the blame – the 'fragmented nature' of their structure was 'no longer appropriate to the type of technological and engineering problems being imposed on it'.[56] He went so far as to say of Dungeness B that the 'problems . . . encountered would have been encountered whatever reactor system had been chosen, because they were of an engineering and constructional nature.'[57] He stated, too, that 'there is nothing technologically wrong with the advanced gas-cooled reactor. . . . The fact that the one at Windscale has run satisfactorily for ten years shows there is nothing fundamentally wrong with the design but the engineering has got to be got right.'

To take the last points first, there were two subtle ambiguities. To say that there was nothing technologically wrong with the advanced gas-cooled reactor did not mean, presumably, that it was economically right, but it seemed to be implied. Low cost was the important object of development. To say that the continued operation at Windscale showed there was nothing fundamentally wrong with the design suggested that Dungeness B was based on the design – not just on the system – at Windscale. Whereas of course it was not – indeed, Hill himself said it followed more closely the design of the later Magnox stations, the first of which had its first operating experience in 1968. The departure from the Windscale design set out broadly in sketch form by the AEA in 1963 was intended to reduce AGR costs and make it competitive. The effort to secure competitiveness had failed; and the experience suggested that the system required costly materials and involved a mass of engineering complexities, making it an inherently high-cost system.[58]

In saying the engineering must be right Hill implied there was a satisfactory basic design which there was not. The engineering, he said, required very high standards of quality control; lack of it had led to 'all these teething troubles on all sorts of bits and pieces of pumps and boilers and generators that have rather plagued us in this country in recent years'.[59] There were instances of bad engineering among the AGR troubles, – such as the distorted steel membranes at Dungeness B. This one involved no overall delay since boiler design problems occurred in parallel. Hawkins rightly emphasised that the main troubles which delayed completion and operation were design problems. He thought the confusion arose from the AEA's habit of confining the term 'design' to their own nuclear work.[60] He warned that we did not yet know what failures in quality control we were facing – until the plants worked. The design problems arose out of the great changes in gas pressures and temperatures, the volume and speed of gas flows and the chemical treatment of the gas. At the outset, some basic data were not established by the AEA: the corrosiveness of the gas, for example, and the metallurgical adaptations it required.

Any prototype development, breaking new ground, is likely to meet with delays. The Wilson Committee, exploring the delays on the whole CEGB power station programme, emphasised this in 1969: 'the greater the advance in technology the greater the chance the difficulties will be considerable.'[61] The advance here was not only great, it was widely thought to be excessive – Guard, the outspoken consultant writing in 1965, thought it an important disadvantage, Jackson, of APC, agreed in 1969 it had been excessive; Hawkins called it 'ridiculous'. Hinton, in a memorandum to the Select Committee, said 'although this trouble had been officially attributed to bad engineering I cannot help feeling that unwise extrapolation . . . is at least equally to blame'.[62] Here was the main source of the troubles.

Hill made much – too much – of the inadequacies of APC as contractors at Dungeness B, to which Stanley Brown had referred in 1967: 'they did not have the resources and they started work before sufficient design had been done.'[63] But although no other AGR plants went so wrong all had severe troubles, arising out of design problems. Why was it supposed the AGR would need less development experience than the LWR? APC was known to be the weakest of the consortia. The other two consortia were spoken of rather flatteringly by the Wilson Committee: 'we feel strongly ... that the considerable expertise in project management of power stations which they have built up is an important asset which should not be dissipated.'[64]

The obvious question to ask is, why was the weakest consortium chosen, when it could not, without a fresh injection of designing and managing skill, do the job? Why did the AEA and CEGB give an air of euphoria when construction was planned and started before there was an adequate design? Palmer thought the Dungeness B troubles started from the 'much boosted competition' between consortia: 'Would it not have been better if there had been a little more time for everyone to consider things?'[65] But it was the competition of the AEA against the LWR that was the key to the situation. This was outside Palmer's idiom. The Second Nuclear Power Programme had been announced in 1964. The AEA had wanted it: they would have liked it bigger. The AEA's only entrant was the AGR; a late developer for the programme, but it had to be nominally ready. The strength of the LWR challenge was unexpected. APC was the only disengaged consortium, not advocating an LWR;[66] the AEA was prepared to assist APC, and to inject some of its own design force into the company. Hill said later that the AEA had not wanted three AGR designs[67] – nor had the CEGB or the other consortia in 1964.[68] It was the AEA who took the step that made it come about. And presumably the AEA experts – and the CEGB – working with the APC believed that the premature start on construction would work well. It was all part of the massive misjudgement. It was not the fault of the stronger consortia.

Since the AGR troubles arose mainly from the prototype nature of the work, Hill's assertion that the same problems would have been encountered, whatever system had been chosen, because they were engineering and constructional, falls down. The LWR was much more developed. Sensitive components would have been imported. The engineering would have had the backing of GE. There may well have been some delay. But not on a Dungeness B scale.

When Hill turned to the FBR he lost some of his defensiveness. He agreed when giving evidence to the Public Accounts Committee of the House of Commons, that they had 'learned the lesson of Dungeness B' when they went from a 15 MW prototype to a 250 MW prototype before going to a full-scale FBR plant for the CEGB. 'We saw that in 1966. We have seen another set of problems and learned another set of lessons.' Furthermore, 'we are not proposing there should be a programme of fast reactors. ... We are proposing that one full scale should be built and we should learn for two, three or four years from the experience of that station before taking another step.' And to crown all, 'we are really proceeding much more cautiously than we did in the early days of nuclear power when we went from a first-off prototype to a full programme of nuclear stations.'[69] Did he now think this greater caution would have led to wiser decisions in 1955, 1957, 1964–65? He was not pressed to say.

### (v) MINISTERS DECIDE

The decision to have the AGR was made by ministers. Under the British arrangements it was one of their prerogatives. In a formal sense the Government had ultimate responsibility. The White Paper of April 1964 made it clear that after competitive tenders to the CEGB for light water reactors and AGRs had been 'judged on a comparable basis' the Government would review the results 'with the supply industry and the AEA . . . in order to decide on the type or types of reactor to be built.' The two ministers most closely responsible for the decision were Frank Cousins, as Minister of Technology, and Fred Lee, as Minister of Power, both formerly trade union officials.

How the Government made their decision, and whether any arm-twisting was needed to get its acceptance with no sign of AEA/CEGB division, is not on public record. No one now doubts that it was wrong (put across as a great national triumph, although the figures, when available, made the claim ludicrous). That it was claimed as a national success by the Government, and then by most of the press, is a clue to part of the motivation.

There was not only no light shed on the factors which determined ministerial actions in regard to Dungeness B, nothing which got under the skin of the *Appraisal*, but there was no public discussion which illuminated the way ministers approached making this kind of decision. Of the two ministers directly concerned, it was noticeable that Cousins would not, in conversation, accept any criticism of the AEA.[70]

There was much in the general nature of the policy, developed under the Conservative Government from 1952 onwards, in association with and inspired by the Ministry of Supply's Atomic Energy Division, and from 1954 by the AEA, with which ministers in 1965 would have felt strongly in accord. It satisfied the taste for 'global' central planning. It paid obeisance to the scientists and technologists, who though they were falling off their pinnacle in the United States had just reached the forefront of Harold Wilson's propaganda as the Prime Minister of the day. For socialists, decisions by ministers and a state agency must be better than scattered uncoordinated decisions by American corporations. The United Kingdom's achievements and leads, as they were presented, nourished both national pride and anti-Americanism. Many of the leading members of the AEA were distinguished and able people. And if very cheap power did become available, and dependence on overseas oil and domestic coal were lessened, these were worthwhile aims.

Nevertheless, ministers in making decisions such as that on Dungeness B were deciding technical matters about which they knew nothing. Sympathy with a few general lines of policy could not change this. Could they contribute anything or were they just rubber stamps, If so, what service were they performing? What, in fact, did they suppose they were doing?

The Select Committee in its 1967 inquiry was much exercised with this question. Both Wedgwood Benn, who succeeded Cousins as Minister of Technology in 1966, and Richard Marsh, who succeeded Lee, also in 1966, as Minister of Power, gave evidence before it and the Committee was disturbed by what it found.

In a remarkable passage at the end of their *Report* they expressed concern that neither of the two ministers 'appeared to have any very effective technical check on the activities of the AEA and the consequent allocation of funds for the Authority's purposes'. The ministers were not, between them,

> adequately equipped to assess the value and significance of what the Authority are doing. This applies to basic research and experimental work which is remote from commercial discipline ... [and] may lead to the pursuit of a particular line of development which is not the best but from which it is impossible to withdraw. It might be argued that this is precisely what has happened in the case of the entire gas-cooled reactor technology.[71]

It must be said that this was not a complete commitment to the view that concentration on $CO_2$/graphite had been a mistake: earlier they had said 'there may be room for argument about the economics of gas-cooled nuclear generation, but there is no doubt that in the main the existing Magnox generating stations have more than lived up to expectations in terms both of output and availability.' (Within little more than one year their output was substantially and so far permanently cut by 10 to 20 per cent on account of corrosion.)

> There is every reason to hope that the AGR stations will do the same or better ... as they represent an improved version ... of a well proved type. ... Your Committee think that in the circumstances it was right for the AEA to have devoted so much effort to the development of the gas/graphite reactor. Indeed ... there is an urgent need to press ahead with the commercial development of high temperature gas/graphite reactors.[72]

The suggestion is there, as often in British writing, that the HTR must follow $CO_2$/graphite development, although the work of the United States and Germany who were both ahead in the field belies this. There was thus an ambivalence in the attitude to $CO_2$/graphite; an exaggerated view of Magnox achievements and misplaced hopes for the AGR.

The force of the Committee's criticism of the ministers' ability to carry out their functions in relation to the AEA is shown vividly in the evidence.

Richard Marsh argued that the Ministry of Power's job was 'not to determine the policy or the technical activities of the AEA or CEGB, but to be able to adjudicate on these', and for this there was 'sufficient technical expertise in the Ministry',[73] including in the Chief Scientist's Division some 'qualified scientists who had worked on nuclear development outside the Ministry'. There were two, both from consortia, but one via the CEGB.[74] 'As far as we are concerned [this was clearly a trump card] the policies of the AEA and their advice get a very powerful check from the CEGB. I have never quite understood why it should be assumed that the CEGB and the AEA have a common interest.'[75] But Marsh did not give illustrations either of the technical activities or policies of the AEA and CEGB on which he was 'able to adjudicate', or of the ways in which the advice of the CEGB had enabled him or his department to 'check' AEA policies and advice or whether the AEA policies had led him or his predecessors to check the CEGB. He did not show how his Ministry's 'technical expertise' had enabled his predecessor to decide that the choice of the APC tender for an AGR for Dungeness B was sound. Nor did he say whether there had been any disagreement between the AEA and the CEGB over Dungeness B, at any stage, and if there had how it had been resolved and reconciled.

Marsh referred repeatedly to the Working Party on Nuclear Power Costs whose report was submitted to the Select Committee too late for any detailed study by them.[76] It was a pretentious statistical exercise (which can reasonably be regarded as a statistical 'folly') based on the premise that the Dungeness B *Appraisal* comparison and the AEA cost forecasts for the AGR were right. The party was composed of representatives of the AEA, the CEGB and the Ministry's Chief Scientist's Department. Its report, Marsh said, represented ten months work of the 'top nuclear scientists of the country' – and there had been 'no real division between them'.[77] There *should* have been division. The main premise was wrong, its validity had been widely and expertly questioned and neither the Minister's technical experts nor the opposed interests of the AEA and CEGB had protected him. Marsh was pressed to say why the NCB had not been added to the Working Party. The answer was that the Working Party's work was merely 'technical' and he would not call on the CEGB to help in analysing future coal-mining costs. But, in the NCB, Leslie Grainger, the Member for Science – no one mentioned it – was a nuclear engineer, as distinguished as any in the Ministry; and he would have counterbalanced the CEGB/AEA axis on the AGR. It is instructive also that Marsh by implication ruled out the possibility that any scientists and engineers in the consortia were among the 'top' British nuclear experts; they certainly were rated so outside Britain and in the AGR/LWR comparison they were right. The Minister was convinced that his limited group provided 'a reasonable amalgam of expertise', and he 'had no reason to assume there is a better qualified body'. And then, the moment of truth: 'if they are wrong, I am not qualified to judge.'[78] The Select Committee agreed.

While Marsh claimed that with his experts and the opposed interests of the AEA and the CEGB he could 'adjudicate', Benn made no such claim. His evidence was the more important because he was the minister responsible for the AEA. Also he gave evidence in 1969 which illuminated the position more: whereas Marsh, like most Ministers of Power (there were four in the Labour Government of 1964–70), was a bird of passage; Harold Wilson, as Prime Minister, moved him to the Ministry of Transport in 1968. 'For an ambitious young Minister', he writes, 'he had been long enough at Power, his first appointment.'[79]

Benn stated broadly that he applied no check on the technical advice of the AEA. Successive ministers exercising their duty of appointing the Authority had ensured it had 'a strong and expert Board led by an outstanding Chairman [and] . . . the corpus of scientists and engineers had always been first rate.'[80] He regarded Penney as his 'principal adviser on atomic energy matters, and it is not thought necessary, right or proper or possible for us to have within our Ministry a complete organisation for the duplication of review and evaluation of the advice given to me by the Authority.'[81] The Ministry was developing a 'techno-economic analysis unit' which would not be perfect for a long time. The Authority prepared its projects with great care, taking technical and economic factors into account; so, 'on matters of this complexity the technical side cannot really be assessed other than by the Authority [and] we take the Authority's word for it.'[82] Thus 'their decision to go Magnox or . . . for the AGR or to work on fast reactors would be one based entirely upon a scientific assessment of what systems were worth further study . . . and it would not be open to me to have a sensible alternative view.'[83] Penney no doubt encouraged Benn in this; he had said forcibly in 1964, after the Powell Committee, that the Government, including the civil servants, should never again be called upon to make technical decisions such as deciding between two reactor types: the AEA intended in future to thrash out all such technical problems with the reactor experts of the CEGB. He would do all he could to prevent a repetition of the committee procedure.[84] Penney was the first chairman of the Authority who was a scientist: a contrast to the civil service administrators who had hitherto held the job.

Asked whether, in view of the reported low prices of American reactors, he looked as favourably upon the possibility of having foreign competition in reactors in Britain, as we hoped potential foreign buyers would look upon our reactors when we tried to export them, Benn replied that the AEA had developed such a wide range of reactors that 'he could not himself see how the use of foreign reactors could bear very strongly on the policy of the AEA itself.'[85] But, the Chairman of the Select Committee pursued, 'since the Authority have great confidence in the excellence of their own products and their own designs, the kind of advice they will presumably give will be in favour of British reactors.' Did the Minister look at international competition within these shores as a kind of check? Benn, characteristically, avoided answering the question: at great length he said that he was informed that a comparison of costs of British and foreign reactors was not unfavourable to the United Kingdom, provided you used the right ground rules. 'Were it, on the basis of such calculations as we can make, to be revealed that the AEA reactor was not competitive then a very different situation would arise, but it has not arisen.' There had after all been the Dungeness B comparison: 'I know there has been some criticism of the way that was conducted, but there was an *Appraisal*.'[86] What more was there to say?

Two points from a memorandum which arose out of Benn's evidence filled out his assessment of the position in 1967. Inevitably they followed the AEA line. First, the general point, partly quoted earlier:

> I believe that this country has been given good value for the money which it has spent in the field of civilian nuclear power. I think the position will be seen more clearly in our favour in a few years time. The present is not a good time for even well informed observers to conclude that the race has gone against us.

Second, a particular one:

> It would be wrong to think that the capacity the consortia had available in the fifties would have been in any way adequate for the development of alternative prototypes, nor would diversion of effort from building Magnox have been of help. ... The study of material under irradiation ... was limited by the available irradiation capacity.[87]

This misconceived the criticism he thought he was answering and the circumstances to which he attributed immutability. It was perfectly possible to deploy Britain's nuclear development resources differently in the 1950s, including those within the Authority – the polarisation of the AEA and consortia was the result of choice, not an act of God; it was quite practical to build more irradiation facilities if that was recognised as desirable. They were costly facilities, but £10m to £20m for instance, although it would have looked a high cost, was trivial compared with the hundreds of millions spent on obsolete Magnox plants, whose design and construction prevented nuclear engineering firms concentrating on projects with a future.

Within two years Benn had seen part of the light. On the general point, he told the Select Committee in 1969 that he would not of course be satisfied 'until we have a major export business in nuclear reactors and systems and licences and parts. I do not think anybody could be satisfied that we have got value for the very substantial sums of money we have invested.'[88]

Benn was mainly concerned to explain to the Committee the principles of the reorganisation he had announced in October 1968 and why he had departed from the Select Committee's recommendations in 1967 that only one design and construction company should be formed. This will be examined later. But it is relevant in this context that the two design and construction companies which he was setting up were to absorb the functions of prototype development from the AEA, and to embrace part of the AEA staff in doing this.

The new companies would be responsible for marketing as well as design and construction, the Minister explained. This was hardly new. But they would 'be able to feed in this marketing experience to research and development so that systems are developed with the market in mind'.[89] This *was* of course new, in the nuclear industry, as organised in the United Kingdom by legislation. The nuclear industry would develop into a 'perfectly ordinary industry' with a 'mixed enterprise operation [by two or more companies] at the heart of it', with research directed by and paid for and increasingly conducted by 'the people whose business it is to secure the pay-off'.[90]

> The Minister of Technology had made a great discovery.
> If you ask me candidly what has been wrong broadly with our whole approach to the nuclear industry or nuclear energy, and indeed shipbuilding or anything else, it is that we have always thought of the scientists as being in the heart of the operation, whereas really the marketing man identifying the market, forecasting, building up relationships, ought to have played a very much larger part. The development of systems to forecast demands for particular products should be very much bigger . . . this particular lesson has been very clearly learned.[91]

So clearly learned that, as Benn put it, even if 'we' have spent a lot of money on a system, neither of the companies would be under a moral obligation to develop the system if 'they do not believe [it] is going to have a market'. 'What has been wrong in the past has been the systems have been developed and then, later, applications have been sought.'

Here then was 'a very fundamental change of emphasis which I am trying to apply across the whole field'.[92]

Had the Minister given due weight, he was asked, to the commercial success of the AEA? Except in its fuel and isotope work, he answered, where it had made some money, the AEA 'had never been a commercial organisation. It had acquired management skills of an enormously high order, as well as its scientific skills', but had not been commercial.

> It has been successful in developing systems which have found favour with home customers, the CEGB through Magnox and now the AGR and later systems ['found favour' was a nice euphemism since the CEGB was forced to order more Magnox than it wanted on two occasions and needed heavy persuasion and pressure at some stage to take the AGR] but there is no doubt that part of the price that has been paid for the old structure . . . was that the industrial firms which would have been acquiring export markets were, by the very fact that the frontiers were drawn where they were, denied the opportunity to extend back into the design and to dictate research, which was open to some of their competitors in the United States.

'That', he added, 'is historically true.'[93]

Thus Benn was now arguing – agreeing with what he had so recently denied – that the 'old structure' reflected a 'wrong approach' and had narrowed development with damaging results, and prevented the growth of firms on the American pattern. Changing the arrangements in order to establish a 'perfectly ordinary industry', however, was causing much friction. 'The transfer of this emphasis into the industrial field', he said, 'does create more serious transfer problems than would have occurred if the change had taken place in 1955 when, looking back on it, it might have been quite a good idea to have done it.'[94] If it had taken place, that is, when the consortia were being created, presumably with much more transfer of AEA staff to the new companies than then occurred, transfer too of facilities and the power to acquire them, including irradiation facilities, with the freedom for the companies to choose their own development projects, and to have state aid, presumably, on the American model, in doing so. This was clearly what Professor P. M. S. (later Lord) Blackett envisaged when in 1967 he said that Britain had taken 'the wrong road' when it decided to rely on its research and development in atomic energy and so on in government stations rather than industry. In America a much bigger fraction of government money was spent on industry.[95]

Benn's second thoughts were better as an explanation of the past than as a guide for the near future, but even in relation to the past they involved some confusion. The initial error in Britain was a failure to have parallel developments of different systems – and different autonomous developers – before any market could express a preference. The 'market' would want what offered lowest costs. The British Government by instituting a state monopoly ruled out the discovery of this by comparisons and the stimulation of it by competition. *Pace* Benn, the engineers and scientists had to be at the heart of the initial development, but you had to sort out the right ones. Both needs were provided for in the American arrangements. Once a competitive reactor system had been strongly established (as the LWR was) there was a new situation. What feedback from their export marketing were the two Benn companies expected to bring back which would influence their development programme – and how quickly? There were unlikely to be any home orders for the two new companies for two years: 'this is having the effect, as you could expect', the minister said, 'of driving the companies to search for business abroad which is necessary to sustain their own capability at home, and I think to this extent will compensate for the economic difficulty of trying to build an export business without a sure home market.'[96]

This was cloud-cuckoo land: the mirage which deluded the Select Committee in the same context. The Minister had 'begun a series of extremely comprehensive reviews ... with the firms' and with all the 'relevant Departments', so that 'we can look at the export field in a more formidable way than we have even been able to do in the past'. It would take a lot of time. But 'the companies are in deadly earnest'. What had they to export? There was the AGR: and *'we just have to get our SGHWR launched'*[97] [my italics]. The future of the SGHWR, indeed, 'like every other project in Britain, not just in the AEA, depends upon the capacity to get export orders for it'.[98] So, after all, this great drive was going to focus on finding markets for reactors which the AEA had already developed, not reactors for which our marketing forces discovered a need so far not catered for by the AEA and which they then produced for the market. This situation is further analysed later. Internal consistency in argument, one must assume, was not important politically. There was, I believe, no chance of export success in 1969–70. Benn's two companies – or for that matter the Select Committee's one, had it been chosen – would have failed in export for the reasons given earlier, because, as the starting point in Benn's thesis stated,[99] the AEA had failed to develop exportable products. The failure would in due course 'call' for further minsterial action, for the one thing never said by minister or Select Committee, was that these were matters with which ministers were not competent to deal. The Select Committee thought the ministers needed more expert advice[100] – the almost automatic response of most politicians, civil servants and economic 'experts' to situations of this kind. The validity of the response is examined later.

(vi) CIVIL SERVANTS ANONYMOUS

Ministers acted with the advice of 'permanent' administrative civil servants. Lacking a Richard Crossman in this field we do not know from the record what advice they gave. We can say that the civil servants' contribution to the Dungeness B disaster was subsumed in the ministers' contribution, but we do not know how far ministers accepted, or disregarded, the advice given to them.

Several groups of civil servants were involved. The principal group was in the Atomic Energy Office, which became the Atomic Energy Division of the Ministry of Technology when responsibility for atomic energy was given to the Minister of Technology. But groups in the Treasury, the Ministry of Power and the Foreign Office were also involved. The first group must be looked at in most detail because its lead was followed by the others.

Benn's evidence to the Select Committee in 1967 showed that the civil servants in his Atomic Energy Division accepted the technical and economic assessments of the AEA and did not check them. This irked the Select Committee who complained that the Minister was not doing his job properly. But this was a situation he inherited, not one he created. It seems clear that the civil servants readily accepted this situation, not as undesirable, or unwelcome, imposed on them by the legislation they were administering, and basically irrational. They did not confine themselves to saying that they lacked the qualifications needed to vet the assessments of the AEA, and merely accepted them for administrative processing, incorporating them into plans, programmes and estimates, because that was what government policy required. They appeared to believe that all AEA judgements and assessments were right, all critics of the AEA wrong, and all AEA answers to criticism right.

This bias seems implicit in the memorandum which Benn submitted to the Select Committee, which his staff would have drafted, and in the phrases he used, which betrayed an administrative tradition whose continuity can be established. It was confirmed by the evidence, slight as it was, of the head of the Atomic Energy Division, Mr M. Michaels, who accompanied the Minister when he gave evidence to the Committee, and by the lack of any sign that the staff tried to assess criticism by objective study so that they understood what it was about.

The principal memorandum submitted by Benn (and quoted earlier) put the AEA's role in a historic setting, bearing the clear imprint of civil service origin. By the Act of 1954, 'Parliament leaves all technical and economic assessments ... to the Chairman and his Board', only the 'ultimate decisions on major policy matters' being taken by ministers. There had been 'every reason for relying on the advice of the Authority' because ministers had made good appointments and 'the scientists and engineers . . . were of the front rank'.[101] Hence, the Minister argued, it was neither necessary, possible, right nor proper to repeat its evaluations or review its advice and impossible to check its economic and technical assessments, which had been prepared with great care as the Authority was concerned to be right.[102]

All this was close to the doctrine expounded by Sir Friston How, head of the Atomic Energy Office from 1954 to 1959, who had been dealing administratively with atomic energy in the Ministry of Supply since 1946. He explained to the Select Committee on Estimates, in 1959, how at the outset it was decided not to duplicate in Whitehall the bureaucracy of the AEA – his office was to be a small one – and how in formulating plans and programmes you had to rely primarily on the Authority, you had to go to 'those who know' and they were 'almost all in the Authority'. And when estimates based on the Authority's proposals were submitted to the Treasury they were normally not modified because the Authority had 'made its case'.[103] Michaels (who retired in 1971 to a part-time post with BNFL) was in How's office from 1954, so there was a remarkable personal continuity in the office, and the How tradition was ostensibly handed down. But how could the civil servants tell that the dominant group in the Authority always 'knew' and 'made the case', that they were not merely clever but right?

The Act of 1954 had given to ministers and by delegation to their civil servants an impossible task. They were to ensure that resources were used in the right proportions for the various applications of atomic energy. This they could only do if they knew what could be achieved by the specific applications proposed and the known alternatives. This was beyond the powers of ministers and civil servants. In the apocalyptic period it no doubt seemed plausible to rely on the Authority (which meant accepting the dominant view within the Authority) for guidance and to develop administrative procedures accordingly, which the legislation required. Although the impossible was asked for, the professional administrator could find a way of making it appear to be done. But as the competing American technology became familiar, and American achievements became increasingly impressive, forming the bases of Continental and Japanese nuclear development, and as the economic and technical bases of AEA forecasts proved weak, and the world status of the AEA declined (when Kronberger died in 1971 he was spoken of as the last AEA scientist with a world reputation,[104]) the procedure was no longer plausible.

Nevertheless, the Benn Memorandum to the Select Committee in 1967, probably drafted by his staff and no doubt expressing their own doctrine, still propounded the view (not in these words) that ministers, who knew nothing of nuclear technology, advised by civil servants, who also knew nothing of the technology, would appoint members of the Authority, who would infallibly get the technical and economic answers on nuclear energy right. The senior civil servants are reported by witnesses close to the affairs to have exerted considerable, though informal, influence over the choice of new members, and to have promoted changes in senior staffing. Ministry officials in rejecting criticisms of the AEA or defending its performance relied happily in 1967 on AEA handouts[105] and, if my own experience was a guide, made no attempt to check the substance of criticisms or even their meaning (which sometimes eluded both Authority and Ministry) by simply enquiring of those who made them.[106]

On the evidence of 1967, the civil servants advising the Minister of Technology in 1964–65 would have dismissed the criticisms of the AGR, and supported the AEA without reservation, in the run-up to the Dungeness B decision. The Windscale prototype would have been adequate, the Magnox operating experience more impressive than LWR experience, AGR fuel at least as advanced in development and testing as LWR fuel, the APC design adequate though only a sketch, the comparison of yesterday's BWR with tomorrow's AGR a satisfactory basis of choice, and so on. The advisers and the Minister were well aware that the AEA was now under criticism, but on the 1967 evidence the advisers would not have protected the Minister against the inevitable bias of the Authority, to which the Select Committee was alert, by informing him of the weaknesses in the Authority's case and the strength of its rival's.

That ministers lacked objective advice in 1964–65 on AEA technical and economic assessments, as the Select Committee concluded they did in 1967, was an important component in the ministerial contribution to the AGR disaster. The ministers may not have been aware of the need for objective advice and may not have asked the civil servants to fill the gap. Even so, the civil servants would presumably have set out to fill the gap of their own volition, had they been conscious of it. That they did not ostensibly change their stance between 1954 and 1967, in regard to AEA performance, is remarkable and puzzling since the environment of atomic development was revolutionised in the period. They were aware of this; and civil servants of this status are sophisticated types.

The explanation is to be found in the form of administrative structure chosen in 1954 and the way in which it was staffed. The responsible minister, who appointed the Authority, retained the duty to promote and control the development of atomic energy. But although the formalities of controlling the AEA were there, the emphasis in the Atomic Energy Office (later Division) was on promotion. Since many members of the staffs, both of the Office and the Authority, came from the former Atomic Energy Department of the Ministry of Supply, they met on a friendly basis, fired by a common enthusiasm. There has always been at least one former senior civil servant among the members of the Authority. As described by Friston How, the relation of Office and Authority looked like a partnership; the idea of the Office supervising the Authority seemed foreign to it. There was no arms-length relation to start with and none developed. 'We are really very close together, much closer than other public corporations are to the appropriate Minister', Benn said in 1967. The Office remained small. Continuity at the top meant that the head of the Office acquired a sort of authority through the sheer mass of knowledge of what had happened in the past and why. The very permanent civil servant may easily present a formidable appearance to a transitory minister. The responsible ministers for the first decade were the Lord President, the Prime Minister and the Minister for Science, hence the Office was not part of a large industrial department where the nuclear industry might have been treated as an ordinary industry (as Benn said in 1968–69 it should be), until it was brought into the Ministry of Technology at the end of 1964. This seemed to produce no effect by 1967, but some by 1969. The impression of partnership lasted unimpaired to 1967 and the Office continued to be positively associated with ministers in the 'unconscious levity' of making decisions which were beyond their competence.[107]

The civil servants, both in the Treasury and the Ministry of Power, appear to have accepted the judgement of the Office (Division) as the sponsoring department, but the records are slighter even than for the Office, and the mental processes more obscure. Within the Treasury some senior officials were perturbed at the Dungeness B decision, but the branch charged with authorising nationalised industries' expenditures – of which for some years Peter Vinter was in charge – ostensibly felt that fears were misplaced and that the Authority and CEGB had 'made the case' advanced for them by their sponsoring ministries. But how, and how far, the 'case' was probed in the Treasury is not known. Marsh made it clear in 1967 that the Ministry of Power relied mainly on the Ministry of Technology for technical assessments of AEA work, but also hoped that the CEGB would be a check on the AEA. The documentation for the 1967 Select Committee proceedings showed that the Ministry of Power accepted in 1966 without reserve both the Dungeness B *Appraisal* figures and the Authority's forward estimates up to 1980 as the bases of elaborate and misguided forecasts.[108] In 1964 high officials in the Ministry of Power were quite prepared to accept that BWR capital costs were much below those of the AGR.[109] But what happened in the interval is not publicly known. What pressures, for example, were exerted on and in the CEGB, and the contribution of the Ministry of Power civil servants to the Dungeness B decision is not, I believe, on public record.

There was an air of logic in the arrangement whereby the responsible ministry's judgement of the AEA was accepted by other departments. But since the CEGB could – if uninhibited – express an expert view which could be educative, there was a flaw in the logic. Since the responsible ministry said it could not assess the AEA's claims but did not try either to make itself able to do so or to free itself from the obligation to give approval and support, there was no logic there at all. The arrangements gave the impression of cumulative assurance – officials of three departments concurred in supporting AEA proposals, projects and programmes – although none had probed what they supported.

(vii) THE SELECT COMMITTEE OPTS OUT

The Select Committee discovered in 1967 that ministers were not adequately equipped to decide whether the AEA's policies and performance were the 'best', and surmised that the Authority's commitment to the gas-cooled technology might reflect this deficiency, but it failed to follow up the problem in later reports. It registered the declining export prospects of the AGR in its 1969 *Report*, and in 1973 reported that 'the choice facing the industry is the successor to the AGR'. But apart from perfunctory questions not followed up, it did not probe why, after the promise of 1964–67, a successor was needed. For a body aware that one of its purposes was to inform Parliament and the public of the facts, and anxious to advise on policies, this neglect of the genesis of the AGR disaster is surprising. One can only speculate on the reasons for it.

Only the first inquiry, in 1967, was by the whole Committee; the succeeding ones were by relatively small sub-committees. The initial Committee represented a wide range of opinions, from the Chairman, Arthur Palmer (who, in the Second Reading debate on the bill establishing the AEA, had welcomed it with reservations, but was content so long as public enterprise was to take the initiative and have the prime responsibility under ministers), to Sir Ian (later Lord) Orr-Ewing, who in the 1954 debate regretted the extent of AEA power and would have preferred it to have had no powers to produce or construct. It was a new committee in 1967, but most of the members had probably reached conclusions before it was formed. The 1967 *Report* reflected the breadth of views represented on the Committee – hence its ambivalence – but its *conclusions* reflected a Labour–Liberal majority. The Committee divided on party lines.

Orr-Ewing left the Committee early in 1969 and David Price, before the 1972–73 proceedings, and the Orr–Ewing point of view was no longer represented. The small sub-committees were less representative; they probably conceived their functions more narrowly as being concerned with *ad hoc* investigations; the members, if not less distinguished, were less experienced; and it was easier to grind an axe without counter-action. The first chairman, charming and urbane, with (almost) unshakeable views, became a more dominant influence. The Conservative presence weakened: sometimes no Tory was there. At the end Airey Neave appeared to be the heaviest Conservative weight. He was not a philosopher in his approach, but he had the advantage, and disadvantage, of being on the board of one of the investor companies in TNPG, so that he understood their official stance well.[110] He seemed, however, to disparage American achievement, without perhaps understanding it. All in all, the changes in the Committee membership did not favour a deep study of AGR problems. The succession of small inquiries by small groups with little documentation, no expert analysis of what documents there were and no expert staff, pointed forcibly to the difference between the Select Committee and the JCAE in the United States.

Though it was known that there were doubts about the AGR in 1967, members would have shown extraordinary foresight had they recognised the full extent of the disaster course on which it was irrevocably set. They pounded away at irrelevant questions, such as exports and size of fuel channels. Realisation of disaster came slowly in succeeding years – if it is yet accepted in these terms. By 1969 the evidence on Dungeness B already gave a disturbing picture of the tender and appraisal episode, which seemed to have been overlooked.

As the indications of disaster became clearer their significance was possibly missed, because, although most members did not go back to the beginning of the AEA, most came to the Select Committee with solutions for the nuclear industry's problems cut and dried. Many had been subjected to the powerful propaganda campaign mounted by the AEA in 1964. The pressure had been kept up. Most of the Committee wanted a bigger monopoly in future. So why should they criticise the monoply in the past? The majority seemed unconcerned to find answers from facts; they knew the answers.

This was reflected in their questioning of witnesses, who were treated, although not by all members, differentially. Thus the CEGB was liable to have a bad time; Stanley Brown in 1967 and Hawkins in December 1973 were treated roughly. The first was against the formation of a single design and construction company; he attacked the sacred cow of replication and showed up the weakness in design and management of APC, the only consortium to support the single-company idea favoured by the AEA, and strongly propagandist. The second supported the LWR and opposed the SGHWR. The AEA representatives were treated with immense respect. It was symptomatic that Palmer asked Hawkins in 1973 why he had been over-confident about the AGR; yet he only asked Hill what part the AEA had played in the Dungeness decision in the light of Hawkins' reference to AEA pressure. When Hill said it was a completely joint agreement with the CEGB, based on a meticulously fair comparison, Palmer did not then ask why the AEA had made the well-authenticated series of monumental misjudgements in assessing its development work and in advising the Government and CEGB, which had been illustrated in earlier evidence to the Committee.

Why the members who came onto the Committee had their minds made up and hence did not probe the AGR disaster, we shall return to later: but it is worth recalling that an element of fantasy had entered into the assessment of British and American achievements which led to incredible assertions in the late 1950s, some by people of distinction. As late as 1964, an article in the *Guardian* claimed that 'only a few years ago Harwell had no rivals as a scientific institution'. It needed a bold journalist to think himself capable of this assessment.

In 1958, at a Conference of the Federation of British Industries (FBI), Sir George Thomson, atomic physicist and Nobel Prize winner, said he was amazed how much American industrialists were prepared to invest in nuclear plant development which they all admit 'holds no prospect of being economic for any time that they can see'. Thomson concluded they were doing it for prestige. It was, for them, like the Sputnik. A utility chairman told him he was planning to build a large BWR – as far as Thomson could see, it was 'just for the fun of it'. It could only pay 'in the very distant future'.[111] He was surprised how many people in America told him how much more advanced Britain was in the practical application of nuclear energy. At the same conference, Sir Claude Gibb, head of Parsons, who had been prominently associated with the early development of Magnox and was a founder member of TNPG,[112] quoted with approval an American utility president who said, 'Shippingport is just a ruddy joke'. 'If [Gibb said] you are willing to throw economics overboard you can at once go for a PWR. The engineering problems are not terribly serious in it.' The cost of power from Shippingport, he said, was roughly 5.40d per unit, compared with 1.25d at Calder Hall. He envisaged that whenever a fuel rod leaked in the PWR you would close the reactor down for a few days. Like Thomson, he supposed the Americans were developing nuclear power for prestige, whereas we were doing it for national *salvation*.[113] How credulous and unperceptive the great can become!

By 1964–65 it was not possible to take this line, although it was argued, strangely, even in 1964, that the United States had only developed nuclear energy for export, not for home needs. This was a thesis impossible to defend in the light of the documents – for example, the reports by Paley, McKinney, Charpie and the JCAE discussions of 1954–55.[114] But a disposition to be contemptuous of American development, to look for what went wrong, to exaggerate difficulties in the United States, to believe and possibly hope for the worst, to treat the American domestic critics of LWRs as reliable and the supporters as misleading propagandists, had been nurtured and it persisted. Several members of the Select Committee were victims. Most of them probably still accepted unreservedly in 1967 the Dungeness B comparison favouring the AGR.[115]

Between 1967 and 1970 one could find facts which on a superficial examination would fit the argument that the AGR, despite delays and redesign, could still ultimately be the winner. In this period the LWR suffered delays, technical hitches and rising prices, which could be made to seem to offset the early AGR troubles, which then seemed limited to Dungeness B; and while LWR prices rose the prices of the AGRs ordered in 1967 were below the Dungeness B prices of 1965. The first big BWR, Oyster Creek, did not operate till autumn 1969 – two years later than promised. One could argue that the rival reactors were level-pegging on delays and the AGR was gaining on cost.[116] The argument could seem plausible even though it was phoney. A typical instance of the angled approach occurred in 1969. The AGR and FBR had both suffered delays due to welding problems. Lubbock asked Moore whether, if the British welding industry was to blame for not advancing the technology fast enough, was not the American industry more to blame, since the Americans also had welding problems. Moore answered yes! No one would have recognised from the exchange that the United States were far ahead of the United Kingdom in welding large components – above all the pressure vessel – for nuclear plant.[117]

After 1970 it was more difficult to sustain an argument which suggested level-pegging for the AGR and the LWR as an increasing number of big LWRs worked, in general well, and the great numbers ordered world-wide, even in France, showed that they were regarded as competitive, while the AGR ran into increasing and now generalised trouble. The conflict over safety in the United States, however, with its sensationalism, and the periodic and normally short spells of derating, provided useful diversionary material. It was nevertheless surprising that in December 1973, Airey Neave argued that the lead time for LWRs was as bad as that for AGRs, as if the planned eight-year lead, of which three years was for licensing, was comparable with the unplanned, unexpected troubles of the AGR which extended the construction time of the AGR unpredictably.

In its 1973–74 Report the Select Committee quoted, as if needing no comment, figures given by Mr F. Tombs, Chairman of the SSEB, to show there had been 200 reactor-years of Magnox experience, but only twenty reactor-years' experience each so far of large (over 500 MW) LWRs. The Magnox, he said, had had a better load factor. Hawkins told the Committee that the average availability of the CEGB

Magnox was less than that of all American LWRs, but this apparently made no impact. What was not pointed out was that only one Magnox plant had reactors of over 500 MW – derated to 420 MW, – that this one had, so far, an exceptionally bad record, and that if the Magnox reactor records were set out in date order the average load factor progressively fell. With each upward step in gas pressure, temperature and unit size, performance became less satisfactory, and the youngest Magnox plants with highest pressures (still substantially below the projected AGR levels) and concrete pressure vessels, had the poorest load factors, well below the LWR average. The Friends of the Earth (FOE) gave the Committee figures from which the right interpretation could have been discovered; the FOE did not disclose it, if they discovered it, nor did the Select Committee. It is shown in Table 10.1.

TABLE 10.1

Load factors in Magnox Plants 1972–73[a]

| Station | Commissioning year(s) (MW) | Total design capacity | Pressure | Load factors | |
|---|---|---|---|---|---|
| | | | | 1972 | 1973 (10m) |
| Calder Hall<br>Chapelcross | } 1956–59 | 447 | 100 | 91 } 91<br>92 | 81 } 86<br>92 |
| Berkeley<br>Bradwell<br>Hunterston A | } 1962–63 | 1668 | 150 | 78<br>65 } 72<br>72 | 83<br>65 } 73<br>71 |
| Hinkley Point A<br>Dungeness A<br>Trawsfynydd<br>Sizewell | } 1965–66 | 2478 | 200–300 | 50<br>68 } 57<br>52<br>57 | 63<br>71 } 60<br>40<br>65 |
| Oldbury | 1967 | 635 | 350 | 52 | 52 |
| Wylfa | 1971 | 1180 | 400 | 26 | 35 |

[All two-unit stations except Calder Hall and Chapelcross, which had four units.]

[a] The table provided by the FOE to the Select Committee is given in its original form in the notes to this chapter.[118]

More recent figures on Magnox load factors for the last ten months of 1975 gave broadly the same picture as Table 10.1.[119] Though drawing wrong conclusions from statistics inadequtely analysed seems surprising at so late a date, one may assume the mishandling was through a lack of understanding; it is fair to surmise that they were used willingly, because the conclusions were welcome.

The apologists for the AGR in the Select Committee also pointed (as Hill did) to the troubles with conventional power stations in the 1960s. As the capacity of fossil-fuel plants was increased, with the use of bigger units and steam at much higher pressures and temperatures, there were serious delays and breakdowns. They arose partly from design failures (the design of large turbine blades for example) partly from manufacturing failures (higher stresses required more precision, more quality control) and partly from industrial disputes. There were analogies with the AGR troubles, but the extent and severity of design deficiencies were by comparison slight, the problems of quality control and minor component failure more conspicuous. The delays were shorter, and the plants when completed came – not always immediately – up to specification. The plants reached the stage of developing faults in operation, which at the time of writing has not happened to an AGR. Troubles with the new larger fossil-fuel plants have occurred in all major industrial countries, but no other country has had an AGR-scale disaster. Members of the Select Committee were not always successful in noting the difference. Thus Dr J. Cunningham, asserting in December 1973 that the Drax coal-fired station had taken ten years to build (the first two units were synchronised in less than seven years after site work started), said Hinkley Point B was no more off course.[120] It already appeared more off course and was delayed for over two years longer. It was expected to be an unsatisfactory plant, difficult to maintain, working, initially at least, below design capacity. The incident suggested a wish to escape from the AGR facts rather than to understand them.

Thus none of the parties concerned with the AGR, not the AEA, nor the CEGB, nor the various government departments concerned with energy policy decisions, nor the Treasury, made a serious attempt to analyse the genesis of the disaster. The Vinter Committee, set up – as seen later – to review reactor programmes could not be expected to do so because of its composition; several of its participants were closely involved in the Dungeness B decision and events leading up to it. They could have been witnesses but not judges. The Select Committee did not fill the gap.

To all appearances it did not try to do so. Many of its members were clearly attracted by 'scenarios' in which there was no AGR disaster. The Committee in effect opted out of serious examination. This failure to understand, or even try to understand, the past did not inhibit the majority of the members of the Committee from advocating with great fervour which form of organisation and which choice of reactor would be most in the national interest in the future.

## NOTES AND REFERENCES

1. *SC Science and Technology, 1973–4*, Q. 284.
2. *Ibid., loc. cit.*
3. *SC Science and Technology*, 1967, Qs 524–5. It was an interesting exercise, Ghalib maintaining that he had not lost faith in the AGR while also maintaining that the TNPG compared like with like 'so the systems' comparison we thought was very fair'.
4. *Ibid.*, p. 98 and Q. 570 for Ghalib's argument that if the AGR were to be competitive for export, design advances were needed which he implied were simple: they appeared to be the use of helium instead of $CO_2$ and the use of coated particle fuel. Neither in fact was easy. (Below, pp. 222–3.)
5. Above, pp. 119, 130.
6. *CEGB Appraisal*, p. 21.
7. Above, p. 119.
8. Above, p. 125.
9. Based on a Memorandum given to me on a visit to Windscale in August 1964.
10. Moore and Thorn, *op. cit.*, p. 101, stated specifically in 1963 that design was not optimised. Vaughan had thought TNPG had optimised it in their design for Oldbury but recognised that the test would come in operation, which was not until late in 1967. Above, pp. 119, 130, 145 n.85.
11. Above, p. 118.
12. For example, see *SC Estimates, 1959*, Qs 2340, 2345 (Hinton).
13. *Nuclear Power*, December 1958, p. 581. Stretch left the AEA to join the NCB as Director of the Central Engineering Establishment in September 1957.
14. *SC Estimates, 1959*, Q. 2346 *et seq.* Also *SC Nationalised Industries, 1963*, vol. II, Q. 1075.
15. The date was recalled by Sir Henry D'Avigdor Goldsmith. See *SC Estimates, 1959*, Q. 2374.
16. *SC Science and Technology, 1967*, Qs 64–70.
17. *Ibid.*, Q. 1234.
18. See below, pp. 103, 193, 265. The *Financial Times* was so impressed with Penney's figures that it was convinced that the design of the 1000 MW FBR station must be started not later than 1968. Penney himself said with regard to his FBR figures: 'I think it is very hard for anyone to challenge the figures we put forward.' *SC Science and Technology, 1967*. Q. 114. This was the monopolist's privilege and the figures were wrong. 'Such boasting as the experts use . . .'
19. *SC Science and Technology, 1973–4*, p. 186.
20. Above, p. 120.
21. *The Second Nuclear Power Programme* (London: HMSO, April 1964).
22. Gowing, *op. cit.*, vol. II, p. 238, quoting a paper by the chief scientist of the Ministry of Fuel and Power.
23. Based on notes of meetings.
24. Above, p. 120.
25. *The Economist*, London, 5 September 1964.
26. Above, p. 119.

27. For example, *Financial Times*, 14 August, 1967: 'Thus stated the Authority's attitude sounds arrogant. When closely scrutinised it is no more so than that of the industry with more to justify it.' This is in a typical article by David Fishlock and Ted Schoeters, still arguing that AGR export orders could have been obtained in 1965–66 and stating that the FBR 'must be launched in the 1960s'. 'There are risks but they are containable.' How did they know?

28. *SC Estimates, 1959*, Q. 2393. Hinton said the senior people were paid rather more than the Authority salaries and were tied by pension schemes. The situation is now, I am informed, reversed.

29. *Nuclear Power*, December 1958, p. 583. One of his colleagues at Windscale told me that Stretch became known as 'The Rebel'.

30. For example, Dr A. J. Little, Principal of the Harwell Reactor School, who told an FBI conference in April 1958 that 'there are two outstanding systems, the PWR and the gas-cooled reactor'. These would be the 'principal contenders . . . for quite some time to come'. *Nuclear Energy*, Report of the FBI Conference on Nuclear Energy, Eastbourne, 10–12 April 1958, p. 99. See also Plowden, chairman of the AEA, quoted in Burn, *op. cit.*, p. 107.

31. The *CEGB Appraisal* concludes that the 'relevant irradiation experience is certainly no greater for the water reactors than for the AGR'. What was 'relevant' was not disclosed. There was clearly a much greater *volume* of experience for the LWR. 'Most of the substantial body of irradiation', however, it says 'has been on stainless steel cans, and the performance of the zirconium alloy can to the irradiation required is not fully proved in a reactor.' Certainly the makers in the United States were still developing their fuels, as they are now, to higher ratings. The most remarkable phrase in the *CEGB Appraisal*, however, was that 'a zirconium can . . . is proposed for BWR fuels' (*CEGB Appraisal*, p. 22). An uninformed reader might reasonably suppose this was an innovation. In fact zirconium can fuels had been developed in the Vallecitos reactor before Dresden I, they were chosen for the first Dresden load in 1959 because of good neutron economy, they were used continuously in Dresden I by the side of stainless cans and chosen exclusively after some years of mostly pre-Windscale experience; they had had 5600 hours of operation with few failures by the end of 1961 – over a year before Windscale began to operate, and the 'coolant environmental conditions' (pressure, temperature and chemistry) were 'virtually unchanged from 1959 to 1969'. By the time of the *CEGB Appraisal* experience was coming not only from Dresden I but also from Humboldt Bay, Big Rock, and BWRs in Germany, Italy and Japan. Reactor experience for AGR fuel was limited to Windscale; it was short, and in order to simulate the Dungeness B conditions of pressure, can thickness was reduced. There had been no experience of the heat exchange characteristics of the fuel bundles chosen for the AGR. Experience in BWR fuel was clearly greater and accumulating faster. *JCAE, Development, 1960*, p. 158 and G. H. White, 'Boiling Water Reactor Progress', paper presented at the Japan–US Atomic Industrial Forum (Tokyo: December 1961) p. 10. A fuller, later survey in 1970 said 'the lead Dresden Unit Type I fuel assembly [with zircalloy can] had been operating in-core for more than nine years by September 1969'. H. E. Williams and D. T. Ditmore, *Current*

*State of Knowledge, High Performance BWR Zircaloy Clad UO² Fuel* (San José, California: GE 1970) Section 4, p. 1. It was interesting that Grainger emphasised how limited in scope and value was the irradiation experience on the Windscale fuel for Dungeness B even in 1967. *SC Science and Technology, 1967*, p. 433.

32. This emerged at a discussion I had at Windscale in August 1964.
33. *SC Science and Technology, 1967*, Q. 1462. Those being ordered in 1967 were of course an advance on Oyster Creek but with much new experimental backing.
34. Above, p. 125.
35. It was interesting that Penney when asked in 1967 what benefit could come from building a BWR said we should obtain some close knowledge of LWR costs. *SC Science and Technology, 1967*, Q. 159.
36. *SC Nationalised Industries, 1963*, vol. III, p. 22 [from a CEGB memorandum]. 'The right programme now is one that brings down costs by technological development and ensures that when nuclear power is justified by its cost *or made necessary by shortage of conventional fuels* there is a sufficient nuclear industry . . .' [my italics].
37. *SC Science and Technology, 1967*, Q. 17.
38. Two examples may be given. The Oyster Creek plant was to start at 515 MW capacity, but GE said that within a year or two the power density would be raised so that it would have a capacity of 640 MW. The utility accepted this and installed a 640 MW turbo generator. The 'stretched' capacity was reached early in 1971 after eighteen months of operation. The second instance was the use of jet pumps for recirculation, which offered several advantages: they had been tested; they were adopted for Dresden II (ordered in February 1965); they worked successfully from the start in 1970.
39. Burn, *op. cit.*, p. 91.
40. I am following in this notes of discussions I had with members of the Board and of the Electricity Council and with directors of the consortia.
41. *SC Science and Technology, 1967*, Q. 1237. This point was before the decision to switch to the larger fuel bundle, and a factor in forcing the switch according to Legge-Bourke.
42. *SC Nationalised Industries, 1963*, vol. II, Qs 3573–95.
43. Burn, *op. cit.*, pp. 91–2.
44. *Ibid.*, p. 92.
45. Deputy Chairman since 1959. He was knighted in 1967.
46. *SC Science and Technology, 1973–4.* Qs 284 and 296.
47. *SC Science and Technology, 1972–3*, Q. 603. Palmer asked Weinstock whether he had any views on that kind of situation.
48. Guard, *op. cit.*
49. *SC Science and Technology, 1972–3*, Q. 150.
50. *SC Science and Technology, 1969*, Q. 698.
51. *Financial Times*, London, 30 June 1964.
52. Cost details for the proposed improved Magnox were given in Vaughan and Joss, *op. cit.* It was claimed that costs per KW in a 500 MW reactor would be 0.5d compared with a forecast of 0.66d for Wylfa.
53. This view was expressed to me early in 1964 by a member of the CEGB. He was speculating on a situation where the utility would not have its

choice subject to a minister's consent, and where it would not itself feel inhibited as a monopoly buyer.

54. *SC Science and Technology, 1967*, Qs 357–63 (Nuclear Design and Construction Limited), Qs 524, 540 (TNPG).

55. The Chairman of TNPG made it clear (*SC Science and Technology, 1972–3*, Q 309) that he had not: he claimed that all these troubles in building a nuclear power station came because no one had realised you did not just design a station and then build it. 'What you have got to do is to order it – and then over a period of two years you have to spend money testing the components in the works. All the troubles we have had have been because the components have not had running tests in the works.' But *pace* Sir Edwin McAlpine, the engineers in TNPG knew all about testing components and models: their own account of the development of the concrete pressure vessel for Oldbury (Vaughan, *Experience with Integral Gas-cooled Reactors, op. cit.*) showed this. An alternative explanation of their Hinkley Point contract put to me was that they expected to gain something from Dungeness B and so put in a competitive, not exploratory, offer.

56. *SC Science and Technology, 1973–4*, Q. 806.

57. Report of the *House of Commons Committee of Public Accounts*, Minutes of Evidence (London: HMSO, 22 January 1975) Q. 37. Hereafter cited as *Committee of Public Accounts, 1975*.

58. A foretaste of the complexities could be gathered from a paper by R. D. Vaughan and D. R. Smith, *Exploitation of the Advanced Gas-cooled Reactor for Hinkley Point B* (London: Institute of Mechanical Engineers, 1966). It contained illuminating diagrams of the gas flows, fuel stringers with gags, the standpipe closure, circulators and insulation.

59. *Committee of Public Accounts, 1975*, Q. 37.

60. *SC Science and Technology, 1973–4*, Q. 353.

61. *Report of the Committee of Inquiry into Delays in Commissioning CEGB Power Stations* (London: HMSO, 1969) p. 26. The chairman of the Inquiry was Sir Alan Wilson, Chairman of Glaxo. Hereafter cited as the *Wilson Report*.

62. *SC Science and Technology, 1973–4*, p. 183.

63. *Committee of Public Accounts, 1975*, Q. 37.

64. *Wilson Report*, p. 35; also pp. 17 and 30.

65. *SC Science and Technology, 1969*. Q. 627.

66. It did at one time contemplate offering a Combustion Engineering LWR.

67. *SC Science and Technology, 1973–4*, Q. 797.

68. Above, p. 162.

69. *Committee of Public Accounts, 1975*, Q. 39.

70. We had met frequently in 1963–64 as members of the economic committee of the Department of Scientific and Industrial Research (DSIR), and I raised the subject with him in the early days of his Ministry.

71. *SC Science and Technology* (Report) 1967, p. xlviii.

72. *Ibid.*, p. xxxvi.

73. *SC Science and Technology, 1967*, Qs 1466–7.

74. *Ibid.*, Q. 1476.

75. *Ibid.*, Q. 1468.

76. *Ibid.*, Qs 1478–1501 and (Report) p. xxxi. Had the Committee had the Report earlier it might have found some of its figures less convincing.

77. *Ibid.*, Q. 1501. Its preparation had involved the production of 100 major papers.

78. *Ibid.*, Q. 1490.

79. Harold Wilson, *The Labour Government 1964–70* (London: Penguin Books, 1971) p. 661.

80. *SC Science and Technology, 1967*, p. 188.

81. *Ibid.*, Q. 970.

82. *Ibid.*, Q. 984.

83. Unless, he qualified, the expenditure was beyond the capacity of the country to pay for it. *Ibid.*, Q. 986.

84. Reported both in the *Financial Times* and *The Times*, 17 July 1964.

85. *SC Science and Technology, 1967*, Q. 990.

86. *Ibid.*, Qs 991–2.

87. 'Comments by the Minister of Technology on the *Political Economy of Nuclear Energy*', *SC Science and Technology, 1967*, p. 411, 413.

88. *SC Science and Technology, 1969*, Q 928.

89. *Ibid.*, Q. 930.

90. *Ibid.*, Qs 952–3.

91. *Ibid.*, Q. 935.

92. *Ibid.*, Q. 938.

93. *Ibid.*, Q. 951.

94. *Ibid.*, Q. 956.

95. Quoted in Burn, *op. cit.*, p. 120. Penney rejected the comment on the ground that the AEA was not a government station but quite an industrial concern, *SC Science and Technology, 1967*, Qs 15–16.

96. *Ibid.*, Q. 940.

97. *Ibid.*, Q. 979.

98. *Ibid.*, Q. 374.

99. Above, p. 173.

100. Below, p. 290.

101. *SC Science and Technology, 1967*, p. 188.

102. *Ibid.*, Qs 950, 984.

103. *SC Estimates, 1959*, Q. 1, *et seq.*

104. *Nucleonics Week*, 8 October 1970.

105. For example Michaels repeated to the Select Committee in 1967 an AEA statement that if so-called 'United States ground rules' of thirty years life and 85 per cent lifetime load were used the latest Magnox plants were competitive with the best coal-fired plants (*Applied Atomics*, Reuters, 3 May 1967, and *SC Science and Technology, 1967*, Qs 1000, 1018). There were no standard United States ground rules, but no utility I knew used an 85 per cent lifetime load factor, and all save TVA had capital charges based on higher annual percentage rates than were used in the United Kingdom (see *ibid.*, p. 357 for my comment at the time). Even if these ground rules had been normal in the United States there was no reason to suppose they were appropriate in Britain for Magnox; it was soon crystal-clear they were not, because the corrosion problem of 1968–69 led to a derating of all the plants, and load factor has been below 70 per cent. It was well known that the CEGB would not (quite rightly) accept the

30/85 basis, that Magnox had been ruled out as too costly by the CEGB since 1963, that Magnox plants could not compete with American nuclear plants on cost nor with American coal-based plants, nor with oil-fuelled plants in Britain even on the 30/85 basis – it was not a correct choice on any economic criteria. The phoney calculation (which was probably not done correctly allowing, for example, for the 20 per cent inflation from 1959 to 1967) could only divert attention from the real problem. But Michaels – and Benn – treated it as though it was significant. Michaels also implied in his evidence to the Committee that competition between members of different consortia in particular export markets had been a serious obstacle to negotiating export of AGRs. In fact, directly the problem arose a new organisation was set up to avoid it; but there were no real prospects of AGR exports to obstruct. The AEA spoke as though they were just waiting for a better organisation. The Ministry accepted the misleading emphasis. *Ibid.*, Q. 1040.

106. To take two instances, the AEA argued that I did not compare like with like when I said Britain had spent 90 per cent as much on civil nuclear power development as the United States had spent by 1965. They said rightly that most of the expenditure in the United States was on R and D, most of the expenditure in Britain on Magnox 'commercial' plants. That was indeed my point: virtually the same expenditures were made ('like with like') but America used the resources productively, Britain used them wastefully. The Ministry repeated the AEA criticism in their memorandum to the Select Committee on *The Political Economy of Nuclear Energy* as their own and dressed it up with new figures without finding out that it was based on a misunderstanding of my argument. I was able to correct the error in evidence to the Select Committee (*Applied Atomics*, Reuters, 3 May 1967; *SC Science and Technology, 1967*, p. 412 and Q. 1685). The Ministry in its memorandum also stated that I had apparently not compared BWR and AGR fuel costs nor comparisons of cost where AGRs and BWRs were offered in export markets, in particular in Belguim. (*Ibid.*, p. 413.) No attempt was made to check this assertion. I had made both comparisons, obtaining data from the firms engaged in the competitive designs, for example, TNPG and AEG, and from various Belgian sources. The results favoured the LWR.

107. L. B. Namier, *Diplomatic Prelude 1937–9* (London: Macmillan, 1948) p. xiv. Professor Namier spoke of the 'unconscious levity' of the British Government in offering the guarantee of Poland's frontier in 1939 and of the Polish Government in accepting it.

108. Above, pp. 170–1.

109. I had discussions at the time.

110. There was a period when Airey Neave thought his TNPG interests were too directly affected by the Select Committee discussions: he then withdrew temporarily from the meetings when the Committee was engaged on nuclear problems. During the Heath Government (1970–74) he was chairman of the main Committee but not of the sub-committee which dealt with nuclear energy.

111. *Report of the FBI Conference on Nuclear Energy*, Eastbourne, April 1958, pp. 163–4.

112. TNPG was an amalgamation: Parsons was one of the firms which formed

the Nuclear Power Plant Company in 1955 – one of the two which amalgamated.

113. *Report of the FBI Conference on Nuclear Energy*, Eastbourne, April 1958, pp. 197, 180, 107.

114. Fishlock in *New Scientist*, 3 September 1964, p. 551. For the United States discussions, see above, pp. 1 and 2.

115. For example, see a letter by Lubbock in *The Times*, 2 March 1967, arguing that because our orders for AGRs were only planned to provide one-third of new generating capacity while the US orders represented 51 per cent of the projected total – 'once again Britain is in danger of yielding a world lead in an advanced technology to the Americans'.

116. See above, pp. 31, 43, 133.

117. *SC Science and Technology*, *1969*, Qs 852–5. Moore did explain that the problem did occur in all countries engaged in nuclear engineering; the nuclear designs required new processing technologies. Naturally people do not develop new processing methods in a vacuum, and there will commonly be a time lag between the demand and the response. The term 'blame' used by Lubbock is inappropriate but typical of much misunderstanding.

118. The relevant extract from *SC Science and Technology*, *1973–4*, p. 152, reads: 'The December 1973 *Nuclear Engineering International* reports the following average percentage load factors for UK Magnox stations through October 1973:

| Authority | Station | 1972 | Jan to Oct 1973 |
|---|---|---|---|
| CEGB | Berkeley | 78 | 83 |
| | Bradwell | 65 | 65 |
| | Dungeness A | 68 | 71 |
| | Hinkley Point | 50 | 62 |
| | Oldbury | 52 | 52 |
| | Sizewell | 57 | 65 |
| | Trawsfynydd | 52 | 40 |
| | Wylfa | 26 | 35 |
| SSEB | Hunterston A | 72 | 71 |
| BNFL | Calder Hall | 91 | 81 |
| | Chapelcross | 92 | 92 |

These figures yield the following averages (or, if we omit the Wylfa station, the averages in parentheses):

| | | |
|---|---|---|
| CEGB only | 56.0 (60.3) | 59.2 (62.7) |
| all UK | 63.9 (67.7) | 65.3 (68.2) |

The averages it will be seen are arithmetical.

119. The figures were: Calder Hall, 68.5; Chapelcross, 86.5; Berkeley, 82.2; Bradwell, 60.1; Hunterston A, I, 92.4, II, 78.9; Hinkley Point A, 58.3; Dungeness A, 67.3; Trawsfynydd, 69.6; Sizewell, 70.7; Oldbury, 51.0; Wylfa 13.8. *Nucleonics Week*, New York, 27 November 1975.

120. The initial work on foundations was started in February 1967. Hinkley Point was started six months later after a delay. Drax I was synchronised in November 1972, Drax II, started a year later, in December 1973. Drax I reached 90 per cent of its full capacity by December 1973. Hinkley Point was then expected to reach full (derated) power by March 1975. In the event it only began to produce power at low capacity almost a year later.

# CHAPTER 11

# Response to Failure: Two Governments Decide

By 1972 it was plain to most participants and close observers, although not perhaps to the general public, that Britain's attempt to establish a nuclear power plant industry had collapsed. There were no exports, no new home orders. The AGR, in any practical sense, was a failure. The CEGB was in no hurry to order more nuclear plants, found it hard to decide what to order, and finally decided it would like to order PWRs designed in the United States. The consortia could not survive with no remunerative business in competitive reactors, and no prospects of such business. The AEA showed no signs of passing out: being financed by the Government it could not fail. It was no longer plausible to pretend that Britain was leading in developing the HTR. Nor could more now be claimed in respect of the FBR than that Britain was on a level with France. Its commercial viability was moreover now seen, despite decades of promises, to be still remote.

Sir Arnold Weinstock, Managing Director of the only large composite electrical engineering firm in Britain, the General Electric Company (GEC), which had absorbed the two others, and who was looked upon by many as the only potential saviour of the nuclear industry, summed up the position succinctly to the Select Committee. 'In the field of nuclear energy we have developed great skills in individual engineers, great capacities in teams of people, and have abused all that in such a way as to have nothing much to show for it in the way of results.'[1] 'We' in this context was an ambiguous concept which suggested a national consensus in error and avoided attributing responsibility to the groups who had blundered, but the result of the long mishandling of nuclear development activity was fairly assessed.

Weinstock remarked to the Select Committee earlier that when policy is decided in a series of lurches, and one great meeting settles all things for all time, then ten years later 'everybody has to go around picking up the pieces'.[2] In 1972–73 the binding decisions, which need not have been binding, but became as it were the ark of the covenant, had been made almost twenty years ago; and now it *was* a question of picking up pieces. Who did the picking up and what pieces were picked up was determined naturally by the strategic strengths and tactical skills of the interests and groups who could exert influence: the AEA, the consortia, the British Nuclear Forum, interested trade unions (especially the IPCS), coal-miners, political party groups, members of the Select Committee and environmentalist bodies. The colour of the Government played its part. It was not a calm rational process, as any reader of the Select Committee proceedings in 1972–74 will observe.

As related above, the first outcome of the conflict of forces was the formation of a design and construction company, the National Nuclear Corporation, which would have the monopoly in supplying the British utilities with nuclear plants. It would, Peter Walker, the minister responsible said, use 'all the skills of the existing industry' – the old, one might have thought played-out, formula. And its management would be directed solely to the company's success. The GEC was to own half the capital, the AEA 15 per cent and the remaining 35 per cent was to be held by a group of private companies collectively represented on the Board. Although the company was predominantly in private ownership, which Walker explained was desirable so that it would be commercially orientated especially for export, the Government retained remarkable ministerial powers. Ministerial approval was needed for the choice of the reactor systems which the company might make; the Minister would have power to ensure that an open-market policy was exercised in purchase of components, so that all potential suppliers should have a 'fair crack of the whip', and the Minister's approval would be needed for any international links – the object being presumably to prevent domination of the NNC by an overseas group. To help the Minister in the exercise of his powers, a new advisory body, the Nuclear Power Advisory Board, was set up with representatives from the AEA, CEGB, SSEB, the National Nuclear Corporation, the BNFL and (initially) two independent members. The plan was, John Davies, Walker's predecessor, had said in announcing the plan, to 'bring together all having a major part to play', so providing the Minister with 'concerted advice on all aspects of nuclear generation policy, and on the Government's role in ensuring the most effective programme in this field'. The Board, he said, would have a major part to play in the decision to be made in eighteen months on the plants to be ordered.[3]

When Walker announced the details of the package in March 1973, Benn (as Opposition spokesman on Trade and Industry) remarked with satisfaction that it was the 'second massive intervention in private industry in a week' by a Conservative Government.[4] Palmer said that he regretted that the state had not retained a majority of the shares, as the Select Committee had recommended, since the country had contributed great resources to the foundations on which the company would rest.

The second major outcome of the conflict of forces came over a year later, when the Secretary of State for Energy, Eric Varley, announced the Government's decision that the Electricity Boards should adopt the SGHWR for their next nuclear power station orders, and should not buy LWRs as the CEGB wanted. This was against the advice of the NNC and against the judgement of most nuclear component suppliers and it looked as though the AEA had two minds about it. It had support – from Hinton, the IPCS and the SSEB.[5]

What was being salvaged by the new arrangements? Would the nuclear plant industry be strengthened? How were the propects of efficient supply of nuclear plants in the United Kingdom, from domestic or foreign sources, affected? Had export prospects been improved? Was a repeat performance of the AGR disaster with another reactor ruled out? What was the effect of the new organisation and new policy likely to be on the cost and use of energy in the United Kingdom? Did it still matter in view of North Sea oil?

These are the questions with which the next two chapters are concerned. The first deals with the emergence of the single design and construction company and the Nuclear Power Advisory Board, its successor with the choice of the SGHWR.

## NOTES AND REFERENCES

1. *SC Science and Technology, 1973–4*, Q. 30.
2. *SC Science and Technology, 1972–3*, Q. 619.
3. *Hansard*, House of Commons, 8 August, 1972, c. 1492 *et seq.*
4. *Hansard*, House of Commons, 22 March 1973, c. 672.
5. *Hansard*, House of Commons, 10 July 1974, cs 1366–7.

# CHAPTER 12

# The Single D and C Company

## (i) THE ARGUMENTS FOR AND AGAINST

Agitation for a single company to design and build nuclear boilers started in 1964, before Dungeness B had been ordered. It was not in response to the Dungeness B disaster, but arose out of the disappointments, frictions and struggles for power of the latter years of Phase I, when orders for Magnox plants were reduced and phased out, when there were no more exports and the selection of the AGR to succeed Magnox was threatened by the LWR. After an AGR was ordered for Dungeness B, it was widely expected that export orders could be obtained immediately. Absence of such orders proved irrevocably that the consortia, established to convert prototypes brilliantly developed by the AEA into commercial plants, were, because of their form, their function and their number, a source of high costs and a damper on export effort. When the second order for an AGR plant was given to TNPG, not to APC, this was seized upon, as recorded earlier, by APC, AEA and Labour and Liberal members of the Select Committee as further proof of the wastes which the consortia system entailed.[1] The case for the single company was thus argued on the assumption that the AGR, as designed by AEA and APC, should be an economic success, but was a success *manqué*, due to the consortia and the consortia system.

This was conspicuous in the 1967 proceedings of the Select Committee when the agitation in favour of a single D and C company reached its first climax. In the second session, Penney set out briefly his case for a single design and construction authority centred in the AEA, which would eliminate the need to pass information about the prototype to three other groups. Towards the close of the hearings he laid before it a memorandum setting out succinctly what the AEA would like.[2] It showed impressively the Authority's zeal for the leading role, and its complete neglect of – or contempt for – the basis of success in other countries. The existing system, whereby separate design and engineering efforts were maintained by the AEA, the CEGB and the

197

three consortia, with a separate organisation the British Nuclear Export Executive (BNX) to coordinate the reactor export efforts of the consortia and the Authority was 'uneconomical and inefficient'.

So a joint company should be set up as soon as possible embracing all these activities, capable of completely designing a station, and having an export organisation. It would design the reactor island for all nuclear plants using British systems, and would act as consultant to the CEGB in placing and executing contracts. The CEGB would place contracts exactly as it did for fossil-fuel plants; industry would continue to be solely responsible for the work of construction and for equipment, and the Authority for the provision of fuel services. The new company, like the consortia, would take over the design and exploitation of systems 'as they became commercially exploitable' – so by implication the AEA would retain its monopoly in developing new systems. In 'the long term' the company *might* take over development resources 'needed in support of commercial design'. The AEA would continue for the foreseeable future to monopolise the supply of fuel: design might be undertaken by the new company but the steady profit business would remain with the AEA. The company would take over such staff as it wanted from the Authority and companies who were willing to transfer. It would be established on the AEA site at Risley.[3] The AEA would preferably have a majority of the shares in the new company.

The spur to efficiency, the Authority said, would come from foreign competition in the home market, if this was thought to be in the public interest. But Penney thought it would not be necessary because you would have 'cut-throat competition' within the single company between teams pushing different reactors, one pushing fast reactors, one the AGR, one the SGHWR. That, he thought, would 'be good enough for a decade'.[4]

The strongest driving force behind the move for a single company was clearly the Authority, whose personnel wanted more dominant power – and security. They were convinced, wrongly, that they were best qualified to design commercial plants[5] and were also shaken by the risk in 1964 that the CEGB might buy LWRs and by the knowledge that the two principal consortia wanted to make one. Some were worried at the size of the design capacity which the CEGB had built up under Hinton. It was not difficult to attract support. The prowess of the AEA was still admired by prominent commentators. Most socialists, like Arthur Palmer, welcomed the idea of a state-appointed body extending its influence.

The attractions of the AEA plan for the Authority were not obscure: it removed the threats to it from the CEGB and consortia completely, would leave no design team immediately capable of supporting a competing development, and would secure for AEA personnel leadership in all foreseeable reactor development and exploitation. The case for it could be made to look seductive and it was fashionable. The market for electricity, and hence for new generating plants, was smaller than expected; the unit size of reactors and generators was larger than expected, the number of prospective orders for nuclear plants in the United Kingdom for some years was low – one a year or fewer. It was to fall lower, but that was not foreseen. The small number of orders meant there could only be a satisfactory load for one concern.

Moreover, if there were several consortia they would all have different designs; these could not all be the best, and subdivision was likely to lower the quality of all the teams and the need to be different would lead to frivolous differences. The existence of several D and C companies would lessen the scope for repetition and standardisation in manufacture and construction. The single company proposed would therefore give the best load, a greater likelihood of high quality, the greatest scope for 'mass production'. It would also eliminate the transfer costs when developments were handed over to the consortia: the AEA teams would just carry on in the new set-up. The new company would have none of the ties to particular component makers which were had in the consortia. The member firms in these had been chosen originally by the AEA[6] because they had complementary capacities, which were necessary for producing a nuclear plant. Naturally the member firms in a consortium expected to get the work which a contract brought in their field. As a result, consortia could not get the benefits of competitive bids.[7] Instead, it was said, the member firms put the prices for their components high.

It was also claimed that consortia partners would have little interest in export orders, since these would normally require that as much manufacture and construction as possible should be by local contractors in the importing countries. Furthermore, the IPCS pointed out, the consortia were not allowed to make fuel, hence had no interest in the fuel export activity derived from exporting reactors which was expected to be remunerative over a long span of years.

There were conspicuous weaknesses in this general case, in logic and in data. To build one team out of five does not mean you get the best. Standardisation can be premature and restrictive in a new technology; development is concerned with solving novel problems – one can take the economics of repetition for granted, but it will not make an uneconomic system or a bad design economic. Experience in other countries does not show that the working together of a stable group of complementary firms is necessarily unsatisfactory, rather the contrary; in Japan for example the exceptionally fast building of plants has been achieved by stable groups of firms. In many countries subcontractors, having established a niche (introducing a structural rigidity parallel to that in the consortia), raised prices steeply in competitive conditions.[8] The IRC, reporting to the Minister of Technology in 1968, believed that the member firms in the consortia were a valuable potential means of getting exports because of their strong export activities.[9] Lack of exports, as we have seen, was not due to the organisation of the firms. The elimination of transfer costs did not lead logically to a single D and C firm – some of the consortia had wanted to achieve such elimination as early as 1959 by being free to undertake early stage development work with state aid on the United States and German pattern. They had wanted an integration backwards.[10] It was only the AEA development monopoly which made a single firm – essentially *integration forward* from the AEA – appear the logical answer.

The British utilities alone could not be expected to provide a market big enough to allow even one firm to achieve the economies of scale achieved by the American firms. Penney recognised this: we could operate the home market 'with a closed circle and keep it going but we cannot do this efficiently'. Many countries were going to try and supply themselves 'but they are going to come unstuck'.[11] This was by now a familiar problem in Europe, not just for nuclear plants. Curt Nicolin, Managing Director of ASEA, had put it graphically in the early sixties: half the turbo-generators of the Western world, he said, were made by two firms in the United States, and the other half by twenty-five firms in the remaining countries. National monopoly did not offer a satisfactory solution, still less the only logical one, because the national market which it offered was too small and because it ruled out the access to alternative reactor systems and competing skills, whose value British and American experience had, in different ways, underlined. Development of international companies would satisfy both requirements. David Price put this – in a stilted form – in the Committee proceedings. Penney thought it attractive in theory but 'practically impossible',[12] as it probably was for the AEA; but, there was no thorough discussion.

The AEA intended to provide the single-company organisation 'capable of the concentrated export effort needed if we are to compete effectively with the Americans'.[13] It was striking that it was completely different from the organisation of the American companies. It was an odd conception. Penney agreed it was an 'architect-engineer organisation' with 'some development activities', and it would design 'in detail the reactor island';[14] but it would manufacture nothing. All the manufacturing based on its designs of prototypes and commercial plants would be done by contractors selected by competitive tendering. Just as in the AEA itself.

In the United States architect-engineers did not develop or design in detail the 'nuclear steam supply systems'; they designed complete power plants around steam-supply systems designed entirely by the

manufacturing firms who first developed them. These firms manufactured a large proportion of the components of the systems: pressure
vessels, pressure-vessel internals, control rods and drives, core components, fuel and accessories, zirconium tubes, instrumentation and
control equipment, pumps, pressurisers, and steam generators. Not all
firms made all these components, but all made a good proportion; two
also made the specially developed turbo-generators and other electrical
equipment – motors, diesel generators, transformers and switchgear.
The manufacturers were also responsible for the erection and testing of
the equipment, for providing facilities for training staff. It was the
manufacturers of reactors and much else who were the driving force
in American development. They had a much greater understanding of
the whole process than the company proposed by the AEA could
possibly have. Whereas Penney said 'I do not think you can just design
. . . you must cultivate some development activities',[15] experience in
the United States showed one must also cultivate extensive manufacturing activities – not a surprising conclusion.

   Penney, coaxed by Lubbock, argued that because the single company would place contracts – on a competitive basis – for the manufacture of all that it designed, there would be *more* competition than in
the consortia system, under which investing firms secured contracts for
themselves.[16] In the new company industrial investors would be welcome, but would not influence contracts in their favour. Nevertheless,
Penney (explaining to Norman Atkinson, MP, why private firms were
wanted) stated that one of the two contributions which industrial firms
as members would make would be a 'feedback of manufacturing costs',
without which 'the main bid is not based sufficiently well'. It was 'very
important that manufacturing costs, real costs, should come into the
organisation'.[17] But by what route, at what stage, in what detail, with
what reservations, with what continuity, would feedback come from
firms who held shares in the company, if they did not get a continuous
business from it? If they did get continuous business how did the
system differ from the consortia system? If feedback could be obtained
from outside firms, who *did* get the business by competitive bids –
why have insiders? Was feedback only necessary for guidance on
costs, as Penney implied? The logic seemed weak.

   The major firms in the United States competed vigorously against
each other. Penney returned frequently to the need for foreign competition in the United Kingdom but protectively found one point
which worried him terribly: 'how are we going to be sure this is not
heavily subsidised?'[18] In principle he would make the new organisation compete with offers from overseas and he thought, reasonably,
that this should apply to turbo-generators too. But what he meant
specifically is not clear; possibly offers from foreign firms directly, not
indirectly through an English licence-holder as for Dungeness B. Yet a

foreign firm would have to employ British subcontractors, and a situation in which a foreign firm could, and a British firm could not, employ them on the same job would have been bizarre. If the foreign system sold well the monopoly advantage of the single D and C company would be greatly reduced. Without avowed or unavowed protection a British D and C company could only succeed against foreign competition if it could provide competitive plants.

This was not the situation. What the AEA had brought into the industry, and would bring additionally into the company, was not competitive. The picture Penney gave of internal competition between the AGR, SGHWR and FBR teams becoming a source of efficiency makes the point impressively. The AGR was on the way out; the SGHWR, too late to prevent the ordering of five AGR plants, required much further development before its economics could be assessed, though it was unlikely to be competitive with LWRs: the FBR was still remote from competitiveness. Few people perceived this at the time: most importantly perhaps, the AEA, in the top ranks, did not. Discussion of a remedy for the British industry's failure, whose premises ignored the main source of failure and would exalt the agent principally responsible to still greater influence, was necessarily a fantasy; but it was illuminating and could be influential.

The penultimate session of the Select Committee hearings at which the AEA proposal was presented was held *in camera*, so there was no possibility of immediate public (or consortia) reaction.[19] The CEGB were invited by the Committee to send comments, and registered substantial disagreements. They agreed that the AEA should retain the monopoly in developing concepts not yet at the commercial stage, but found unacceptable the plan to set up a monopoly for design and construction of power stations at home and abroad. 'The Authority are the repository of specialist nuclear knowledge', but are 'far from being so as regards design, manufacture, erection and operation of commercial plant. They are seeking to enter into spheres in which their natural advantages are no longer paramount.' They found their judgement reinforced by the mishandling of the Dungeness B design with all the AEA involvement. The Board regarded the abolition of technical competition as undesirable (they appeared to mean competition between two independent commercial designs); competition was 'the best spur to efficiency', and 'stimulated responsiveness to market requirements'. To deny industry responsibility for development must lead to stagnation with little incentive to cut costs or improve performance.[20]

They had thus, effectively dismissed the engineering claims of the AEA but not yet recognised the danger of AEA monopoly in developing new concepts.

(ii) THE SELECT COMMITTEE DIVIDES

The 1967 Report of the Select Committee showed the same ambivalence over these conflicting submissions by the Authority and the Board as over the assessment of the gas/graphite system,[21] and again it reflected mainly the party divisions, although also some confusion. Remarkable inconsistencies were presented with great aplomb.

The Report favoured more effective competition. One of the six main current problems of the industry, it said, was the need to maintain competition within the home market and in face of rapid advances by foreign competitors.[22] The 'turnkey' system, whereby the CEGB bought whole nuclear power stations, should be abandoned. The Board should treat nuclear power plants as 'normal industrial jobs'.[23] acting, that is, as its own architect-engineer and ordering components (boilers, turbo-generators, transformers, and so on) from competing firms. The Report recommended that the AEA concentrate on their primary task of research and development. Reorganisation of the industry should have as its aim 'the more effective integration of the Authority's effort on research and development with more *competitive* industrial activity than is now the case'.[24] The Report noted that American makers, in seeking exports, had taken fuller advantage than the British of the existence of indigenous firms capable of undertaking nuclear installation work and argued that 'British manufacturers [noticeably plural] were likely to be most successful in exporting British reactors in close association with Continental makers and suppliers of heavy power plant'.[25] Thus in Germany GE was linked with AEG, Westinghouse with Siemens, and now APC had linked with BBC – three competing German firms. 'It is much in the British interest to break down nationalism in the Continental and world development of nuclear technology'. Hence generating boards in the United Kingdom 'must be free to shop overseas, if it is in the national interest to do so'.[26]

Along with this, the Report favoured the single company with the AEA participating strongly on the Board to design and make nuclear boilers. Within this company a major part of the reactor research and development activities of the AEA were to be included. The single company would design and make all reactors based on British development, for domestic utilities or export, and would be responsible for the export effort.[27] It was close to but not identical with the AEA proposal.

This recommendation of more centralisation and monopoly was supported by all but one (David Ginsburg) of the Labour members of the Committee and by the one Liberal, Eric Lubbock – the Conservatives voted against it. This was not made plain in the Report itself: a result possibly of a decision to make the first report of the new

Committee an agreed document. It was in the Minutes of the Committee's Meetings,[28] which most readers would probably not consult. It was later referred to in debate in the Commons, but it was the Report, not the debate, which was subsequently quoted.

The Report also recommended that the new single D and C company should design and develop fuel and research on it but not manufacture it, and there should be a new monopolist fuel supply and manufacturing company consisting of the 'AEA and others'.[29] A subcommittee which visited the United States had recommended that 'competition in this field, though difficult in the UK, is desirable' in the interests of scientific development.[30]

The Report set out the standard arguments for the single company; size of market, myths of exports and replication forgone, the effectiveness of competition between teams working on rival reactors in one institution, and the powerful potential effect expected if the CEGB showed a willingness to 'invite' (not necessarily to accept) foreign tenders. It also produced a new, essentially *ad hoc*, argument. Because the AEA was the repository of much essential knowledge, it would naturally be 'strongly represented in the proposed single organisation'. It would be impractical to contemplate a second company in which the Authority did not also play an important part. Competition would not be realistic between two companies in which the AEA participated and it would be difficult for two companies to have equal access to the research and development facilities, such as irradiation testing and sodium technology.[31] The possibility of AEA staff going into different autonomous companies, without AEA participation in management, did not figure in the Report.

Unlike the AEA Memorandum, the Report recommended that a large part of the AEA reactor research and development should at once be transferred to the proposed single company. It pointed to what it called *the* 'striking feature of American structure ... the integration of basic research, development and prototype development within the commercial organisations which have to sell the products'.[32] This came about, it was said, through the AEC acting as a controlling and licensing body only, and placing large nuclear contracts with private industry; a system which 'for historical, constitutional and political reasons could not be transplanted into this country'. The Committee (that is the Labour-Liberal majority) was eclectic in its reference to American practice, and failed to refer to other not less striking features which were vital in its success. Each company made only one type of competitive reactor and developed it intensively. All made fuel for their reactors, all competed vigorously with one another, and all had to sell to highly commercially-minded utilities who were numerous, free to choose, applied commercial criteria, could not be 'bent' by government pressure or subsidies, and were prepared themselves to invest in order to support the development of plant whose future value they recognised.

The Conservatives were unable to see how the majority proposal could be reconciled with the need to retain competition in the home market or with the granting of freedom to the CEGB in ordering nuclear plants as a 'normal industrial job' or with the recommendation that the AEA should concentrate on their primary task of research and development where they were expert. Without legislation to forbid it, any British company could act as licensee to a foreign plant manufacturer, although they were forbidden to have a licence to build any AEA reactor system – making the single company meaningless unless foreign competition was restricted. Foreign manufacturers would be able to have in export markets licences for British reactors which were refused to British makers for home or export. This would look, and would be, ridiculous.

If domestic manufacturers other than the single company were banned by law from having foreign licences, direct foreign competition would be a weak check on the efficiency of the single company. It would always be possible to argue that competition was 'unfair'. There were for example restrictions on entry into important foreign markets, above all the United States, so that, if foreign sellers were successful, agitation to prevent it could be based on arguments which sounded plausible and the Government would be pressed to secure an adequate return on the nation's great investment.[33] Aiming at a wider national monopoly was not the most persuasive way to 'break down nationalism as a means to more international trade'; this doctrine was perhaps for export only.

The Conservatives 'preferred the market solution'. They accepted the CEGB's low estimate of the AEA as designer of commercial plants, and the high value the Board placed on competition; they emphasised the AEA's lack of export experience, and the uncertain size of markets for nuclear plants – Europe's demands might follow the American explosion. Britain's electrical engineering, they pointed out, was undergoing radical changes which would produce fewer but stronger firms: cross-frontier links were possible between British and foreign firms, which would help exports.

The Conservatives did not appear, however, to challenge the view expressed in the Report 'that the Authority have done their basic task, that of civil nuclear research and development, with success'.[34] They, as much as the Labour-Liberal majority, failed to see the overwhelming case against the proposal for the single company. The AEA had backed the wrong horse, and ridden the horse badly. Available gas-cooled reactor systems were unexportable: Magnox was a dead end, the AGR, as handed over, was underdeveloped and a strain on consortia resources. There were conclusive reasons for not extending the area of centralisation in general, and for not giving the AEA dominant power in a new field where it was not a leader. Nor did the Conservatives pick up the contrast between the utility structure in Britain and that in the countries more successful in developing competitive nuclear power – notably the United States and Germany – where, there were several utilities large enough to buy nuclear plants and free to make their own assessment of the economic advantages of particular systems and suppliers.[35]

### (iii) BENN REJECTS THE SINGLE COMPANY

In the Autumn of 1966 Benn, when he became Minister of Technology, favoured the single design and construction company, and set out his views to the consortia early in 1967. But, to the chagrin of the Labour-Liberal majority of the Select Committee, he changed his mind, and in July 1968 announced his decision in favour of two companies. The Minister had been advised by Sir Frank (later Lord) Kearton, Chairman of his new Industrial Reorganisation Corporation (IRC) and on Benn's instructions the IRC proceeded to negotiate the forming of two companies. The reasons he gave for changing were (i) that the IRC said there was enough business for two companies; (ii) the shareholding members of the consortia had valuable overseas links to help in exports;[36] and (iii) the companies in the existing consortia would not voluntarily all form one company. He seemed averse from passing a law forbidding British companies to become licensees of foreign companies perhaps because it would have brought retaliation, and the IRC favoured the overseas links.[37] The Minister specifically said he had not been influenced by the CEGB's wish for competition – his reasons he said were sufficient, and he was not responsible for the Board. As seen earlier, the two companies were to absorb the functions of prototype development from the AEA, and therefore part of the AEA staff. They were to feed their marketing experience into research and development so that systems were developed with the market in mind.[38]

In addition to these two companies there was to be a new fuel company, as the Select Committee had proposed: initially all the shares would be held by the AEA, but private firms might be asked to take some shares later. This fuel company was to have a substantial minority holding in the two design and construction companies, who would design, but not manufacture, fuel. Until the fuel company was formed – it needed legislation – the AEA would have its holdings in the D and C companies.[39] The two could thus hardly be regarded as autonomous and independent. Benn also planned to set up an Atomic Energy Board, including representatives of the Authority, the CEGB, the design and construction companies and the fuel company, which was to 'concern itself with the composition and financing of the R and

D programme, the coordination of activity in the export field, and major matters of policy'. This too required legislation, which the Minister thought was still a year in the future when he gave his evidence to the Select Committee in 1969: in the interim he intended to strengthen the 'existing coordinating machinery'.[40] The Labour Government had been replaced, before legislation to set up the Board had been introduced, and the precise functions Benn thought it should perform were never spelt out.

Benn, as remarked earlier, had seen only half the light, and his solution was muddled, with no bearing on the central problems. Of the three consortia, APC went out, as it would have in 1964–65 but for the AEA. The two new companies were, in their industrial composition, virtually identical with the other two consortia, save for one significant change which happened almost by accident. TNPG remained TNPG, NDC became BNDC. Both had the AEA and IRC as additional shareholders, which hardly added to their industrial or exporting strength.[41] The accidental change was that, after NDC (the former English Electric – Babcock & Wilcox – Taylor Woodrow group) had been reconstituted (as BEEN),[42] GEC absorbed English Electric, which brought Sir Arnold Weinstock into the group. He was widely looked on as the future strong man in the industry, but at this stage he had larger immediate preoccupations, and his conspicuous interventions were holding and defining, not creative, operations. Weinstock played a large part in deciding that BNDC would not have its headquarters at Risley but remain at Whetstone, and would not take on contracts to complete the Dounreay FBR project.[43] He obtained a price for Heysham much above previous levels, contrary to IRC expectations.[44]

The transfer of AEA functions took the form of a transfer of teams, the SGHWR and FBR teams, but not of real responsibility to the new companies to plan development in response to market experience, which was, in Benn's rhetoric, to be the fundamental innovation about the new companies. The transfer of teams, both in the end to TNPG, was a means which ensured that AEA projects went forward under the existing AEA project leaders within the new companies, and so conceded what the AEA had been seeking since 1964. The Minister's object, was to 'allow those who worked in the AEA full scope to carry their work forward into the exploitation and sale of the systems they had developed'.[45]

The trade union of the transferred teams would have preferred this extension of function to be within the fully state-owned sector: the AEA itself. In explaining to the Select Committee why this would not be, Benn recognised that the '*status quo* was really rather comfortable and pleasant' for the research staff, and gave a complete security which no firm could give, but 'it would be wrong' for the AEA research staff to suppose they could 'remain in their present employ-ment in their present numbers doing their present job' when the research function was reducing. A research organisation with a single function does necessarily work itself out of a job. Atomic scientists, who by their work greatly disturbed the coal miners' prospects, should not themselves expect to escape the consequential changes of their success. The best teams are teams brought together for certain pur-poses, not teams held together for no purpose. The industrial end may require qualities not always found in the same person who is a success-ful researcher. Nevertheless, the Minister hoped the teams would trans-fer to the new mixed enterprise company.

But the process was not envisaged by the AEA as a handover: the company would nominally 'manage', but, to take the case of the fast breeder, the AEA intended to 'retain the overall responsibility to come to a successful conclusion of the prototype'. Some of the work con-nected with its conclusion 'will be done for us . . . under contract, by TNPG', who had not taken over the prototype: 'they are merely doing more work on contract' which the AEA would otherwise have done for itself. TNPG had taken on 150 FBR staff from the AEA,[46] but, as the proceedings described it, it remained an AEA cell within TNPG, and the contracts on which TNPG worked were financed by AEA. The AEA went out of its way to emphasise how little was changing: TNPG were 'trying to take over some of our design and engineering effort' but the generic work on systems remained with the Authority. The AEA would get 'a little more help' from the companies than in the past. But the AEA would retain exclusively the irradiation facilities: it would be quite uneconomic for the companies to be able to test fuel designs.[47]

Of the two new companies, TNPG seemed prepared to play as the AEA wanted: but not so BNDC under Weinstock. So two AEA teams –FBR and SGHWR – went into TNPG, none to BNDC. BNDC had no contract to develop any newer system, although when TNPG was occupied on the FBR and the SGHWR, BNDC was regarded as the focus for HTR work.[48] They did no development work on their own on it: there were no resources.

The two companies, as created, bore no resemblance to the great new departure which Benn proclaimed. The Select Committee recog-nised this position in its 1969 report. The new companies were doing 'fractionally more on existing reactor types whether commercial or not' than previously. 'But official influence over the way that future development is directed is still wielded by the Authority.' The days of market-oriented development had not dawned. The Authority 'showed little sign of a serious withdrawal'.

This, the Committee said, 'we can hardly criticise in the circum-stances'.[49] They remained wedded to centralisation in the form in which it had failed. The nuclear industry must be dominated by the

AEA and the Government. The Committee had envisaged a division of the Authority into four segments in 1967, but within each segment it would have been dominant.[50] Now, the Committee argued, the Authority should continue to dominate while the Government failed to make decisions, and when the Government made decisions these must back up the Authority. The AGR was unexportable, but the Authority had developed three new reactor systems which the industry could 'offer for commercial operation' in 'the next two or three years'; the SGHWR, HTR and FBR.[51] The Government must underwrite these and ensure, by subsidisation, that the utilities ordered them, so that they could provide the basis of export orders.

These were grotesque misjudgements: but the Select Committee still took AEA promises at their face value, although they knew that two reactor types had been backed by ministers on AEA advice, and forced on utilities in larger quantities than they chose, which turned out to have no export prospects and were more costly than alternatives. Detached observers now recognised that the absence of exports was not due to the organisation of design and construction. *The Times* made the point forcibly in 1968 and 1969: 'Impressive as the AEA's achievement has been it has certainly not been enough to justify handing over the whole industry to its control'... 'No matter how many times the structural changes are rung this will not of itself help to sell power stations.'[52] But the Select Committee majority did not – perhaps could not – follow the simple logic of events.

(iv) CONSERVATIVE FORMULA

The two companies proposed by Benn, dubbed 'denationalised' – within the AEA, the IPCS, among MPs and in the press by frustrated seekers of supreme power for the AEA – were declared a failure almost before they were formed, which was premature, but they were inevitably failures by the criterion adopted. They received no export orders because they had nothing saleable to export. They had no home orders: the CEGB's needs were smaller than was expected, and the Board deferred ordering the Sizewell AGR in 1970. With the Vinter Committee appointment, it became clear there were doubts whether more AGRs would be bought. The Board continued to order oil-fired plants, and announced formally, in autumn 1973, that it would order no more AGRs, at least until one had worked. Oil-fired plants were a better bet. So the problem – what to do with the nuclear industry – had to be faced again by the new Conservative Government, to whom Benn could pass (unwillingly no doubt) the poisoned chalice.

They focused initially on the right problem, if not the right way of handling it: they set up the Vinter Committee late in 1970 to review the reactor development programme. The key problem was, what should the industry now make? It must have a product with a market, must get orders. The Committee set up two working parties, one to study the fast-breeder position, the second the thermal reactor work. This was not to be another Powell Committee, it was said; but Peter Vinter, who chaired it, had headed the Treasury department on nationalised industries and had handled atomic energy problems there: he was at this time a deputy secretary in the Department of Trade and Industry. The other two principals were the chairman of the AEA and CEGB, Hill and Stanley Brown, respectively – not the team to inspire great confidence since their institutions were the source of the disaster. E. C. Williams, appointed Chief Inspector of Nuclear Installations in February 1972, also played a central role. The names of the working party teams were not to my knowledge published: those on the thermal reactor party included – according to the grapevine – at least two of the AEA's principal representatives on the Dungeness B *Appraisal*.

The Committee took long to reach any conclusions, and on thermal reactors the conclusion was that further studies and development work – of varying intensity – should be carried out on the SGHWR, AGR, HTR and LWR. On the latter 'our objective is to achieve assurance about the questions that have arisen as to its safety'. Arthur Palmer, when John Davies, Secretary of State for Trade and Industry, announced this decision to defer a decision, in August 1972, made plain the obstacle to agreement: 'Does the statement mean that there is no danger of this country making the gigantic blunder of turning to an American reactor system?' The fast breeder was to remain the main element in the long-term programme: the 250 MW prototype was now expected 'to run next year' (1973).

So the question – what should the industry do, make, and sell? – was shelved. But the Conservative Government decided to proceed with reorganisation into a 'single strong unit' as soon as possible, before deciding what the industry should do. All the parties to the Vinter Committee appeared to have concurred on this though it was not their initial problem. It took eight months to agree the bases of organisation and much longer to get the organisation working.

Palmer naturally claimed that the Select Committee Labour-Liberal majority report of 1967 was vindicated and made much of this.[53] Those who had opposed now accepted, or were resigned to, the single D and C company: both the Benn companies, the CEGB, the Conservatives on the Committee – all except R. A. Morton, who had headed the IRC exercise which advised Benn.[54] This acceptance was not due to the persuasiveness of the majority's arguments: no one seemed to return to these – their irrelevancies and illogicalities remained. The acceptance was almost certainly due to the poor financial shape of both companies as the decline of their activity progressed,[55] inevitably, and due to the failure of the gas-cooled reactor development imposed on them by the AEA and ministers, from which they had not been allowed to escape in 1958–59 or in 1964–65. Palmer did not recognise – or did not acknowledge – the source of the decline. From all he said the unwary might have supposed that the single company could have stopped it, and he continued to advocate the AEA as majority partner in a single company.[56]

The Conservative Government turned this down. They had gravitated steadily since early 1972 – perhaps earlier – towards solving the organisational side of their problem, perhaps the whole problem, by persuading GEC, which meant Weinstock, to undertake the salvage operation. Conservative members of the Select Committee hoped in 1967 that the reorganisation going forward in the heavy electrical industry would throw up strong firms who could take over the nuclear plant industry. GEC had come out of the process unexpectedly as the one dominant company. Many politicians had become increasingly dazzled by GEC management skills, and the company clearly had the strong financial background and international trading connection needed to export nuclear plants – supposing there were nuclear plants to sell. Weinstock's personal reputation rose steadily; in April 1971 the Government appointed him a director of the nationalised Rolls Royce. Cecil King recorded at that time that he was 'growing in self-confidence and authority all the time'; by 1974 he was 'surely the leading industrialist of our time'.[57]

To ensure that Weinstock and the GEC would undertake the salvaging the Government had to find a form of organisation mutually acceptable to them and Weinstock, acceptable to the other interests involved and capable of being sold politically. Weinstock had enemies: some feared he would want to buy American, some would lose jobs if he won.

The form agreed upon was aptly described by David Ginsburg as 'an odd animal'. The opening passage of Weinstock's evidence to the Select Committee in March 1973, just before agreement was reached, provides an appropriate preface to its analysis. Asked why Britain was doing so badly in export markets he said, first, that 'the light-water reactor was a better choice than a gas-cooled reactor', and second that 'in the US the whole range of activity was concentrated into single companies, General Electric and Westinghouse, who handled the development programmes, dealt with the choices, offered reactors to utilities on a more commercial basis than was the case here, where we set up a totally different system to deal with this industry'.[58] This was as close as could be, in a truncated form, to the thesis of *The Political Economy of Nuclear Energy*.

It is of major interest to observe in the light of this diagnosis how far the terms which formed the basis of agreement in 1973 moved towards the American model and away from the 'totally different' British model. The main elements in the new arrangements are analysed in succession from this standpoint.

The capital structure was odd: GEC would have 50 per cent of the shares. A large number of other companies would together hold 35 per cent indirectly through the new company, British Nuclear Holdings, which would be *one* shareholder of the new company. The Government would have 15 per cent, to be held by the AEA. This was linked with a two-tier management structure unusual in Britain, with a main board which would appoint an operating board. Half the members of the main Board would represent GEC, and all members would need approval by both GEC and the Government. GEC with its 50 per cent holding and board membership would lead and supervise the operating board; the GEC would 'be managing this company by the agreement of the whole board, which is not in doubt'. It would receive a fee for its contribution to management services. The position was defined thus by the Minister for Industry before the board was formed. The involvement of GEC in the company was 'in order to provide the leadership, and by leadership this means supervision and management together with financial strength'.

Why, Ginsburg asked, should GEC not have a majority share-holding, 51 per cent? Because, Thomas Boardman, Minister for Industry, answered, the company would then become a subsidiary of GEC, and directly accountable to GEC, whereas on the fifty-fifty basis the operating board would be answerable to the main board, which would have an authority and influence which it would not have had if the company were a subsidiary. But with GEC so firmly established in executive management the role of the board was not crystal clear: Weinstock made clear in his evidence to the Select Committee his view that policy must spring from day-to-day experience and be formed by executives handling day-to-day commercial and technical business, entirely by persons in day-to-day touch.[59] These would be GEC persons. By what criteria would the main board, remote from day-to-day business and heavily GEC-weighted, check the efficiency of GEC management in this business where the outcome of important decisions was usually remote in time? It is difficult to see why the bunch of companies, most of them previously in consortia, should wish to be shareholders since no shareholder, not even GEC, was to get prefer-ential treatment over sub-contracts and there were no early profit prospects, or why this arrangement should be a good one for the appointment of external directors acceptable to GEC, but supposedly able to detect any plunge into inefficiency.[60] Needless to say this capital and management structure was remote from that of the great American companies.

So, too, was the competitive status of the new company. Weinstock made it clear that he supposed it would have all the domestic orders.[61]

No influential voice was now raised against this. The fears expressed by the Select Committee, and even Penney, in 1967 of the monopolistic dangers to efficiency and cost if the utilities were not free to invite tenders from abroad for the supply of nuclear boilers[62] had been apparently forgotten. No one said the dangers had been removed. In the United States, and most other countries where nuclear plant building was conspicuously successful, there was strong domestic or foreign competition – or both. The defence of this changed attitude towards monopoly would have been that demand for nuclear plant was even smaller than expected.[63]

Turning to the scope of activities envisaged, again there was a sharp contrast between the new British and American companies. The great American firms manufactured components for nuclear boilers, designed and made fuel, carried out the development work on the reactors they chose to make and sell, and made only one type of competitive reactor. There was no sign that the new company was to make components: the subject was not, I believe, referred to in debates. Lord Aldington told the Select Committee that the NNC intended immediately to get facilities set up in Britain to make components for LWRs if these were ordered, but whether they would be operated by NNC or established by a subcontractor had yet to be decided. 'We have an open purchasing obligation and that has to be dealt with.'[64] The universal assumption appeared to be that the open tender system which could not be contemplated for nuclear plants was best for all components, and *might* be operated on an international basis.[65] As to fuel it was decided BNFL should not have shares in the new company, but there should be a 'commercial link', perhaps a joint marketing company, in any case some 'very close links in a variety of ways both in management and marketing'. The company was established without this vital relation being defined: the possibility that the NNC should become part owner of the fuel company was not raised. For American companies fuel development and production was a logical part of reactor development, the two had to be developed together and mutual adaptation was likely to be simpler under one control.[66] The cost of development would be covered by the sales of manufactured fuel. In exporting, the inability of consortia to give an overall guarantee for a plant's performance because it did not supply the fuel, was among the unwelcome features of British export tenders; and when a firm *could* offer both plant and fuel it could adopt a price policy giving little profit on the plant but good profit from the fuel.[67] If the new British company was to develop and design fuel but not make it, its development costs would not be covered in the usual way.

In regard to development, the dividing line had again not been worked out when the agreement was announced. Boardman said categorically: 'The AEA will continue to have the principal role in research and development.'[68] The new company would be 'responsible for its own R and D' on commercial projects ('those projects for the construction of which it will be responsible'): 'it will be closely involved with the AEA . . . in the longer term R and D which the AEA will wish to carry out and associated with that very fully.'[69]

The Government would have the last word, Boardman said, because it was spending the money, and the dividing line between company and AEA would be the subject of advice from the Nuclear Power Advisory Board to the Government.[70] It seemed clear however that the 'integration of basic research, development and prototype development within the commercial organisations which have to sell the product', aided by large direct Government development contracts, not via the AEA, was not to be a feature of the new company. The Select Committee had recommended the transfer of a large part of the AEA reactor R and D to the new single company it proposed: Benn appeared to concur on behalf of his two new companies, which were to base the choice of development programmes on their commercial evaluation of reactor types in international trading.[71] BNDC (whose management was supervised by Weinstock) in evidence to the Select Committee treated the transfer from the AEA as of the utmost importance. Palmer realised that 'there is something of a resistance movement' on the part of the AEA.[72] Boardman's statement confirmed that the resistance had succeeded.

Thus the new company would not have the scope of operation which characterised American companies and in Weinstock's view explained their success.

It was equally clear that the reserved powers retained by the Government were capable of being extremely restrictive and further differentiated the new British company from the American pattern. The power to intervene, to ensure that the new company – which meant GEC here – should give all potential component suppliers equal opportunity by the open tendering system, can be regarded as a typical anti-monopolistic measure; but it could inhibit the development of component manufacture by the company itself in the American manner. It reflected the consortia myth – now repeated almost universally – and the shrinkage of business.

The right to determine which international links the company might form, because they 'could have wide national implications', could prevent a company adopting a policy which was based on its assessment of the market, the kind of policy Benn had advocated. What was in ministers' minds possibly was a wish to be able to mould the company's policy in a European rather than an American direction: again it was a radical divergence from the American pattern.

The right reserved by the Government to decide which types of reactor the company should make was the most revealing, and most conspicuously in conflict with American practice. Ministers broadened the range of advice on which they would formally rely, but, said Boardman, 'the responsibility must ultimately be that of the Minister'. The Minister still thought he was competent to take this responsibility. Asked by Palmer why the customer (the CEGB) should not decide on the product he required, Boardman said his voice should be the loudest, but there are cases where if a customer 'would be prepared to take something a little different' – a different kind of shackle-bolt was his example – 'it might sell in all parts of the world'. Even the Minister was constrained to say this was a 'stupid analogy' and proceeded 'we are determined to see that the technology we have established in this country is given the opportunity to be exploited throughout the world, and I think the CEGB will be very anxious that that should be so'.[73]

Weinstock had identified the mistaken choice of the gas-cooled reactor as a main source of British failure, and his company – BNDC – in a memorandum to the Select Committee in March 1973 said, 'decision on the choice of reactor ... should be made ... on business grounds.' Hence it was puzzling that he undertook the job on the Government's terms. He had also argued, however, that much had gone wrong by 'lack of direction' in the consortia. How better direction would have offset the disadvantages of a high-cost system and an immature prototype was not explained. Whatever the reasoning, he did accept the restrictions, and there was a prospect of seeing how management mystique could transform a situation if a bad reactor choice was made.

The British market for nuclear plants was discussed almost exclusively in terms of scale: fewer were wanted than was expected. But the United Kingdom domestic market as seen earlier differed sharply from the American market in two other respects: it was dominated by an almost monopoly buyer, and the buyers were not free agents to buy what they wanted. These distinguishing characteristics of the British market were kept intact by the Conservatives in the early 1970s. Two of the small British utilities – first the NSHEB then the SSEB – fancied the SGHWR: the CEGB did not wish to buy it. The NSHEB decided it could not afford to buy it in isolation.[74] The substantive arguments about the reactor systems are analysed later; but the lack of balance between the two Boards was important. The second distinguishing characteristic of the British market was spelled out by Boardman who, in discussing what would happen if the CEGB was disinclined to choose a reactor favoured by the Government, said: 'The overall Government control of the CEGB is through the capital investment approval.'

In the late sixties the desirability of change in these features of the market had been discussed. It was a topic in which the Select Committee showed interest. I had found support within the Electricity

Council for the view that because of the complexity of the CEGB purchasing alone, leaving on one side the value of competition, a division of the CEGB into three would be desirable. In 1968 there were rumours that a subdivision of the CEGB into six regional bodies was contemplated.[75] This was not vigorously discussed, and by 1972–73 not discussed at all. It was symbolic of British handling of nuclear problems that when the potential usefulness of competitive purchasing of plants had been illustrated the topic should pass into limbo, and that the centralisers should be unchallenged in 1975 in the conclusions of the Plowden Committee on organisation of the electricity supply industry (referred to more fully later), whose chairman was the central planner whose feet 'never left the ground'.[76]

The new company launched by Walker was thus contrasted in all main features – including market environment – with the great American companies and with the nuclear power plant manufacturers, in varying degrees successful, in Germany, Japan, Sweden, Switzerland, Italy and France. It was easier to interpret its features in terms of political convenience than of industrial dynamics. It remains to consider what effect the Nuclear Power Advisory Board, part of the Conservative package, might have.

### (v) NUCLEAR POWER ADVISORY BOARD

Its terms of reference were 'to provide continuing and concerted advice on all strategic aspects of civil nuclear energy policy'.[77] It was possibly a response to the criticism that ministers had been inadequately advised in the past. The Select Committee had recommended forming a technical assessment unit. Benn's plan to set up an Atomic Energy Board was also presumably a response to the Committee's criticism, but not what they suggested; he would have brought together representatives of the AEA, the CEGB, fuel company, and the two D and C companies. His Board might have had some executive function. The Conservatives' Board was only advisory. Its membership was wider than Benn's. AEA, BNFL, CEGB and the new National Nuclear Corporation (NNC), the Electricity Council and the SSEB were represented – each by its Chairman – and AEA and BNFL each had an additional representative.[78] Two scientists were added: Penney, formerly Chairman of AEA, and Sir Alfred Merrison, Vice Chancellor of Bristol University, a distinguished nuclear physicist, who had had a spell at Harwell. The BNFL representative, J. R. S. Morris, a merchant banker, was on the Board of Courtaulds. The AEA present or past had in one guise or another five out of nine seats. The minister responsible for civil nuclear development was chairman.

This was not a well-balanced body to give technical advice. Two members only were nuclear engineers, two were nuclear scientists; all of these were past or present members of AEA. It could not serve the purpose of the Select Committee's proposed technical assessment group which was to help the Minister judge AEA technical and economic assessments. No non-AEA member chosen could give views based on close recent experience of the technical problems. The Minister said the Board would give him the 'best' advice. The members between them could have access to most British technical experts whose views were significant, and could call in experts for consultation. But this did not make them individually able to assess the experts' views or collectively able to provide the best advice that could be based upon the views.

The Minister said that apart from being the people 'best capable' in their 'individual capacity' to give advice, the group would 'bring together the whole range of interests which are concerned with this industry'.[79] This was far from true. Among the interests not included were makers and exporters of nuclear plant components (collectively organised with the D and C company and the AEA in the British Nuclear Forum), foreign plant makers – the best experts on LWRs among other things – the coal and oil industries, trade unionists and environmentalists.

The live questions were, what could be achieved by bringing together regularly on the Board representatives of the chosen interests, and, if the Board managed to produce 'concerted advice', what would be the basis of agreement? The Minister gave no answer. The Minister could get advice from all the sources represented on the NPAB separately, possibly in a less inhibited form. Why have a Board? It could be argued that the Germans had a much more elaborate committee system: but the minister in Germany did not regard himself as competent to tell a utility what reactor to choose. Boardman clearly thought that the first task of the NPAB would be to help him make this decision. If he wanted to turn down the CEGB-NNC choice of an LWR he would like to have the Board's unanimous agreement – he would like, that is, to have the CEGB and the NNC (who were members) give up their preferences so that he would not need to use his power to direct the CEGB's investments. Perhaps he thought that as Chairman of the NPAB he would be well placed to exert his influence as the Minister who knew the national interest. Two questions were posed. Was it realistic to suppose the CEGB-NNC would be diverted? Even more important, was this going to make the Minister more competent to make the kind of decisions which ministers had made so disastrously in the past? Would there be any basis for a decision against a utility–manufacturer agreement except anti-American and pro-AEA bias?

Boardman showed how easy it was to fall into this bias when he was asked why in his remarks favouring international collaboration his emphasis was all on Europe. Europe, he replied, 'offered the greatest opportunity for collaboration'.[80] If Britain wanted to use the LWR the United States offered the greatest opportunity.

The proposed NPAB like the rest of the Conservative package showed the failure to understand why atomic policies had ended in disaster. The first results of the Board in action, under Varley as Chairman, confirmed pessimistic assessments of its value: they are discussed in the next chapter.

## NOTES AND REFERENCES

1. Above, pp. 125, 127.
2. *SC Science and Technology, 1967*, Q. 122 and app. 8, p. 387.
3. *Ibid.*, Q. 1637.
4. *Ibid.*, Q. 1636.
5. *Ibid.*, Q. 1223 *et seq.* For a later statement see *SC Science and Technology, 1972–3*, Q. 266 *et seq.*
6. See Lord Mills, *Hansard*, House of Lords, 10 July 1963, col. 1420. 'The AEA – very ably – took steps to bring into being the consortia of manufacturers so that this programme could be implemented.' (He was the Minister of Fuel and Power at the time.) *Ibid.*, col. 1443, Lord Carrington: 'the Authority was originally closely involved in their formation.'
7. Penney for example advanced this argument: *SC Science and Technology, 1967*, Q. 1588. Atkinson on the Committee assumed it in the same interchange.
8. This was stated for example in Germany and the United States.
9. *SC Science and Technology, 1969*, Q. 15.
10. Burn, *op. cit.*, pp. 86, 107, 114.
11. *SC Science and Technology, 1967*, Q. 1593.
12. *Ibid.*, Qs 1634–5. Price suggested there might be two architect-engineer organisations on a permanent basis with British and continental participation (no reason given!) operating in the European community which 'in theory should be one market'.
13. *Ibid.*, p. 387.
14. *Ibid.*, Qs. 1591, 1604, 1628.
15. *Ibid.*, Q. 1604.
16. *Ibid.*, Q. 1601.
17. *Ibid.*, Q. 1587. The other contribution was that the new company 'must use the large numbers of persons these concerns have' in promotion overseas.
18. *Ibid.*, Qs 1593–1600, 1641.
19. *Ibid.*, p. lxxiii: 'Ordered that strangers be not admitted.' In retrospect I had reason to regret this personally since giving evidence later on the same day I would have welcomed the possibility of relating what was said to what I had to say. The reason given for considering the Memorandum, and presumably the discussion on it, confidential was that the AEA had had informal discussion on it with the Ministers of Technology and Power and the CEGB had not sent the document to the consortia.
20. *Ibid.*, pp. 403–4.
21. Above, p. 169.

22. *SC Science and Technology, 1967*, Report, p. xvi.
23. *Ibid.*, p. xli.
24. *Ibid.*, pp. vi, xlii, xlvi. The italics are mine.
25. *Ibid.*, pp. xxxv–xxxix.
26. *Ibid.*, p. xxxv.
27. *Ibid.*, p. xlii.
28. *Ibid.*, pp. lxxix–lxxxii.
29. *Ibid.*, p. xlvi. This was ostensibly unanimous.
30. *Ibid.*, p. lxi. The main report envisaged that at some stage the single UK company might go to foreign suppliers for fuel: so presumably could the British utilities.
31. *Ibid.*, pp. xliii–iv.
32. *Ibid.*, p. xli (my italics).
33. The Conservative Members' argument is well set out in *ibid.*, pp. lxxx–lxxii.
34. *Ibid.*, p. xlv.
35. I thought this important and referred to it in Burn, *op. cit.*, and in my memorandum to the Select Committee, *SC Science and Technology, 1967*, p. 362.
36. 'We are not guided', the IRC witness told the Committee, 'by the practical possibilities. What we could perceive was that a great many companies in this [nuclear] industry like International Combustion, Babcock & Wilcox, English Electric, all had close relationships with companies of considerable strength in Europe... We were aware that there were very few companies in the country that did not have possibilities for very good friends in Europe.' *SC Science and Technology, 1969*, Q. 15.
37. The IRC envisaged links with American firms in time: Morton, who headed the IRC work, discussed the problem with firms in the United States and the AEC, but details of what was said, though they appeared to be promised, were not produced. *Ibid.*, Q. 22.
38. Benn's reason's were set out at length in *ibid.*, Qs 925–6.
39. *Ibid.*, pp. 175–7.
40. *Ibid.*, p. 177, Q. 985.
41. J. C. C. Stewart, formerly Member for Reactors in the AEA, whom many had expected to become chairman after Penney, became Deputy Chairman of BNDC, and since 1975, Deputy Chairman of NPC. He was not, however, a representative of the AEA: he saw the future of development in the companies, and believed that there was a serious intention to pass R and D functions to the new companies. Initially this consortium, before GEC absorbed EE, was going to have the AEA fast reactor team, and be sited at Risley.
42. Babcock English Electric Nuclear (BEEN).
43. *SC Science and Technology, 1969*, Qs 98, 344–50.
44. Morton in his evidence surmised that BNDC would make a 'damn good try to get the [Heysham] contract: they will pitch their price very tight'. *Ibid.*, Q. 32.
45. *Ibid.*, Q. 965.
46. *Ibid.*, Qs 108–11.
47. *Ibid.*, Q. 125.
48. The German group BBC–Krupp (from which Krupp withdrew, to be replaced later by General Atomic) were surprised to discover after a period

in which their contact with the British was with TNPG that in future it would be with BNDC. Their managing director asked me in 1971 if I could explain how it happened!

49. *SC Science and Technology, 1969,* p. xix.
50. *SC Science and Technology, 1967,* p. xivi.
51. *SC Science and Technology, 1969,* p. xiv.
52. *The Times,* 23 July 1968 and 10 October 1969. In 1968 it emphasised that the export business had gone to light water reactors because of their lower costs: 'can the blame for this be laid at the doors of the consortia?' It thought commercial pressures should help to keep design groups in touch with markets. In 1969 it made the cryptic assessment that even if some cost reductions were achieved, the product would remain more costly than a light water plant, or that the price of the product in the protected market would be made more profitable through the lessening or removal of competition.
53. *Hansard,* House of Commons, 8 August, 1972, c. 1496; 2 May 1974, c. 1379.
54. *Financial Times,* 14 February 1973.
55. TNPG cut its headquarters staff by one-quarter when the CEGB cancelled the Sizewell project: BNDC said to the Select Committee that it could not afford staff for new reactor development.
56. *Hansard,* House of Commons, 22 March 1973, c. 676.
57. Cecil King, *The Cecil King Diary 1970–1974* (London: Jonathan Cape, 1975) p. 102 (21 April 1971) and p. 376 (16 July 1974).
58. *SC Science and Technology, 1972–3,* Q. 570.
59. *Ibid.,* Qs 618–19.
60. The AEA in a Memorandum hoped that the large heavy electricity and boiler firms would assist and guide executive management and allow facilities in overseas offices to the company; but the first point at least was ruled out by GEC's position, and the suggestion gives no indication of what the participating companies would gain, hence what motivation was assumed. *Ibid.,* p. 100.
61. *Ibid.,* Q. 604.
62. Above, pp. 198, 201, 203.
63. It was envisaged that foreign competition in supply of components might be encouraged. It was significantly referred to as 'international collaboration or international purchasing'; but this appeared to exclude foreign tenders for complete nuclear boilers.
64. *SC Science and Technology, 1973–4,* Q. 672.
65. *SC Science and Technology 1972–3,* Qs 646–9. The AEA in a Memorandum had envisaged that ultimately a single D & C company might wish to make components for itself which were not satisfactorily obtained through competitive tenders; but this did not seem to influence the decision in 1972–3. *Ibid.,* p. 100 (1973). Aldington envisaged specialisation between European component makers, *SC Science and Technology, 1973–4,* Q. 672.
66. This was so, not only in the period when fuel was being designed and tested but in subsequent operational development of 'arrangements for getting the best results . . . in the use of fuel by shuffling etc.'.
67. The point was made in the Select Committee proceedings (above, p. 100); some possible (not likely!) purchasers raised the point in discussions I had with them: for example, BASF.

68. *SC Science and Technology, 1972–3*, Q. 668.
69. *Ibid.*, Qs. 662, 664.
70. *Ibid.*, Qs. 662, 668.
71. Above, pp. 173–4.
72. *SC Science and Technology, 1972–3*, Q. 544.
73. *Ibid.*, Qs 700–6.
74. Below, pp. 227, 258 (n. 47).
75. Palmer drew my attention to this in 1968 and agreed that it was consistent with public ownership. I have been told that Sir Ronald Edwards supported the idea of forming six independent regional electricity authorities for all functions including, therefore, generation.
76. The Committee submitted its report late in 1975 but it was not published until after this section was written. For a further reference see below, pp. 284–5.
77. Report of the NPAB, 1974, p. 1.
78. The chairmen were respectively Sir John Hill (AEA and BNFL), Sir Arthur Hawkins (CEGB), Lord Aldington (NNC), Sir Peter Menzies (EC) and F. L. Tombs (SSEB). The additional AEA member was R. V. Moore; and J. R. S. Morris represented the AEA subsidiary, BNFL.
79. *SC Science and Technology, 1972–3*, Qs 698, 705, *et seq.*, Boardman's evidence in March 1973.
80. *Ibid.*, Qs 713–14.

# CHAPTER 13

# LWR versus SGHWR

## (i) LOW-COST CONTENDERS IN 1969

In May 1969 the CEGB told the Select Committee that the HTR and SGHWR both 'showed promise of significant economic advantages on the AGR'.[1] The great development potential of the AGR was no longer talked of. The CEGB preferred the HTR; it promised lower costs, greater long-term economies, could be a direct source of high temperature process heat for industry, and, being gas-cooled, would use familiar technologies. They thought the remaining technical problems could be solved quickly, and that it could 'probably be ordered in the next year or two'.[2] In 1970 the CEGB called for tenders to be submitted in April 1971 for an HTR to be built at Oldbury.[3]

The CEGB thought the SGHWR could be ordered soon too, but they did not want to take on another technology – such as water cooling – for no economic gain.[4] Most of the Board were resolutely against taking another look at the LWR but one member told Ernest Tremmel of the AEC that he regretted the world's largest utility did not have an LWR if only for experience.

Although one more AGR was ordered after this, the CEGB clearly hoped to turn quickly to a new reactor type developed by the AEA, and thought this practicable. They were encouraged in this by the consortia. The full impact of the AGR disaster was still to be felt. The CEGB still referred to the Windscale AGR as a 'closely representative prototype plant' for the 600 MW plants under construction.[5] The development process was not yet comprehended.

By April 1971 when the Oldbury tenders were submitted light was dawning. All the AGRs were now in trouble. The AGR became progressively less attractive. But from the slow progress of HTR development work in the United Kingdom and abroad, the belief that its outstanding problems would be quickly solved was dispelled. The plan to order an HTR for Oldbury was soon dropped.[6]

On the other hand the LWR had to be looked at. World orders amounted to 41,000 MW in 1971, and 11,000 MW were operating. 'It could be ordered in quantity at once, though recent controversies about their safety would first have to be resolved.'[7] The SGHWR was halfway between the HTR and the LWR: 'after further detailed design and component development' it 'could be ready for a commercial order in 1974' ... 'Further orders could follow as confidence was established'.[8] These assessments were not accompanied by any cost data.[9]

## (ii) THE ATOMIC ENERGY AUTHORITY HAS THE BEST REACTOR

Once again the choice of the next reactor for the British utilities lay between an American-designed reactor and an AEA development. But the competition was more than ever unbalanced – in favour of the LWR. Against the massive constructional and operating experience with large LWR units the only SGHWR plant was the 100 MW prototype which had worked, with some discontinuity, since the end of 1967. But the AEA claimed again that they had the best. Moore took this line with the Select Committee in August 1972: it was unfortunate, he said, they had christened their reactor the SGHWR because it gave the impression 'that the plant is new and novel ... a wrong impression because it is basically like a boiling-water reactor ... a version which has several important advantages.'[10] (Moore chided engineers who said the use of saturated steam turbines which was involved was a reversion to Victorian technology;[11] in 1964–65 it had been a bull point in AEA propaganda for the AGR that you did not use these old-type turbines![12] There was brilliant opportunism in the propaganda.) It was more accurate to say it was a cross between Candu and a BWR, which had been deemed not economically attractive in America.

The SGHWR belongs to the family of reactors (of which the earlier Candu has been most successful) in which the fuel rods are cooled within a large number of pressure tubes, instead of all within one large pressure vessel as in the LWRs. The moderator, heavy water, is contained in a large low-pressure tank, the calandria, which is penetrated by a large number of tubes. The pressure tubes are inside these calandria tubes, close to but not in contact with them, separated by spacers. Each pressure tube includes a bundle of fuel rods, and the coolant flows through the tubes. In Candu the coolant is heavy water; in the SGHWR it is ordinary water which, as in the BWR, is converted to steam and used in a direct cycle to drive a turbine. In the Candu reactor the coolant flows into a heat exchanger to produce steam: Candu is an indirect cycle system.[13] The point of asserting kinship rather than novelty was to establish that early maturity was possible. 'The HTR', Moore said in 1972, 'is a high risk technical problem. By comparison the SGHWR is a proved and established technique, and subject to doing the necessary work to produce a fully detailed design – which might take nine to twelve months – one could immediately order one.'[14] Sir Edwin McAlpine and Mr S. Ghalib, Chairman and Managing

Director of TNPG, which after reorganisation under Benn included the SGHWR team of the AEA, and moved headquarters to Risley, took the same line: no research and development was needed for the SGHWR, it had been ready since 1971, but it would be wise to develop some components for a year or so before starting site work.[15] They were as enthusiastic and certain about the SGHWR as the old TNPG had been in 1968–70 about the HTR. To choose an LWR when you could have the SGHWR would be 'tragic'.

### (iii) HOW THE INTERESTED PARTIES LINED UP

Before the Government would support the LWR the CEGB had to be in its favour. There always had been some advocates among its engineers, but in 1969 the chairman made it clear that it was not an option the Board was considering. During 1971, as the Vinter Committee proceeded, obviously the Board accepted it as a strong contender; but when Vinter told the Select Committee in summer 1972 that no reactor type came out on top according to *all* the criteria applied,[16] this no doubt reflected the Board's position early in 1972. The American companies – the 'siren voices' as one MP called them – still had much persuading to do. By early 1972 BNDC had opted for the LWR – informally, but the news got around: by the summer it was being said the CEGB and Weinstock were ganging up against the AEA, but the new Chairman of the CEGB, Arthur Hawkins, was unaware of it.[16] Weinstock thought Hawkins did not 'intrinsically' want *any* nuclear reactor.[18]

The change at the top probably made the full conversion easier. For as well as Stanley Brown, whose period of office ended, other dominant figures went, E. S. Booth, D. R. Berridge and Leonard Rotherham (who had flanked Stanley Brown when he gave evidence to the Select Committee) and Owen Francis, the Vice-Chairman, who had given evidence with Hinton to the Select Committee on National Expenditure.[19] All were associated with Dungeness B optimism.

But when Hawkins appeared first before the Select Committee in August 1972, a month after he had taken over, he said he did not have to make a choice. Growth of demand for power had fallen far below the 8 per cent per year forecast when the National Economic Development Organisation (Neddy) plans were drawn up. Much surplus power had resulted from the ensuing programmes, and it would be more embarrassing when the new plants worked well. For the next ten years growth of demand for electricity was forecast to be between $3\frac{1}{2}$ and 5 per cent a year. Additional plant in the pipeline would supply half the extra power which the 5 per cent would require; the further orders needed ranged from 16,000 MW on the 5 per cent to a mere 4000 MW on the $3\frac{1}{2}$ per cent hypothesis. The nuclear part would be four plants on the higher figure but only one on the lower figure, the total of new nuclear plant orders to be placed before 1980.[20] None was needed immediately. 'The demand for new reactors in the next few years ... does not exist.'[21]

It was not necessary to discuss the LWR at length in this context (though the CEGB gave its opinion that LWR safety problems would be overcome): but the SGHWR had to be considered, because if no order was expected work on it should stop. Hawkins, pressed to give the Board's attitude, said the SGHWR could be a convenient stopgap, but they did not need one. It was out-of-date as a technology, with less development potential than the HTR: it would be better to use all British resources on the fast breeder where we had a lead. To go down the SGHWR alley 'would not lead us into the future'.[22]

When Hawkins next gave evidence to the Select Committee, the CEGB was converted to the LWR and proposed to order eighteen LWR reactor plants in the next decade, starting with two in 1974. They also wanted an HTR plant ordered at once so that the British system could be developed to take over from the LWR as soon as possible. For the present the LWR was the only reactor available of proved engineering design, with reliable components – and the cheapest.[23] Hawkins expressed a preference for what he called the third generation PWR – the Zion plant of Commonwealth Edison.[24]

Not surprisingly the Select Committee saw this as a *volte-face*. The scale and specificity of the proposal were challenging – unnecessarily since as Hawkins himself said a ten-year programme is not immutable.[25] It could well be argued that in relation to the Electricity Council's forecasts of load increase Hawkins' second programme was too large and his first too small[26] and both were proved excessive because demand fell far below the Council's expectations.[27] But if probable costs made it desirable to build nuclear power plants there was no need to link the choice of the LWR with the size of the programme for a decade of ordering, which envisaged that the next 45,000 MW of nuclear power, virtually all the 'competitive' plants to be put in up to nearly 1990, would be based on American technology. The sheer scale of this prospect underlined the failure in British development, and it probably stirred more opposition than if a more modest start had been proposed.

Hawkins said it was a temporary departure – as soon as the British HTR was satisfactory in a large demonstration plant the Board would go back to the British gas-cooled tradition, and it envisaged moving quickly to the British FBR.[28] Was not this then, Palmer asked, a gap that needed stopping, and had not Hawkins originally said the SGHWR would be a useful stopgap although it was obsolete technology? No, Hawkins said, the SGHWR was 'obsolete' in the sense that it was only a 100 MW plant; he could not meet his needs for 45,000 MW with 100

MW units, hence it could not stop his present gap. If the SGHWR was to be built in larger sizes it must go through the demonstration plant process – it would not be proved satisfactory to order in large sizes for eight to ten years, during which he wanted to order eighteen large twin-reactor plants.[29] It was not surprising these explanations caused irritation.

But Hawkins put the case for the LWR and against the alternatives well. The rest of this chapter is concerned with the objective factors in the assessment, and the reactions of interested groups.

It is convenient to summarise the group attitudes first. Broadly, when the CEGB decided, the NNC made it clear it also favoured the LWR. Despite the BNDC attitude Weinstock was not prepared to commit himself in March 1973: he had argued that the AGR might still be turned into a winner. But by December he had come down forcibly for the LWR.[30] The bulk of the British nuclear engineering industry outside the AEA had supported the choice of the LWR since 1972 because this alone offered a prospect of substantial exports – of components. TNPG (while it existed) and some of its constituent firms did not take this line: nor did Faireys (whom Rotherham had joined). But almost all the British Nuclear Forum firms were pro-LWR.[31]

The AEA seemed poised in January 1974, after the CEGB had decided, to jump in either direction. Hill thought, understandably, that the period of indecision should be ended. He felt the Government and CEGB should not just aim at a strong British nuclear industry but should continue with British technology; and he thought they both wanted to. Complete development of the FBR was the universally agreed *terminus ad quem*. In his view we should not overload our development programme, so apart from the FBR the United Kingdom should develop only one of the alternative new systems – SGHWR and HTR: he favoured the HTR, as the CEGB did. It had more promise. Hill would not call the SGHWR obsolete, but it should have been taken up five years ago (at the very latest two years ago) if it was to be accepted on a world basis. It 'would not be possible to get such close international links on this route as on the one chosen by the CEGB'.[32]

Whichever type was chosen, Hill argued, only one should be ordered immediately – a demonstration plant. A second could be ordered three or three and a half years after ordering the first.[33] The SGHWR and HTR would each require a big design effort and run into teething troubles, so Hawkins would have a gap to fill, for which Hill looked on LWRs as appropriate. They would have to be approved by the Safety Inspectorate,[34] and there would be no need to provide for more than a two or three-year gap – after that you could survey the situation again. Hill had talked to Hawkins about all this, and apart from the ten-year ordering Hill seemed to accept the former's line. He did not think the CEGB was committed to more than requiring 45,000 MW in *nuclear* plants ordered by 1982: they did not *all* need to be LWRs.[35] Palmer was surprised that Hill was so indulgent to the suggested purchase of LWRs.[36] Hill agreed that a gap could be filled by building more AGRs (but a new design would be needed – it did not exist), or more Magnoxes (again a new design would be needed), or by buying in Candu reactors from Canada: if he had to choose between Magnox, Candu or SGHWR (an unrealistic range) for closing a gap he would plump for the SGHWR.[37] At this point the proceedings conveniently closed. He was not pressed to say how large a gap or how long it would take to fill it by any of these routes. Hill's original thesis had been by now lost sight of in untidy but tendentious questioning.

Of the witnesses before the Select Committee in its 1973–74 sessions only Mr F. L. Tombs, Chairman of the SSEB, gave the SGHWR unreserved support. Some AEA witnesses had done so formerly – Mr F. J. Doggett (Deputy Chairman of UKAEA), and Richard V. Moore did so in August 1972. Unlike the HTR which was a 'high commercial risk' they said the SGHWR involved no risk and would be ready to build within a year.[38] The IPCS had supported it from 1969 as the best prospect for the British industry. In a Memorandum in March 1973, which Airey Neave found 'impressive', they argued that if an American system was chosen now 'in five years' time there would be no British industry capable of exploiting a British fast reactor' and selling it abroad. The only way to avoid this was to choose the SGHWR now and make the CEGB buy it. This would secure exports in undeveloped countries and give credibility to the British industry.[39] Neither the Institution nor Airey Neave saw any need to explain why most British nuclear plant manufacturers assessed the position differently.

Tombs, flanked by D. R. Berridge, formerly Chief Generation Design Engineer of the CEGB and now his Deputy Chairman, had considerable but not any recent experience in the British heavy electrical industry (in GEC and Reyrolle-Parsons). He argued that it was desirable to turn to water reactors because the capital cost of gas reactors would continue to be unfavourable even when the immediate AGR difficulties were overcome; he was disenchanted with high-temperature reactors – he implied that everyone was. Experience with the LWR was unsatisfactory (their load factor, he claimed, misinterpreting the statistics, was lower than for Magnox), and building LWRs with safety assured in Britain would take a long time, so it was wrong to suppose a bigger gap could be filled by LWRs than by SGHWRs. The SGHWR, being a 'modular system', could be converted quickly to large sizes and could be called a proven reactor. He also argued that the SGHWR had a good export potential, whereas there was little for the LWR. He repeated the assertion that the SGHWR would have sold for export in the past had CEGB ordered one. This had been denied by

BNDC, and accounts of particular negotiations indicate that where it was claimed a sale was likely (but did not occur) the AEA was offering a special inducement. In Finland not only was most of the work to be done locally but assistance was to be given for the Finns to export SGHWRs; in Australia, the British centrifuge know-how was offered; in Greece payment would be taken in tobacco, if the cigarette manufacturers would cooperate. (They would not!) In Australia the chief utilities voted against the SGHWR and wanted an LWR, preferably from the United States – it was $50m cheaper.[40] Hinton and Rotherham set out their reasons for preferring the SGHWR in memoranda to the Select Committee, adding no new reasons.

Several members of the Select Committee had shown a fondness for the SGHWR since 1967. They saw it a near export prospect in 1967 and 1969, and were disappointed the CEGB were not prepared to buy it but appreciated the reasons. In spring 1973 they decided that because the CEGB still would not buy it and there was no longer a 'potential overseas market for it . . . it might well be there is little point in continuing it, sad as is such a conclusion.'[41] When they reported little more than six months later, they had returned to their old love, 'that poor lonely reactor that goes round and round', as Palmer called it. They referred to the evidence for, but not against.[42] They concluded that on the evidence publicly available no proposal to build American light water reactors in Britain should be approved by the Government; that common sense would indicate that until the HTR and FBR were available on a commercial basis the way forward would be to use one of the British technologies already proven; and in this connection 'we note the enthusiasm of the SSEB for the SGHWR'.[43]

We turn now from the line up of the interested parties to the balance of evidence – starting with cost estimates.

(iv) THE BALANCE OF EVIDENCE

*(a) Costs*
Cost computations given to the Select Committee by the CEGB and SSEB showed LWR costs – both capital and generating costs – approximately 10 per cent lower than AGR costs. Magnox costs were appreciably higher. The SSEB gave higher figures than the CEGB, basing them on smaller plants, assuming a shorter life (twenty years instead of twenty-five), and giving what they thought would be March 1974 prices, while the CEGB gave March '1973' prices. But the comparison between the costs of different reactor systems was very similar. They relied on the same sources for the basic information, including the British plant manufacturers and the AEA. The figures referred to 'second-of-a-design' (in 1974 the CEGB said *fourth* of a design) plants, so transitory development costs were eliminated. Here are the figures.

TABLE 13.1

(A) Estimated Capital Costs in Britain of Nuclear Plants[44] of
Different Systems 1973–74 (£/KW)

| | System | Magnox | AGR | SGHWR | LWR |
|---|---|---|---|---|---|
| Construction cost, including | CEGB 2 × 1320 MW[a] | 313 | 222 | 195 | 172 |
| interest during construction | SSEB 2 × 600–700 MW[a] | 306 | 250 | 221 | 198 |
| Initial fuel | CEGB | 15 | 19 | 18 | 14 |
| | SSEB | 23 | 18 | 17 | 13 |
| Total | CEGB | 328 | 241 | 213 | 186 |
| | SSEB | 329 | 268 | 238 | 211 |

(B) Estimated Generating Costs in Britain 1973–74[a] (p/KWh)

| | Load factor (%) | Magnox | AGR | SGHWR | LWR |
|---|---|---|---|---|---|
| CEGB | 64 | 0.72 | 0.57 | 0.51 | 0.46 |
| SSEB | 60 | 0.89 | 0.75 | 0.67 | 0.60 |
| | 70 | 0.78 | 0.65 | 0.58 | 0.53 |
| | 80 | 0.69 | 0.58 | 0.52 | 0.47 |

[a] The figures given for Magnox by both Boards refer to plants of 2 × 500–600 MW. For CEGB they are at '1973 price levels'. For SSEB they are prices 'expected to obtain in March 1974'.

These figures looked straightforward to interpret. But this was only so if they all had equally solid and realistic bases and if whichever system was chosen plants could be ready for use in as short a time and in equal numbers. This comparability was not present.

Several large PWR plants had been made and were operating, and a great many were being made, in the United States and elsewhere, their designs could be repeated and their cost of manufacture be related to experience. But they had not been made in Britain nor examined by the British Safety Inspectorate. The CEGB and the NNC believed that only small design changes which could be made quickly would be needed to satisfy the inspectorate, and that American construction costs and construction time schedules in the United States, Germany and Japan (and elsewhere) would be a reliable guide to costs and times in Britain. Hence they believed that after a few minor changes they could undertake production of the PWR in any numbers required by CEGB programmes. A few components would initially need to be imported, but representing only about 10 per cent of total cost. These judgements were all disputed: some opponents were convinced it would take several years to establish PWR safety.

The SGHWR existed only as a 100 MW prototype. Great changes in design and specifications were required in extrapolating it to 1300 MW capacity. The CEGB and NNC said bluntly the SGHWR was not 'proven', and the Board insisted that it must see the operation of a large demonstration plant before ordering several as 'commercial', which would need upwards of eight years. The SGHWR had not been accepted as safe and could not be until a full-scale detailed design was available for a specific site, but there was a disposition to expect no difficulty.

The two gas-cooled reactors were in an intermediate position: Magnox had operated in medium-large sizes (*very* large sizes were not practical), and the AGR was almost operational (though further off than was recognised in January 1974) but both required considerable redesign in order to be acceptable for further orders and the designing had still to be done. Their probable costs were sufficiently above the others, especially the LWR, to make them uncompetitive.

The cost comparisons in Table 13.1 could thus not be accepted at their face value; one had to establish lead times, repeatability, and safety prospects.

*(b) Lead Times and Repeatability*
Tombs said in December 1973 that if a power plant were ordered in 1974 the lead time would be a minimum of six years for a fossil-fuel plant and a minimum of seven for a nuclear plant,[45] and that for a nuclear plant ordered in 1976 an SGHWR would be finished sooner than an LWR, and a gas-cooled reactor would take longer than an LWR. This would happen because the SGHWR would have the largest proportion of factory-built input, whereas the gas-cooled reactors would have the least.[46] This implied that one was comparing like with like, and that by 1976 the work needed to establish a repeatable design of an SGHWR would have been done, so that several could be ordered in 1976 with as much confidence in expecting an uninterrupted manufacturing and construction programme and as competitive a result as there would be if an LWR were ordered.

This was not the position and Hawkins raised the discussion to a more realistic plane by spelling out some of the changes of design which had already been sketched out or would be needed in extrapolating from the 100 MW unit operating at Winfrith to a 600 MW or perhaps 1300 MW *demonstration* plant. The changes, it will be noticed, were needed not only because of the increase in size but for safety, durability and cost reduction, and this underlined that the reactor though described by Tombs, Ghalib, Doggett, Moore *et al.*, as 'proven' was still in the development phase.

> The fuelling system now proposed [Hawkins told the Committee] would be quite different from the one in use at Winfrith. It would be very different. It needs a pond to be constructed over the reactor that is not at Winfrith at all. Containment is quite different from the Winfrith plant to meet safety requirements. . . . Also with regard to seals at the top of each channel, which have given a great deal of trouble, they will need to be redesigned to prevent leakages during operation. The cooling circuits are austenitic at Winfrith Heath. In order to get down cost . . . they were tendered in ferritic steel.[47] That is a change. It may be a good change. You have got to prove it. . . . The geometry of the pressure circuit itself has been changed to reduce the number of tubes that will be in trouble if you have a leak. That again is a change. The calandria will be of stainless steel and zirc-alloy instead of aluminium. Again that is a major change. . . . All right, if we need to go ahead we build a single demonstration large SGHW and you wait eight years to get experience before you build any more. On that time scale I do not believe – and my colleagues would agree with me – the SGHW is at this stage something worth developing.[48]

The much-emphasised modularity of the SGHWR in which the pressure tubes were uniform 'modules' and plants could be made larger or smaller by having more or fewer modules, was of no practical value until all the development problems were solved; and there remained many crucial problems to solve. Hill's evidence confirmed Hawkins' judgement on this.[49] The development programme undertaken after the Minister decided in favour of the SGHWR showed they were right.

The LWR was, as we have seen, a developed reactor – once a design had been accepted by the Safety Inspectorate it could be used straightway for several plants. So safety became the key issue. Opponents of the LWR made much of this. They also argued – and it is useful to deal with this first – that the LWR would take a very long time to build. They based this both on American precedents and British weakness: Tombs said the LWR would take seven years to build. Others suggested more. Aldington said that NNC could do it in five to five and a half years from the start of construction, once through the safety hoop; assuming he had the support of the American licensor.[50] He was fully justified by world experience, including that in the United States, as reference to Figure 4.5 shows.[51] After the early period of ordering beyond the industry's capacity, both to make and develop, the long lead times in the United States reflected the extraordinary licensing delays. Most of the plant – all the nuclear part – could be made and erected in a little over four years. It was the same in Germany, Japan, Switzerland and Sweden.

There was, however, it could be argued, a special obstacle to be overcome in the United Kingdom, of which Aldington was well aware, that the engineering required for much of the LWR was exceptionally refined, above the standard normal in most British mechanical and electrical engineering, though possibly not above some in jet-engine manufacture. 'First of all there is the pressure vessel' – to which everyone referred. 'Then there are the internals which require very high quality engineering. The rod drive mechanism also requires very high quality engineering. There are the steam generators and the pressurisers which require slightly higher quality engineering than in ordinary boilers'.[52] Aldington's approach to this was that it would be a mistake to try and establish the manufacture of all these components in the United Kingdom – the scale of orders would not justify that – but it should be possible to develop an international specialisation within Europe; Britain would do some of the work, for ourselves and Europe, and European countries would specialise on other parts, for themselves and for us. Lifting the level of British engineering would be advantageous. The balance of payments from these interchanges would in all these countries be nil, and we should become quickly free from any import requirements from the United States.[53] He possibly was planning at this stage an initial joint operation with the Westinghouse operations in France.

Before all this happened the NNC would import the most sensitive items from the United States and have at the same time Westinghouse's technical support. Aldington did not explain, nor did anyone ask, why there should be a discrimination against importing from the United States.

Some were sceptical of the time promise because they were against the LWR, or against the United States. There were some who were simply sceptical because they thought the CEGB and British manufacturers would, because of bad traditions, on both sides, make the LWR a more expensive reactor in Britain than it was in any other country. This seemed to me a real risk – but a risk to take in this situation.

## (c) Safety

We turn now to the crucial issue. The Select Committee, reporting hurriedly in January 1974, in order to get a message to the Minister before he decided possibly in favour of the LWR, stated in heavy type that 'in view of the conflict of opinion on the safety of the LWR it is in our opinion for the proponents of light water technology to prove its safety beyond all reasonable doubt, rather than for their opponents to prove the contrary.'[54] It was recommending for the LWR rather grandiloquently, as though it were exclusive, what every country insists upon for every type of reactor. For this purpose it was usual to appoint an inspectorate, composed of experts, engineers and scientists, who must collectively be, or become, familiar with all the problems in as up-to-date a guise as possible and in the light of the most advanced current technology, and must examine all the relevant details of proposals and work over all the calculations on which they are based against that background. When the chips are down this judgement can only be made by experts.

In Britain the Chief Inspector of the Nuclear Installations Inspectorate was a ministerial appointment. At this date the responsible minister was the Secretary of State for Trade and Industry; in 1967 it was the Minister of Power. It would be for the Inspector, at this time Dr E. C. Williams, to decide whether the LWR was safe; he would begin to investigate this when the CEGB had asked the Minister for approval, which the Minister could not give, except conditionally, without the Inspectorate's agreement. The Select Committee seemed to pre-empt the Inspector's function in its assertion that 'no proposal to build American light water reactors should be approved by the Government on the basis of the evidence publicly available.'[55] Though Williams' decision would be 'crucial to the issue'[56] they did not stay for an answer; the way forward should be to choose at once a British, that is an AEA, technology.

The opponents of the LWR, including apparently most members of the Select Committee, argued that Williams would take two years or more to report on the LWR, and would require big changes in design which would amount to requiring a new British prototype LWR, so that the delays and ultimate cost would not be less than for an SGHWR.[57] What was the 'evidence publicly available' on which this conclusion was based?

This is the problem with which this section is concerned. Was there good reason for believing that a proper assessment of the 'safety' of the LWR would overturn the evidence that it offered the lowest-cost electric power then available and offered this low-cost nuclear power in larger quantities sooner than any alternative type of nuclear reactor? Was there anything in the evidence to justify the Committee's pre-empting the Inspectorate's function?

The balance of evidence given verbally to the Select Committee was strongly in favour of the view that LWRs were safe. Only one witness supported the critics, and not very strongly. Tombs thought that 'concern about critical fracture of the pressure vessels arose considerably after the start, when people were committed', and it had not yet been answered. 'If we had a dozen light water reactors we would probably take a different view, because of familiarity and commitment about safety problems.' But 'we can take an uncommitted view'. People who had doubts about LWR safety were not 'cranks'. He had complete confidence in our safety inspectorate, and thought there would be no gain through public hearings: the issues were complex and difficult to expose in a public hearing, because 'what people tend to do at public inquiries is to seize on out of the way aspects and blow them up out of proportion.'[58] Tombs was thus not passionate on the issue, believing it could safely be left to Williams' organisation.

Williams, in giving evidence, said properly that until he had been asked to assess the safety of a specific LWR project, he could not possibly have a view, and it would take a long time. He would like to do it in less than two years but might not succeed. He had had some of his staff examining the evidence given at public hearings in the United States: he knew the problem areas – who did not? – vessels, ECCS, scram failure. But he 'pleaded with the Committee not to get into a slanging comparison of the relative safety of different reactor types. This is a glass-house situation, believe me, which I have to live with.'[59]

Hill, to Palmer's obvious annoyance, refused to call the LWR inherently more dangerous than other reactors. All reactor systems have their problems – the right question to ask is can they be made safe in the eyes of the licensing authorities *and* obtain public acceptance. Foreign licensing authorities were most responsible people and in accepting the LWR were not being irresponsible. Whether the greater density of population in the United Kingdom would require special treatment was for the Nuclear Installations Inspectorate (NII) to consider; so too the questions whether the steps taken to ensure integrity of the pressure vessel were adequate. Hill had made it clear he was prepared to see LWRs used by the CEGB as a 'fill-in', though not in such large numbers as Hawkins proposed, but this restriction was not for safety reasons.[60]

All other witnesses who gave verbal evidence, Weinstock, Stewart, Hawkins, Matthews (CEGB's Chief Safety Officer), Aldington, in the December 1973–January 1974 sessions argued strongly on the basis of their own or their colleagues' minute examination of American design, manufacturing and licensing, that the LWR was safe, and that modifications required by the inspectorate would be small, minor additional redundant safety equipment possibly, such as a fourth spare diesel unit, involving little extra cost or time.

The greatest volume of evidence given to the Select Committee verbally against the LWR on grounds of safety was in the questions of members – with Ronald Brown, leading, Palmer, Neave and Edward Leadbitter, MP, giving aid and comfort. We return to this. There was support for the anti-LWRs in memoranda from Hinton, Rotherham and the Friends of the Earth, and a memorandum by Sir Alan Cottrell, then Chief Scientific Adviser to the Cabinet Office, was used by the 'antis'.

Cottrell had been Deputy Head of the Harwell Metallurgical Division from 1955 to 1958 before becoming Professor of Metallurgy at Cambridge. He was asked to submit a memorandum because Williams had consulted him – he had, also like Williams, been asked to meetings of the NPAB. The memorandum was basically reassuring. Rapid fracture is possible from large cracks, above a certain 'critical' size, in large pressure vessels. It is essential therefore to avoid such cracks, which must be done by exceptionally careful manufacture and exceptional inspection – this during and after manufacture and also during the operating lifetime of the vessel since areas might lose fracture toughness gradually during operation. Ultrasonic techniques will identify cracks well below the critical size and 'seem to offer a good margin of safety'. There were parts of the vessels where theoretic determinations of the critical size are uncertain, and here judgement is important. Theory was however supplemented by proof tests to 125 per cent of design pressure, which should reveal any weaknesses, and the AEC was carrying out a large research programme, not yet completed, which included testing to destruction of large models with large slots. The programme had so far confirmed the theory. The parts which are most liable to lose fracture toughness under prolonged low radiation in a vessel are not the parts most under stress. Two problems Cottrell concluded needed more scientific investigation, the gradual growth of small cracks in highly stressed regions through ageing and corrosion, and the effect of thermal shock from emergency core cooling water in a loss of coolant accident.[61]

There were those who argued that the great care in manufacture and inspection was beyond human capability; Cottrell said nothing in the memorandum which suggested this though he may have said it privately. The crucial questions arose over the last two problems. These were not matters on which nothing was known. They were matters on which a layman's judgement had no value. Palmer pressed Hill to comment, hoping, it appeared, he would say they implied an unacceptable risk. Hill sturdily refused to say what Palmer wanted. The manufacturers of the vessels would say they already carry out the work in the meticulous form required. Scientific work must of course always go on in all areas of advanced technology, and further scientific work on pressure vessels must continue. The need for such work was thus no proof of unacceptable danger. It would be for the Nuclear Inspector to decide whether what is done in the manufacture and operation of vessels is adequate. The vessel problem did not arise in connection with the SGHWR, but 'all reactor systems have to go through their own safety analysis, if not for the pressure vessel, for other features.'[62]

The two memoranda by Hinton and Rotherham, which treated the

LWR as – in Hinton's words – 'of current reactors the one whose safety should be most strongly questioned',[63] were less scientific, informative and precise than Cottrell's, but fully committed to a view. Neither had been familiar with LWR design, manufacture or operation. But the members of the Select Committee might reasonably think the views of such eminent nuclear figures justified them in insisting on assurance from experts currently familiar with the situation in detail, as Williams could be if required. These were not matters in which members' own judgements could have any objective value.

The same problems were posed in a higher degree by the memorandum submitted by the Friends of the Earth.[64] Palmer said it was 'very extensive and well arranged'.[65] It repeated the tactics and was based mainly on the literature of the intervenors in the ECCS marathon.[66] Its thirty-five pages made it an exceptionally long document for the Select Committee, but it was trivial in size compared with the sources it used. The record of the ECCS hearings was said to cover 54,000 pages. As we have seen, the purpose of the hearings was to find out to what extent advanced knowledge of some uncertainties required more conservatism in LWR design and operation: the answer arrived at was – very little. This important conclusion was tucked away in a note in the FOE document: the AEC's conclusions 'as expected embody many minor cosmetic changes but no change of substance'.[67] Records and quotations out of context were all devoted skilfully, in the tradition of adversary legal proceedings, to present one case; they were supplemented by disturbing 'statements', attributed to unnamed persons supposedly holding high office.[68] The quality of the statistics in the discussion of Magnox load factors was parallelled in discussing the capital cost of LWR and Magnox plants.[69] It was characteristic that the 'six semi-scale tests at Idaho' were referred to more than once:[70] 'this series yielded six failures out of six and was forthwith declared irrelevant.' At no point was it mentioned, though it was well known when the memorandum was sent to the Committee, that the tests had been repeated under more stringent conditions establishing that they *were* irrelevant.[71]

'We are not ourselves experts on all technical details of reactor safety', the authors say, but 'we can claim a sound acquaintance with the technical issues.' This did not emerge from the document. (They were physicists, not engineers.) Here again the problems involved were technical problems concerning heat transfer rates, metal deformation, behaviour of fuel cans and pellets and their mutual reactions at different temperatures, metal-water reactions, crack propagation and the like, where the layman's judgement has no objective value. It is frivolous for a legislator or minister to pretend to solve such issues.

Ronald Brown gave evidence on the same lines as the FOE in his questioning both in the February and December-January sessions of the Select Committee. For him the choice of the LWR in America had been 'very clearly wrong', and he appeared to visualise the AEC as covering up the past mistake. 'It was a confidence trick that got the LWR into Europe', he told Weinstock, 'for most of us [who were 'we' – the whole committee?] the AEC's information is now not acceptable'; its conclusions are 'challenged by almost everyone' in the United States: public hearings showed that 'no one has any faith in AEC research'. The AEC suppressed inconvenient evidence and witnesses. LWRs gave 'very limited protection' to public safety. The ECCS were 'concocted by computer' and the sub-scale tests (still with us!) 'show that none stands up'. 'There is a vast amount of accidents happening. A really bad one would cost 40,000 lives.' The CEGB cost calculations were wrong because all big PWR reactors were derated by 30 or 40 per cent. If subjected to the usual standards the makers of the LWRs 'could not get away with the present standards of manufacture'.[72]

This and much more. Brown said he had fitted himself to ask the right questions.[73] All these questions could have been derived from a few memoranda submitted by Interventionist bodies such as the FOE and the Union of Concerned Scientists to various Congressional Committees and Regulatory hearings. Whether Brown had read some of the other relevant documents in the enormous literature – which no one person can read entirely, still less absorb – his questions did not show.[74] Brown expected those whom he questioned to accept statements of the kind set out above as the premises for their replies. How could the witnesses in view of these 'facts' justify the purchase of LWRs which would not be safe or cheap? He had a duty to protect the public from such things. Since they were unprepared to accept his premises, in his view the proponents of the LWR 'failed to answer'.[75] In fact they treated most of his premises, rightly, as mere fantasy,[76] and showed the extremely solid foundation on which the American Inspectorate, with their strong background of independent expert advice and multiple sources of research data, based their licensing. The CEGB submitted a memorandum setting out in detail their own exhaustive investigation into the design and manufacturing processes on which safety rested.[77]

Airey Neave also like Ronald Brown contributed what he presumably thought evidence of LWR danger in stating that 'the present American philosophy is that the failure of a pressure vessel is incredible'. Williams said he would not accept that, he would examine all the evidence.[78] One might suppose from Neave the Americans did not examine the evidence in respect of every individual vessel. Their actual approach was, as Cottrell's memorandum showed, to do all they could, in a very intensive activity, to *make* failure incredible. The handling of steel research in this connection, and the development of sampling to trace changes in steel qualities from year to year and of new methods of testing in addition to those referred to by Cottrell have been mentioned earlier.[79] And while the American developers and licensors believed they were being successful in preventing a sudden rupture, a belief shared by the CEGB and, within a few months, by the Director of Harwell, they nevertheless took steps to minimise the effect of one should it occur and cause some failure in the containment. Initial research on this was reported on in 1968 at the Oak Ridge Nuclear Safety Centre. 'Power reactors are now considered to be so carefully designed', the preface says, that a 'significant release of radioactivity... is extremely unlikely.... However it is in line with the Commission's safety policies to investigate all possible avenues... to better protection of the public'.[80]

Another misconception of the American position crept into the Report. 'The proponents of the PWR [it appears this refers to British proponents] are all convinced that it is a safe reactor. The Americans themselves are sufficiently doubtful to be building a small PWR in the Idaho desert which they will test almost to destruction to prove the efficacy or otherwise of the safety systems.'[81] The intention of United States licensing was to establish criteria which assured safety, given current knowledge, which included the lack of the information which would be obtained from the Idaho LOFT experiment, initiated in 1967. This, it was expected, would allow the introduction of less than current conservatism into design and operation.[82] The CEGB believed current conservatism was, on the data already available, excessive.[83] This conclusion, like all other evidence of LWR safety, did not rank for inclusion in the Select Committee's *Report*.

When Palmer was trying unsuccessfully to persuade Williams to say the LWR was inherently less safe than other systems – both Hill and Williams told him this was the wrong question to ask – he said that 'while there is a certain worry everywhere about all reactor stations... to be fair they operate in this country with a very fine record of safety.'[84] It was a revealing remark; for it would surely have been even fairer and more apposite to say that everywhere where the LWR reactor system he was attacking was used, and this included an enormously greater volume of activity than in the United Kingdom, they had a superb record of safety. It was a measure of his sense of the 'evidence publicly available'. Mr Ian Lloyd, a recent recruit to the Committee, said with refreshing candour in a subsequent Commons' debate that 'the scale of the evidence was utterly disproportionate to the complexity of the issues' – no comprehensive evidence from the United States, nothing for Japan, Western Europe or Sweden – and the experience of some witnesses was 'narrowly based in a geographical sense'.[85]

An incident during the Committee's proceedings made one wonder whether all the 'evidence publicly available' would have been welcome. Following on a complaint by Leadbitter that he had been asked to meet Westinghouse representatives, Palmer said 'there are American reactor interests in this country and they have been trying to put a certain pressure on members of the Committee. But I have not met them and I have no intention of meeting them. I think this is a matter for Parliament to decide by its own machinery in the usual way.'[86] There seemed no similar reluctance on the part of MPs to meet representatives of Atomic Energy of Canada Limited (AECL) who wished to see Britain adopt the SGHWR – if not Candu itself – and expected to gain some export advantage. Like Westinghouse they would provide know-how, but presumably not for nothing. A Canadian minister came to London to press the case for AECL. No doubt a different attitude was called for towards AECL by the fact that it was a state-owned enterprise.

In 1967 Palmer had been one of the group of three members of the Committee who visited the United States to assess the American nuclear development and he had visited GE and Westinghouse as well as the AEC and JCAE. It was a wise decision; the Group's report which recognised the impressive qualities of the firms, was enlightening, though only a little can be learned in one short visit. When Ronald Brown in February asserted flamboyantly, in questioning Williams, that the Americans had 'farmed this stuff [the LWR] out to the rest of the world', but 'with all their razzamatazz cannot find enough people who know anything about the problem to advise them' on the 'possibilities and probabilities' (presumably he meant the safety risks). Williams replied: 'They have a very large core of absolutely superb engineering expertise in the United States, both in the Atomic Energy Commission and in the big companies such as Westinghouse and General Electric. You may argue they are biased, but they are responsible engineers.'[87]

The 'comprehensive evidence' of which Ian Lloyd spoke would have had to include direct evidence, however acquired, from the American manufacturers and the AEC and their independent advisers, from the 'responsible' and 'superb' engineers to whom Williams referred, and from the vast available records. This would have helped to put the odd snippets seen in the distorting mirrors of intervenor groups in perspective. Many members appeared unaware of the inadequacy of the data they were using – even Palmer rejected Aldington's plea that he should get evidence of construction times saying they had got it already,[88] and members, as seen earlier, were uncritical of statistics served up to them.[89]

The Select Committee, to summarise, could not assess 'the evidence publicly available' on the LWR since it only had a fragment of what was available, 'utterly disproportionate', as one member put it, to the task the Committee undertook. It had no direct evidence from anyone

who designed, manufactured, operated or licensed LWRs. The balance
of what it did have, did not support its conclusions. Evidence of witnesses who were directly familiar with the United States performance,
who all favoured the LWR, was not set out in the report, which also
did not explain this rejection. The one witness – Tombs – who treated
the LWR as a risk worth avoiding, when asked whether 'the costs of
redesign' of LWRs for British conditions would 'considerably narrow
the margin between water reactors and gas-cooled reactors', replied:
'No, I would not have thought so.'[90] Cottrell's memorandum was
quoted without reference to the reassurance it gave on major points,
and although Palmer 'doubted whether Cottrell's "evidence" would
have been uncovered but for the Select Committee'[91] (which was
incorrect since Williams and the NPAB had obtained it), no attempt
was made to bring to light evidence from distinguished metallurgists
who had knowledge of vessel manufacture and testing at first
hand. The evidence in the CEGB memorandum, which was based
on direct examination of the United States vessel manufacture, was not
referred to, nor the evidence of Matthews who probably was the one
witness who had had direct contact with the United States manufacture.[92] The cavalier treatment of evidence not supporting the Committee's conclusions, which Weinstock surmised were arrived at before
the evidence was given,[93] matched the readiness which has been illustrated earlier to use uncritically bad statistics which seemed favourable to their conclusions. While the Committee persisted in advancing,
as though established, the argument that making the PWR safe would
make it uneconomic, and disregarded in its Report evidence which
made this conclusion improbable, it also left out any reference to evidence which suggested there would be safety problems for the
SGHWR: the problem for example of preventing trouble in one pressure tube generating trouble in others, to which Hawkins and Williams
referred, and Hill's more general point that the SGHWR did not present the vessel problem, but every system has its own problems.[94] It
would have been possible for an open mind to read into this remark
specific uncertainties.

Although the Committee emphasised safety so much in its January
1974 report, this was not the chief motivating factor: the safety inspectorate was there, all the witnesses before the Committee thought it
reliable, and only Ronald Brown among the members questioned its
capacity.[95] There were signs that a few members were unaware of the
technical nature of the decision on safety and fancied they could make
the judgement themselves. Ronald Brown told Hawkins that he had so
much evidence he could 'talk to your nuclear man on equal terms'.
There were signs that some members feared Williams might come out
in favour of the LWR without expensive changes – he had advised
against a 'slanging comparison' of different reactor types, spoken with

admiration of American engineers and licensors, and described an anti-LWR 'World in Action' television programme as 'grossly biased.[97] He was perhaps not a safe bet for the anti-LWR lobby. There were members, too, who seemed to suppose that where there was uncertainty it was absolute – 'Would you conclude', Mr Christopher Tugendhat asked Aldington, 'that the question of the LWRs cheapness is an open one in the sense that we do not yet know what the Nuclear Inspectorate will demand?'[98] This is the stance of the person to whom 'the man of judgement is the man who suspends judgement'.[99] As Aldington's reply showed it was a matter of probabilities.

The motivating force behind the Committee was probably the desire that nuclear plants in the United Kingdom should be based on United Kingdom technology. Palmer put the background in lyrical form to the House of Commons – 'For the last 25 years Parliament and the country have come to regard Britain as a leader in nuclear technology. A whole generation of British infants has grown up imbibing that faith almost with their mothers' milk.' The Report argued that if the CEGB purchase of LWRs were allowed, 'the country would be virtually abandoning the long established British nuclear R and D effort'.[100]

After the Select Committee had reported and the Commons had debated the choice of reactor in May 1974, but before Varley had decided, Cottrell, disturbed at support for the LWR, wrote to the *Financial Times*[101] setting out publicly the view which had been widely attributed to him, though he did not state it in his memorandum to the Committee. 'I hope', he wrote, 'the safety of the public in this country will never be made dependent upon almost superhuman engineering and operational qualities', which he believed 'necessary for the integrity of LWR vessels'. He praised the 'fail safe' characteristics of the SGHWR. The public statement allowed a public rebuttal: it was delivered by Dr H. J. C. Kouts, Director of the AEC's Division of Reactor Safety Research, also in a letter to the *Financial Times*.[102] In discussions with research scientists, he wrote, he had met no opinion that would support Cottrell. It had been shown that construction of vessels according to standards 'much more casual' than those actually used, would 'guarantee a remarkably low probability' of failure. 'Even if vessels were made by these older and less careful methods we would not expect such a vessel to fail catastrophically over the entire period during which the world will have to depend on nuclear fission as a source of energy.' In recent tests at Oak Ridge National Laboratory, in which pressure vessels were deliberately flawed with cracks of enormous dimensions, no vessel failed at a pressure less than ten times the normal operating level of a PWR. The technologies of building and inspecting vessels were not only 'transferrable but exportable', and Britain had recently been provided with 'full United States research data on reactor pressure vessels'. This was to be incorporated in a report for the NII by a team led by Dr Walter Marshall, AEA board member for research. A provisional summary of this team's conclusions was favourable to the LWR vessel, as Patrick Jenkin had already told the House of Commons before Varley chose the SGHWR. Two further elements in the safety discussion call for a brief reference.

First, that Williams had taken no steps to make a close examination of LWR safety as late as January 1974 was remarkable, since by the end of 1971 it was clear the LWR had to be considered seriously as a candidate for use in the United Kingdom. Proceedings for licensing could only proceed when a utility presented a specific project. But, according to a DTI memorandum, the Inspectorate, 'by agreement had been moving for some years now into direct discussions with British designers at a much earlier stage than was customary in the past', which 'helps to provide the Inspectorate with an effective insight into the weight that has been given by designers to safety aspects of their reactor concepts'. But according to the Department, 'similar opportunities of direct discussion in the formative stage do not exist for designs developed overseas, and in particular for those which are largely finalised'.[103] Put simply, the Inspector had been encouraged by the Department to look at AEA systems as they were being developed but was not to look at the LWR. Six – out of sixty-seven – of Williams' qualified staff were on new British reactor systems early in 1973, and six were going to be on the FBR.[104] Williams did have 'some of his chaps' looking at the evidence at public hearings in the United States and experiments at Oak Ridge and Idaho, but none looking at the detailed design and manufacture of LWR reactors at the works. A German, closely involved in nuclear development, hearing Williams' evidence at this date, commented on the extraordinary naïveté and nationalism of a policy of discrimination which left the United Kingdom badly informed on what was now to most people, even in Britain, the most important commercial reactor system.[105] Responsibility for this rested presumably with the ministers, the civil servants and Williams himself.

It was remarkable secondly that with all the Committee's show of interest in safety it showed no interest or awareness in the political dilemma which it posed. If government is to ensure the safety of a technical development, which must depend on the judgement of experts, and if the object is to obtain advantage from the development and not just say 'no', how are the experts to be chosen? How is the quality of their work to be judged and their vigilance sustained?

The American and British safety organisations provided a remarkable contrast. The British organisation was a closed and, to all appearances, unitary system; the Chief Inspector was appointed by the responsible minister, there were no published guides to the standards and codes of practice to which proposed nuclear plants must conform, the Inspector's staff could, and did, call upon the AEA for assistance but there was no public record of the transactions, the detailed reports of the Inspector on plants were not published.

The American system on the other hand was markedly open, and contained countervailing forces and diverse, potentially rival and even competitive initiatives – checks and balance in the American tradition. As we have seen while the Director of Regulation was appointed first by an Executive Agency, the AEC, and from 1975 directly by the Executive, it was flanked by an institution created by the Legislative branch, the ACRS, established by Congress by a statute drawn up by the JCAE. Both bodies conducted separate exhaustive analyses of utilities' nuclear plant proposals extending over about a year, and further exhaustive examinations of plants before operation. The ACRS members chose their successors independently, though the appointments were formally by the AEC. The work of the Regulatory Branch and the ACRS had to be integrated. There was an extensive AEC (later ERDA) safety research programme, not organised by the Regulatory Branch. This resulted from joint discussions between all the groups involved and was subject to revisions and additions at the instance of the regulators, the ACRS or others, and it was carried out in various research centres, some in AEC research institutions, some in non-profit-making private research institutions, some in corporate research bodies, and some by leading reactor or component manufacturers. These were not free from rivalries and rancours, and there were signs of strain between some of the AEC research centres and headquarters, such as when the centres felt that favourite projects, the molten salt reactor at Oak Ridge for example, were not sufficiently supported. Research results were circulated to all research centres; to the Regulatory Branch, AEC Research headquarters, and the JCAE. There were many competitive centres of initiative and criticism. This non-centralised system was unlikely to be dominated by one strong personality, and was likely to be self-correcting.

The openness of the system existed at all stages: conditions which were to be satisfied by an applicant for a licence were set out in public documents (on this and analogous bases, the judiciary, as well as the executive and legislative branches, could take a hand), hence could be discussed, and were often amended. Consequently the preliminary safety analysis by an applicant was publicly available, the reports of the Regulatory Body and ACRS were publicly available, important ACRS committee meetings were open to the public, the terms of licence agreements were public, there were public hearings, as described above, before licences were issued so that dissentients could be heard, records of performance on important safety parameters were published, records covering even very minor accidents were published, and research results were sent to specific interested groups and publicly available. At first making available was in a sense passive: the documents were in the public records office of the AEC but increasingly attention was drawn by the AEC to salient points. It was astonishingly different from the British situation.

Despite the great emphasis of the Select Committee on safety, no interest was shown in the United States system of regulation apart from a slight interest in public hearings. This may have been partly because, being overpersuaded that LWRs were unsafe, they could not imagine that the American organisation was good, but still more because a system which relied on the interplay of experts from different groups, in which the experts sorted themselves out, with ample publicity as a check, was probably too far outside recent British corporatist and socialist traditions. The Committee was preoccupied, so it seemed, with the inappropriate task of choosing the right reactor itself.

### (d) The NPAB Divides

The Nuclear Power Advisory Board's report on reactor choice, published several months after the Government's decision and undated, represents presumably the last statement to Minister and Government of the arguments over the choice before it was made. As seemed likely, the divisions between those members who gave evidence to the Select Committee remained, with no detectable change resulting from discussions in the Board. Some factual errors were removed in the Report (statistics on experience with large reactors were more accurate, load factor figures were not corrected) and some attitudes were sharpened. Tombs – depersonalised in the Report and referred to as the SSEB – not merely supported the SGHWR but suggested a programme for the whole country: orders should be placed for eight reactors of 660 MW capacity (totalling 5000 MW) in the next four years, with series orders following (say from late 1978). The 5000 MW should all be completed by 1985–86: thereafter plants could come into service at a rate of 2500 to 3500 MW per year. And no LWRs.

How the members who did not give evidence to the Select Committee – Menzies, Merrison, Moore, Morris, Penney – sided, we do not know. In the closed tradition of British practice this was not revealed. The Report says the SSEB programme had 'support from some other members', but does not say from whom. The AEA appeared, as Hill did before the Select Committee, poised to jump according to the decision. With the CEGB and NNC it thought the HTR should be supported 'for the longer term', and favoured an early order for a 1300 MW demonstration plant.[106] Tombs did not agree. All members concurred, with advice from the AEA, that 'we would not have the resources for a substantial effort on the HTR if we have a programme of SGHWR and are maintaining our effort on the Fast Breeder'.[107] However, 'the AEA regard[ed] the SSEB's proposal as practicable but they recognise[d] the case for the LWR'.[108] Which they preferred was not stated.

The Report was more reassuring than the Select Committee on LWR safety – it referred to the encouraging provisional result of the AEA team's visit to study vessel design, manufacture and testing in the United States – and slightly less reassuring on the SGHWR.[109] It mentioned, possibly for the first time in public discussion, that the large steel drum on the primary circuit presented a risk to be guarded against.[110]

The Report included a calculation that, if the CEGB's assessment of electricity consumption and of the relative rate at which LWRs and SGHWRs could be obtained were right, fossil fuel requirements would be £500m a year higher by 1985/6 if the SGHWR programme were adopted, and £1000m higher by 1990/1, due mainly to either a higher cost of oil imports or a reduction of oil exports. Most importantly it would mean a large increase in energy costs.

Palmer in August 1976 said 'the Advisory Committee reported in rather a guarded way, but coming down broadly in favour of the SGHWR'.[111] If the published report is the real report, as I suppose it is, it did not come down at all, and Varley knew, as one would have supposed Palmer knew, that the members most familiar with the LWR supported it against the SGHWR, and that the AEA's recommendation of early support for a large HTR prototype was not possible given an SGHWR programme.

(v) THE CHOICE AND AFTER

*(a) The Basis of Choice*
'The period of uncertainty is over.' Thus announced Varley on 10 July, after announcing the decision of the Government that the utilities must buy the SGHWR. But was it over? Was it more than a choice of a new uncertainty? Was Tombs so certainly right and the NNC and CEGB so certainly wrong in the LWR/SGHWR assessment? It seemed unlikely, though experience had shown there could always be surprises over nuclear energy.

One of the most breathtaking had occurred quite recently in the May (1974) debate. Palmer, well launched it seemed on the predictable speech – satisfaction that the need for the single company using all the best talent together was recognised, the need to support British technology, belief in Britain's nuclear lead, support for Mr Tombs – suddenly struck a dissonant chord. If, he said, the CEGB insists on taking the LWR it should be allowed to do it, on a more limited programme than they had drawn up. He was not too sure about the much canvassed argument that Britain could not sustain a mixed programme of reactors. After all the Swedes had done it. If the NNC did not want to make both LWR and SGHWR, and Weinstock was disenchanted, it would be reasonable for American firms to make LWRs for the CEGB, so long as they gave most of the work to British sub-contractors.[112]

This was so breathtaking that nobody took any notice of it.[113] What was its motivation? Had he developed doubts about SGHWR prospects? Safety he would leave obviously, and rightly, to the NII; possibly Marshall now outweighed Cottrell in his judgement, which was right. But there was no light on these questions.

Much that Palmer said had been said before, more than a decade before, by *The Times* for instance, and in part by the CEGB.[114] What was remarkable was that it was he who now said it. It was not necessary or desirable to be committed to a big reactor programme extending over many years; it was quite practical to mix reactor types in the United Kingdom, there was no reason at all why design and construction should be by one firm or by an indigenous firm, there was no need to exclude American types – by implication the LWR could be safe – there was no need for a minister or select committee to instruct utilities what they should do or from whom they should buy. Palmer's ability on occasion to rise above the habitual and automatic response was refreshing.

The Government was not deflected from making the choice that compelled the CEGB to buy the reactor it did not want to buy and the NNC to make the reactor it did not want to make. The decision resulted in dropping for the time being and probably for good the development of the HTR. We had no resources, Varley said, to go ahead at the moment with a demonstration HTR plant, though he also said his aim was to support British technology. Hitherto gas-cooled technology had been rated as the great United Kingdom contribution. The defeated parties were asked not to mention the LWR for some years,[115] though the NNI was instructed to complete a generic safety study of the PWR.

What led the Cabinet to turn down the proposal of the principal utility and act against the judgement of the exclusive D and C firm, one can only surmise. It was suggested that Varley's affiliations with coal-mining may have had some influence. 'Will he confirm', Winston Churchill asked the Minister, 'that the full programme suggested by the CEGB was unacceptable to the National Union of Mineworkers?' To which Varley replied, 'I do not know what connexion that has with the matter under discussion.'[116] *The Times* in an outspoken leader gave the answer: 'the present government seem determined to provide extra demand for ... the coal industry. A decision in favour of the SGHWR effectively puts off the introduction of a new nuclear generating capacity for the better part of a decade.'

The Government were known to have been heavily lobbied by other trade unions, the IPCS and the Electronic Telecommunications and

Plumbing Union, who represented staffs engaged on the SGHWR and other British development projects. These reflected the second strand, in what *The Times* thought were the three 'suspect and dangerous' forces which determined the Government attitude – the 'nationalistic theme', that the SGHWR should be backed because it was British technology. The third strand was safety: *The Times* said straight-forwardly, 'so many technical problems of the SGHWR remain to be solved' that it is impossible to claim it is 'safer'. Varley had said in the debate in May that, 'to me and my department safety is a fundamental factor', as though this distinguished him from the CEGB and NNC and all nuclear engineers and operators the world over. 'We must consider possible public antipathy', he also said, revealing a more political motive – collecting environmentalist votes.

In the May debate he had shown himself remarkably aware of the drawbacks of the SGHWR – but these were not mentioned after the decision. 'We can proceed to order it quickly', he said, but *The Times* said, you can order any plant quickly but the SGHWR could not be built as quickly as some other systems, and 'it remains to be seen whether they work reliably and how much, in the end, they cost'.[117]

Within two years there were strong indications that *The Times*, and those who shared its fears, were right, that another disastrous choice had been made by ministers. The AEA joined in sounding the alarm in 1976. 'Why are we doing it?' – that is, developing the SGHWR – the deputy chairman asked, saying: 'No one's been able to give me an answer';[118] and Sir John Hill recommended stopping the development and adopting the PWR. It had not been possible to order an SGHWR quickly: the industrial experts, the NNC and CEGB, who said a long period of development was needed and that safety hurdles had to be overcome, were – not surprisingly – right. They had been right too in asserting that the LWR was safe. Obtaining new nuclear generating capacity had been put back for *more* than a decade, and prospects of participating in LWR development had been damaged further. The significance of delay in getting new cheap nuclear power was obscured by the continuing slow increase in the consumption of electricity, bely-ing the forecasts of the Electricity Council, so that there was no press-ing need to order new generating plant to add capacity. Ministers and Select Committee members acted as though oblivious of their creative role in the disaster.

The end of this section deals first with some salient points in the SGHWR development, then with the possible changes of course, and the probable economic consequences of the 1974 choice.

## (b) SGHWR Development, 1974–76

Those who said that, contrary to the assertions of McAlpine, Ghalib and others, the SGHWR still required much further development, including redesign and R and D, and that a prototype commercial-scale reactor could not be ordered quickly, were vindicated. This was not surprising because the chief buyer, the CEGB, had set out a long list of specific needs, some for economy some for safety, and the NPAB had identified, additionally, a major further safety problem in the steam drum. (Each reactor would have two steam drums, into each of which steam from 300 pressure tubes would enter through forty nozzles, which would each be welded into the drum, and the welds could suffer corrosion, fatigue and fracture, with a risk of 'missiles' entering the drum.)[119]

This and the other problems raised by the CEGB, and some others (it was decided for example to have a different fuel assembly, again partly for safety), had to be sorted out before a reference design could be accepted as the basis of a detailed, priced design. A considerable R and D effort was immediately mounted by the AEA with whom the NNC worked closely in several centres, though mainly at Harwell and Winfrith, whose objects were described briefly in the Authority's annual report for 1974-75, requiring substantial recruitment of new staff and withdrawal of existing staff from other work. [120] A reference design was agreed between NNC, CEGB and SSEB in July 1976. It was reckoned that the earliest date on which an order could be placed, after the reference design had been turned into a detailed price design, would be the latter part of 1977, and the earliest date of operation 1984-85. It would, the deputy chairman of NNC said in 1976, be ten years at least before any meaningful feedback for design of commercial plants could be obtained from operating experience, and possibly not until then or later that component production could be organised on the mass production basis used for LWRs. Meantime redesign had led to great increases in cost. In March 1976 it was estimated that a plant would cost – including IDC but not the first fuel core – approximately £500 per KW. Allowing 60 per cent of the increase above the cost estimates of 1973-74 to be due to inflation, this figure represented a rise in real cost of 30 per cent above the SSEB's estimate for March 1974, and 15 per cent above the CEGB's estimate. It was about 25 per cent above the estimated cost of an LWR in the United States in September 1976, assuming the exchange to be £1=$1.6[121]

Promoters of the SGHWR argued that delay in agreeing on a reference design was due to the CEGB and NNC dragging their feet – Benn said there had been no obstruction.[122] Tombs argued that the cost had been increased needlessly by setting higher safety standards for the SGHWR than for the LWR, and by 'reported applications of additional design margins at all stages of the design'.[123] In view of Tombs', and his associates', attitude on LWR safety, this defensive line provoked derisory comment. As Benn unlerlined, the layman could not judge whether American or British reactors are the safer.[124] Data on the exacting standards required of American LWRs, and their remarkable safety record are available; there were no comparable data available publicly for the SGHWR, and no performance in large sizes to record. Since, as Hill had said, each reactor system has its own safety problems, direct comparison was ruled out. All promoters and designers of reactors tended to find that regulators set standards too high – as the United States manufacturers often thought the plants were made 'safer than safe enough'. It could well be so, especially since they could be made 'safe to any degree of concern'. It was not only manufacturers who thought safety got overdone. Speaking in 1970, H. J. Dunster, now Deputy Director General of the Health and Safety Executive, and responsible for the NII, and at that time an AEA expert and ICRP committee member, remarked at a United Nations conference on nuclear development and the environment at the height of the Gofman / Tamplin campaign, that 'if we were to do a true cost-benefit analysis, we would find that far too much expense has gone into safety, considering the trivial aspect of some radiation controls.'[125] It was the price paid, a British safety expert said recently, for adopting standards everywhere designed to make the risk of serious accident less than one in a million years. There was no published reason to suppose that the SGHWR design suffered particularly: what was remarkable was that the most conspicuous large risk it presented had not figured in public discussions before Varley's announcement of the Government's choice. He had himself been alerted by the NPAB.

By spring 1976, before the reference design was complete, an air of pessimism was spreading about the SGHWR, and it was widely believed the project would be dropped. The high cost was not due just to safety measures but, as with the AGR also, to inherent engineering complexity. The Marshall Report establishing the safety of PWR vessels was reported as nearing publication. The SGHWR had emerged as a high-cost system, which 'can be built and will work' as Hill put it in a letter to Benn, 'but . . . it is not possible to see any specific feature of it which gives it a clear-cut and unambiguous advantage over other systems',[126] and it was still some way off. It became known in July that Hill was advising Benn that the SGHWR programme should be abandoned, and recommending that in its place the United Kingdom should adopt the PWR for its thermal reactor programme, but failing that the AGR. Some thermal reactors must still be made in order to provide sufficient plutonium for the fast breeders when they were made in quantity.

Benn discussed the substance of Hill's letter to him on this with the Select Committee who began to review SGHWR progress in August 1976. He gave copies of Hill's letter to him on this, but not for them to publish. Palmer found it extraordinary that the Chairman of the AEA should 'after two years of not very intensive work' suggest that on balance 'his own child, if you like, should be abandoned and that we should turn to the American technology'.[127] Hill was presenting the consensus of the whole Authority. Benn said the recommendation 'is bound to impact on the credibility not only of British technology' but on the credibility of other AEA systems and specifically the FBR.

Hill was reported as attributing what looked like a *volte-face* to two sources. The large demand for generating capacity which the CEGB assumed in 1973–74 would have justified the cost of developing a new system, but the demand had evaporated. And in 1973–74 neither industry nor the CEGB had been able to show the LWR was safe, whereas now we had the Marshall Report to establish it.[128] This was a myth, though useful perhaps for face-saving. Hill had himself accepted the PWR as a possible stopgap in 1974, knowing then that the safety would have to be proved to the NII and that it would take two years, but he clearly believed it would be established. The two years were up in 1976, and the proof, which the CEGB and NNC were confident in 1973–74, would satisfy the NII, had been accepted in regard to the most vital area – the pressure vessel. Marshall had provisionally indicated the probability of this in spring 1974 after his team's visit to the United States and before the SGHWR decision. Kout's reply to Cottrell's criticism of pressure vessels at that time had been crushing. It had obviously not been desirable in 1974 to continue developing at great cost a high-cost system which had no advantage over existing developed systems and whose development would delay the use of nuclear power, raise its cost and not provide exports, when as it seemed in 1973–74 cheap nuclear power was wanted quickly. He had in 1974 emphasised that the SGHWR would require much design work and a prototype stage, and that the AEA preferred the HTR but could not develop both.

The NNC and CEGB had been right in 1973–74 in judging that the LWR was safe, and cheaper than the SGHWR could be, that it would be available in numbers sooner, that the SGHWR was 'not something worth developing'. But they got no credit for being right.

A decision in July 1976 to make further cuts in Government expenditure was seen by cynics as likely to provide a further aid to face-saving, and so it proved. No orders for the SGHWR, it was announced, would be placed in 1977–78, none therefore before spring 1978; and a proposed expenditure by the utilities of £40m on SGHWR components, which were to be developed before large plants were built, was cancelled. AEA research on SGHWR would be curtailed. 'The chances of survival' for the steamer were now, the *Financial Times* said with splendid anticlimax, 'diminished'.[129]

A Belgian Ministerial Commission's comment on the SGHWR episode early in 1976 offered the perfect epitaph: 'Il n'est jamais bon d'avoir raison tout seul.'[130] Would the epitaph be needed? What would happen to the nuclear programme next?

## (c) *Reluctance and Delay,*

Though technical opinion was now virtually united against proceeding with the SGHWR as not economically justified, the politicians were reluctant to agree. They had been shown with inconvenient rapidity to have chosen badly again. Palmer asked whether, since the Government had willed the end, it should not will the means – even, presumably, if the end was discredited.[131] Ronald Brown appeared indignant that the AEA and NII had investigated the safety of the LWR since that was 'absolutely contrary to the decision taken in 1974'.[132] Dr John Cunningham, shortly to become a Minister of State in the Department of Energy, was shocked that 'once again the British nuclear industry has failed to deliver the goods on time'[133] – as though oblivious of the fact that it was the choice of system, first by the developers then by the Government, which was at fault. Wedgwood Benn said he 'would be very reluctant to be driven off' the decision taken in 1974, he had never cancelled anything! If the SGHWR was abandoned it might be replaced by a coal-fired not a nuclear programme. At another point he argued, we must have more thermal reactors for the FBR; but he was sure the FBR programme would slip, and 'if you knock out a system on which you have worked for so long you will raise the question whether the whole nuclear programme needs to be proceeded with at such speed'. He was responding to the environmental pressures. He seemed to argue that if you did not abandon the SGHWR, the credibility of the AEA – which had decided against it – would nevertheless be saved. And he revived the Tombs' argument that, if you adopted the LWR, you might have all the same troubles with it that you had with the SGHWR. He thus kept his options open.[134] It was widely assumed there would be no decision until 1977. Lord Aldington suggested in October that the NNC should study for six months the relative costs of the LWR, SGHWR and AGR systems; this was agreed and provided a mechanism of delay.[137]

## (d) *Economic Impact*

Britain would get nuclear power from the 1974 programme, if at all, later than was intended, and if the SGHWR programme was continued the power would cost more – in real terms, not just through inflation – than was expected. The disadvantage, *vis-à-vis* other industrial countries with the cheaper LWR power, would thus be greater in two ways, and the advantage over coal-based power in Britain less.

How much did it matter? Benn was not alone in speculating whether perhaps the whole nuclear programme should go slower. (Of course the British programmes should not have gone at all!) Putting aside the environmental opposition, the arguments were that the United Kingdom was in no need of more nuclear power quickly, all nuclear power costs were higher than had been pretended or supposed, the advance of nuclear power in other countries was being delayed by lack of demand for electricity and by environmentalist opposition (so if Britain continued to slip it would be slipping in company), and finally Britain had North Sea oil, and plenty of coal, so the sensible thing was to depend on her own resources.

On the other side of the account, the importance of having more nuclear power quickly had been increased by the phenomenal rise of oil and coal prices. Between 1973–74 and 1975–76 the cost of coal and oil delivered to the CEGB power stations rose by 130 per cent. The delivered cost of coal, £4.80 per tonne in 1969–70 rose to £6.74 in 1973–74, £15.41 in 1975–76 and £17.70 in spring 1976. The average working costs of all fossil-fuel power stations in 1975–76, 0.95p per KWh, of which 0.78p was for fuel, was more than twice the estimated total generating cost for an LWR in March 1973 – 46p. Working costs were spread widely round the average according to location and the efficiency of plants. If the 1974–75 distribution of costs is used as a guide, the 16 per cent of output with highest costs in 1975–76 would have had average working costs of almost 1.5p, the next 30 per cent an average of 1.05.[138] By 1975–76 estimated LWR costs would have risen far above the 1973 level, but if the rise had been as much as 125 per cent, the conditions would have been satisfied for building LWR plants to replace much high-cost coal-burning generating plant.

No estimate was available of the cost of building an LWR in Britain in 1976; the best guide was probably in American prices. There had been big increases in construction costs elsewhere, in excess of inflation rates. In the United States the estimated cost (excluding escalation) had risen, by September 1976, 90 per cent above the level in January 1973 (above, page 31), partly because the lead time had stretched from eight to nine years on account of licensing procedures. To the extent that LWRs built in Britain would depend on imports of materials or components or knowhow, the sharp fall in the value of sterling (from almost $2.5=£1 in January 1973 to $1.7 in September 1976) would have pushed costs up faster than in the United States. There was also everywhere a big, difficult to quantify, rise in the cost of nuclear fuel, affecting equally all converter reactors. But while these cost changes may have lessened the economic justification for replacing much existing coal-fuelled capacity in Britain – the answer to this is not known – the American cost estimates showed that LWRs at the higher prices had lower generating costs than coal-based plants with coal at substantially lower costs than in Britain. It still remained true that, to the extent that competing industrial countries extended their LWR capacity, they would have increasingly the advantage of cheaper energy than the United Kingdom. There was nothing to say, therefore, on this score in favour of slowing down the expansion of nuclear capacity – provided it was of the right kind. It was not sensible to slip voluntarily because others slipped involuntarily. It was invalid to recommend greater reliance on coal because reserves were plentiful: they were also costly to mine. Coal prices were likely to continue their steep rise, and at their high current level were subsidised. The fact that much cheaper coal could be used successfully in parts of the United States and elsewhere could be misleading. It was invalid too to regard North Sea oil as a reason for slipping; it would not be cheap oil and would not provide the basis for power as cheap as LWR power. There were plenty of other uses for North Sea oil. It was of course illogical to decide because Middle East oil was, immediately, more costly than coal that more coal should be used for power generation but no more oil; this was another confusion to which Benn was committed[139] – convenient perhaps in political strategy.

Supporters of the LWR forecast that choosing the SGHWR would damage economic prospects by stopping development of the HTR and postponing the introduction of highly refined repetitive quantity production of components. The HTR was stopped – and the introduction of new component manufacture was delayed indefinitely, more seriously than was foreseen.

In 1973–74 almost all experts except Tombs thought the HTR had a more significant future as a reactor of international status than the SGHWR. The Varley announcement spelled the end of the international Dragon project, where the main British activity on the HTR was centred and other R and D on the HTR was largely eliminated. Symbolically and sentimentally this was important, because it meant an end of intensive work on gas-graphite reactor systems which had been the distinguishing feature of British development.

By 1975 it became natural to ask whether this was a real loss. The HTR was suddenly in trouble. Were we seeing a partial or total eclipse? This is not the place for lengthy discussion. Britain had already lost the lead in the development to the United States and Germany. Both had had small prototypes which were much larger than Dragon operating for some years: both were building 250–300 MW plants – of which the American, Fort St Vrain, had been supposed to operate in 1972–73 but only just started at low capacity by 1976. Both obtained in 1973–74 orders from utilities for large reactors of 800 to 1000 MW – thus General Atomic in America secured orders for ten, Hochtemperatur Reaktorbau (HRB) for one in Germany. But as mentioned earlier in 1975 most of the edifice collapsed, all orders for big plants being cancelled with great financial losses in the United States to the main firm involved.[140] In the United States there were faults in the development engineering, troubles due to changes in licensing policies, and contracts had given inadequate protection against inflation. But technical doubts spread: few people now expected the system to produce cheaper electricity than the LWR. The use of a direct cycle with a gas turbine instead of a steam turbine seemed to offer little advantage, and markets for high-temperature industrial-process heat were generally deemed remote. Was it worth developing the HTR for at most small gains? Two other potential advantages had no early economic interest; first, the system could use uranium and thorium together as fuel (which seemed to interest the Germans and Americans much more than the British); and second, a fast breeder variant could be developed which some experts thought preferable to the almost universally chosen liquid sodium FBR.

The time for high process heat (to make hydrogen for example) would probably come, the uranium thorium cycle was of interest, and many thought an alternative FBR desirable: so for Britain to lose the HTR, probably without even the small compensation of the SGHWR, was a net loss.

The slow and then suspended development and the high cost of the SGHWR obviously delayed the creation of a sophisticated nuclear engineering industry. The 1974 choice was a further step in destroying Britain's nuclear industry though the trade unions persisted in saying that an AEA reactor must be retained to keep the industry alive. In the countries which used the LWR specialised component, manufacture had been encouraged or carried out by main contractors within the last five to ten years, not only in the United States but in Germany, France, Italy, Japan, Spain, Sweden, Switzerland, the Netherlands and Belgium. The analogy with the consortium arrangement in much of this development has been pointed out: where it has occurred it would be difficult if a British firm tried later to break in. The opportunity created by GEC of linking with Westinghouse and the French PWR development, with the intention of establishing international exchanges of specialised components and acting jointly in exporting, was lost irrevocably. Britain had earlier lost a possible opportunity of becoming a major centre (perhaps *the* centre) for GE nuclear and turbo generator manufacturing in Europe. Even if the SGHWR was continued it would be several years before its design would have reached the stage when sophisticated production of components in numbers could be organised and set up, and the international link envisaged, with Canada, was with the producer of the analogous but not identical reactor (Candu) so that scope for repetition would be limited.

The Nuclear Power Company (NPC), the operating company under the NNC, had still some years of work ahead completing the AGR programme, but with the stop on SGHWR development it had no major design projects in sight unless a large FBR was ordered. (It had some development contracts arising out of the FBR project.) In order to strengthen its basis and presumably bolster its income, it was reported in spring 1976 as wishing to buy the Whetstone plant of GEC which made nuclear plant internals, some for export. This was a move towards the structure normal in other countries, but faced opposition from other suppliers of AGR components who feared loss of business, several of these suppliers being members of British Nuclear Associates which held 35 per cent of NNC shares.[141] GEC had reduced its holding to 30 per cent, but the position of GEC in the managerial arrangements was not, it seemed, altered. However Cecil King recorded in July 1974 that 'Weinstock had recently withdrawn somewhat from atomic energy matters', disagreeing with the SGHWR decision.

> The American reactors are of a proven type, they will work; the engineering problems have been solved, the time it takes to build them is known. The heavy water reactors we have chosen are prototypes that may or may not work, and even if they are satisfactory no one knows when they will be operating. However the programme is a small one, and the Government's decision means in effect we are opting out of atomic power for the next twenty years, relying on coal and oil which will be much more expensive. Atomic energy has wasted more money than Concorde, though it is Concorde that gets all the criticism.

We were not after all to see that management mystique could transform a situation where a bad ministerial decision had been made.

When GEC reduced its holding in NNC by 20 per cent, this was added to the AEA holding which became 35 per cent. The distribution of offices in the NPC showed how immensely successful the AEA had been in securing positions of power – J. C. C. Stewart, Deputy Chairman, went early to BNDC from the AEA, Dr N. L. Franklin, Managing Director, was from AEA via BNFL, R. Campbell, one of two Deputy Managing Directors, came from the AEA, via TNPG Risley, Dr T. N. Marsham, a director with special responsibility, was still Deputy Managing Director of the Reactor Division of the AEA. Two of these had been conspicuously associated with promoting the AGR, which seemed the route to success. No one who had been right on the AGR was brought to the fore.

As the debate over the reactor choice, still the Minister's prerogative (Benn said that if need be he would override the National Inspectorate in the interests of safety),[142] got under way again, the AGR itself was brought to the fore. It worked, one was told, with remarkable smoothness. The same had been said by all users of LWRs: as reported of the AGR, however, one seemed supposed to believe it exceptional. Hill said that after all it had been cheap; and bad statistics were circulated to establish this. Hawkins said the fuel cost was low, as though this was an unexpected bonus not to be achieved (possibly improved upon) by LWRs. The inherent high capital costs of the system, increasingly emphasised from 1969–74[143], and the need for extensive re-design seemed forgotten or forgiven.

By the time the Minister got round to choosing between the LWR, AGR and SGHWR – or in favour of more coal – the personal setting would be notably different. Benn announced in August that Tombs, supporter of the SGHWR, would be the next Chairman of the Electricity Council from April 1977, and in November that Mr Glyn England would be the Chairman of the CEGB from July 1977 – which meant that Hawkins, protagonist of the LWR and opponent of the SGHWR, was not being reappointed, as he had hoped, for a second term of office.

## NOTES AND REFERENCES

1. *SC Science and Technology, 1969*, p. 122.
2. *Ibid.*, Qs 706, 708.
3. *CEGB Annual Report, 1969–70*, p. 41.
4. *SC Science and Technology, 1969*, Q. 703.
5. *Ibid.*, p. 122.
6. *United Kingdom Atomic Energy Authority Annual Report 1972–3*, p. 15.
7. *SC Science and Technology, 1972–3*, p. 2, in a memorandum by the DTI, submitted on 26 July 1972, which summarised the assessments at which the Vinter Committee Working Parties had arrived some months earlier.
8. *Ibid.*, p. 3.
9. The Vinter Report was not published, but the memorandum referred to in the previous two notes was based on it. How much detail was omitted, we naturally do not know.
10. *SC Science and Technology, 1972–3*, Q. 232.
11. *Ibid.*, Q. 247.
12. For example a brochure on the AGR published by the AEA after the completion of the prototype in 1963 lists among the advantages of the AGR that 'advantage can be taken of modern and cheaper engineering design for the steam cycle'. The point was made in the *CEGB Appraisal*, p. 1.
13. There are useful brief descriptions of SGHWR and Candu in the *United Kingdom Atomic Energy Authority Annual Report 1973–4*, p. 21 and a more detailed one of the SGHWR in *ibid.*, 1974–5, p. 20. A much fuller analysis, though a few years older, is in *An Evaluation of a Heavy Water Moderated Boiling-Light-Water-Cooled Reactor*, Wash. 1086 (Washington: US Government Printing Office, for the AEC, December 1969). This analyses the AECL project to use natural uranium fuel and, as Milton Shaw says in his introduction, in some parts it 'reflects their enthusiasm'. The costing is entirely out of date. But it is an interesting guide to some problem areas.
14. *SC Science and Technology, 1972–3*, Q. 241 (7 August 1972).
15. *Ibid.*, Q. 309 (30 January 1973).
16. Peter Vinter set out 'six or seven issues': cost, performance, known performance in the past, hoped-for performance, development potentiality,

safety, the introductive costs of a reactor in a particular stage of design, international collaborative opportunities, and perhaps most important the CEGB's requirements in getting the right mix of plant in relation to their skills and experience. *Ibid.*, Q. 24.

17. *Ibid.*, Qs 4, 114–15.
18. *SC Science and Technology, 1973–4*, Q. 31.
19. Booth was the Member for Engineering; Berridge, Chief Generation Design Engineer; Rotherham, Member for Research. Francis moved into the Board from the Ministry of Power.
20. *SC Science and Technology, 1972–3*, Qs 103–4.
21. *Ibid.*, Q. 181.
22. *Ibid.*, Qs. 181–91 and p. 26 (1972).
23. *SC Science and Technology, 1973–4*, pp. 42–3, Qs 248 *et seq.*, 284.
24. *Ibid.*, Qs 301–5.
25. *Ibid.*, Q. 417.
26. I made an estimate in 1972 on fairly similar assumptions and my results were half way between the upper figure of the CEGB in July 1972 and their December 1973 figure.
27. *CEGB Annual Report, 1974–5*, p. 9.
28. *SC Science and Technology, 1973–4*, Q. 284 and *passim*.
29. *Ibid.*, Q. 291 *et seq.*
30. *Ibid.*, Q. 1; *SC Science and Technology, 1972–3*, Q. 627.
31. There were naturally instances where a firm had a good prospect of component work for the SGHWR; equally a boiler maker might have felt that Weinstock's interest in a link with Framatome and Le Creusot might rule out the possibility of work on the PWR, and this was an interesting sidelight on the possible unsatisfactory impact of monopoly.
32. *SC Science and Technology, 1973–4*, Q. 750 and Qs 764–5.
33. *Ibid.*, Qs 748–9.
34. *Ibid.*, Qs. 722 and 770.
35. *Ibid.*, Q. 725.
36. *Ibid.*, Q. 723.
37. *Ibid.*, Qs 783, 807, 812–13.
38. *SC Science and Technology, 1972–3*, Q. 233.
39. *Ibid.*, p. 106 (1973).
40. *Financial Times*, 28 March 1966, 10 May and 11 September 1968 (Finland); 2 May, 9 May, 19, 24 June 1969 (Greece); *The Times*, 9 November 1970, *Financial Times*, 3 June 1971, *Nucleonics Week*, 20 May 1971 (Australia).
41. *SC Science and Technology, 1972–3*, Report, p. xiv.
42. Their use of evidence was instructive: they quoted Hill for example agreeing under pressure with Tombs in calling the SGHWR a 'proven' reactor without referring to (1) his preference for the HTR, (2) his emphasis on the great amount of further development needed for the SGHWR, (3) his view that Britain could not do both.
43. *SC Science and Technology, 1973–4*, Report, p. ix. Palmer has emphasised in retrospect that they did not *recommend* the SGHWR.
44. The full figures with notes on the basis of calculations are given in *ibid.*, pp. 188, 192. The only comments I have come across on specific figures in the tables are in a memorandum by two members of the Friends of the Earth. They confessed they did not understand any of the figures and claimed that

seven economists and two engineers shared their inability; two were professors, two were readers, some came from the CEGB. None were given names. The figures were quite normal discounted-cash-flow estimates similar to those used in the Dungeness B discussions. If different assumptions were made on discount rates or load factors the LWR/SGHWR ratios would not be much altered. The FOE quoted United States figures (*ibid.*, pp. 148–9) to give a more correct cost basis, as they argued, for the LWR costs. These, however, were typical American figures including estimated escalation which British cost figures always excluded and which, with inflation at 6 to 8 per cent a year for eight years, made a high figure. J. C. Stewart (BNDC) examined the FOE figure for the Select Committee and found the base figure was 'consistent with those that both the CEGB and BNDC have used, and leads directly to the total present worth cost of £233 per KW quoted by Mr Hawkins'. The FOE also gave a figure for Magnox costs which 'was not correct' and to get the right figure you had to multiply by two. (*Ibid.*, p. 119.) They had presumably used a figure for a plant ordered much earlier, failed to allow for inflation and not known that the older British figures normally omitted the cost of interest during construction.

45. *Ibid.*, Q. 141.
46. *Ibid.*, Q. 146.
47. For a tender to the NSHEB for a plant at Stake Ness.
48. *SC Science and Technology, 1973–4.* Q. 295 (December 1973).
49. Above, p. 226.
50. *SC Science and Technology, 1973–4*, Q. 706.
51. Above, p. 50.
52. *SC Science and Technology, 1973–4*, Q. 672.
53. *Loc. cit.*
54. *Ibid.*, p. viii.
55. *Loc. cit.*
56. *Ibid.*, p. vi and ix.
57. *Ibid.*, p. vii.
58. *Ibid.*, Qs 124–30.
59. *Ibid.*, Q. 444.
60. *Ibid.*, Qs 770–93.
61. *Ibid.*, pp. 122–4.
62. *Ibid.*, Qs 776–83.
63. *Ibid.*, p. 183.
64. *Ibid.*, pp. 143–78.
65. *Ibid.*, Qs. 699 and 703.
66. Above, pp. 65–8.
67. *SC Science and Technology, 1973–4*, p. 176, note 33 to the FOE Report.
68. See, for example, *ibid.*, pp. 165, 166: 'a senior AEC official has told one of us that a disastrous LWR accident within the decade would not surprise him.'
69. Above, pp. 184, 258 (n. 44).
70. *SC Science and Technology*, 1973–4, p. 159 and 176.
71. Above, p. 67. This was not an isolated instance of omission. Thus, for example, P. Rittenhouse of Oak Ridge is quoted as naming twenty-eight experts who shared his reservations about ECCs effectiveness. But the

FOE did not refer to a devastating cross-examination at the hearings in which it was reported that Rittenhouse was shown to have found in his laboratory experiments a 70 per cent reduction in the passage for ECCs water through a core due to deformation of fuel cans, and then argued it would be 100 per cent in a real reactor. He then admitted that he had not seen the real reactor and did not know its design. He was led to say he favoured a time plus temperature limit rather than an absolute temperature limit, which was what the manufacturers were recommending. *Nucleonics Week*, 27 April 1972.

72. *SC Science and Technology, 1973–4*, Qs 26–8, 78–80, 227, 348, 419, 695 and *passim*; *SC Science and Technology, 1972–3*, Qs 442–3.

73. *Hansard*, House of Commons, 2 May 1974, c. 1422.

74. For serious analysis and judgement it was essential to cover a range of sources; for propaganda it might be a disadvantage.

75. *Hansard*, House of Commons, 2 May, 1974, c. 1422.

76. To take up a few points: the subscale test is a good instance of factual error – after the error has been publicly established. The suggestion that the evidence was suppressed, which was quite baseless, has been dealt with earlier. As to the reputation of the AEC, it will be noted that Cottrell quoted its research results obviously with complete confidence, and Williams and Hill both spoke with confidence of the regulatory authorities. The claim that big reactors were all derated by 30 to 40 per cent was quite wrong: Zion I (1050) was initially derated by 15 per cent because of a valve defect, which was soon corrected; the Brown's Ferry units (1065) came in at full power.

77. *SC Science and Technology, 1973–4*, pp. 194–6.

78. *Ibid.*, Q. 559.

79. Above, p. 77 (n. 3). In particular I have in mind the acoustic emission devices which might be used for continuous monitoring for crack growth through high-frequency sound waves emitted by flaws. See AEC, *The Safety of Nuclear power Reactors* (*Light Water Cooled*) *and Related Facilities*, 1973, ch. 7, p. 9.

80. *Review of Methods of Mitigating Spread of Radioactivity from a Failed Containment System* (Oak Ridge, Tennessee: Oak Ridge National Laboratory, September 1968) p. 1.

81. The Loss of Fluid Test Facility (LOFT), a 55 MW reactor, was proposed earlier for a different object but reoriented to investigate ECCS in 1967. AEC, *The Safety of Nuclear Power Reactors* (*Light Water Cooled*) *and Related Facilities*, 1973, ch. 7, p. 17.

82. Aldington referred in his evidence to the fact that a large part of AEC enquiries were, according to one of the Commissioners, to make less strict the precautions that are taken. Ronald Brown said that the information came from a tainted source. *SC Science and Technology, 1973–4*, Q. 690 *et seq.*

83. *Ibid.*, p. 196.

84. *Ibid.*, Q. 540.

85. *Hansard*, House of Commons, 2 May 1974, c. 1399.

86. *SC Science and Technology, 1973–4*, Qs 432–3.

87. *SC Science and Technology, 1972–3*, Q. 443.

88. *SC Science and Technology, 1973–4*, Q. 634. He referred to written and

verbal evidence; the written evidence was presumably the FOE selection.
89. Above, p. 184. One further instance is illustrative. Williams gave evidence that to date (end 1973) operating experience with large PWRs from 500 to 900 MW capacity cumulatively was seven-and-a-half-years for five reactors. It was a good figure for anyone who wanted to say the experience was slight. The basic figures were easily available; but there was a question of interpretation. Which figures should count as relevant experience? There were twelve candidate reactors, whose total reactor years experience since first producing electricity was $32\frac{1}{2}$ years, or $28\frac{1}{4}$ years if commercial operation was taken as the starting point. Here is the list:

| Plant | NSSS maker | Capacity (MW net) | First electricity A | Start of commer. operation B | Reactor since A | Years since B |
|---|---|---|---|---|---|---|
| San Onofre | W[a] | 430 | 7/67 | 1/68 | $6\frac{1}{4}$ | 6 |
| Haddam Neck | W | 575 | 7/67 | 1/68 | $6\frac{1}{2}$ | 6 |
| Ginna | W | 490 | 11/69 | 3/70 | 4 | $3\frac{3}{4}$ |
| Robinson | W | 700 | 9/70 | 3/71 | $3\frac{1}{4}$ | $2\frac{3}{4}$ |
| Point Beach I | W | 497 | 11/70 | 12/70 | 3 | 3 |
| Palisades | CE[b] | 700 | 5/71 | 12/71 | $2\frac{1}{2}$ | 2 |
| Point Beach II | W | 497 | 7/72 | 10/72 | $1\frac{1}{2}$ | $1\frac{1}{4}$ |
| Surry I | W | 788 | 7/72 | 12/72 | $1\frac{1}{2}$ | 1 |
| Turkey Pt. 3 | W | 693 | 10/72 | 12/72 | $1\frac{1}{4}$ | 1 |
| Maine Yankee | CE | 790 | 10/72 | 12/72 | $1\frac{1}{4}$ | 1 |
| Turkey Pt. 4 | W | 693 | 6/73 | 67/3 | $\frac{1}{2}$ | $\frac{1}{2}$ |
| Zion I | W | 1059 | 6/73 | 6/73 | $\frac{1}{2}$ | $\frac{1}{2}$ |
| | | | | | $32\frac{1}{4}$ | $28\frac{3}{4}$ |

[a] W = Westinghouse.
[b] CE = Combustion Engineering.

It could be reasonable to exclude San Onofre because its capacity was 14 per cent below 500 MW – though when it had a slight accident the intervenors counted it as relevant. Haddam Neck used steel fuel cans; in other respects its experience was certainly relevant. Ginna and the Point Beach plants were just below 500 MW. An unsophisticated statistician might exclude them, but wrongly. I cannot personally see why any others had a claim for exclusion – except that two were not made by Westinghouse! The figures are given in detail to emphasise the need for proper definition of statistics, which the Committee did not call for. These figures to my mind give a very different picture from those the Committee used and quoted. Varley used figures comparable to the Committee's in the debate of 2 May 1974; *Hansard*, c. 1395.
90. *SC Science and Technology, 1973–4*, Q. 168. He was pressed to concede that it would make *some* difference and agreed. (*Ibid.*, Q. 169 *et seq.*).
91. *Hansard*, House of Commons, 2 May 1974, c. 1381.

92. The CEGB memorandum recorded among other things the results of direct examination in the United States by CEGB engineers concerned with safety, and Matthews, who gave evidence, was in charge of the safety operations.

93. *SC Science and Technology, 1973–4*, Qs 35, 38–9. Some members had expressed their views to the Minister, but these Palmer said were the views of individual members rather than the opinion of the Joint Committee. Six members had joined a deputation to see Walker early in November: they included Palmer and Neave. *Financial Times*, 9 November 1973.

94. Above, pp. 231 and 234.

95. Ronald Brown did appear to imply that, since the United States regulatory organisation with its enormous numbers failed (an assumption not substantiated), Williams with his smaller numbers must fail also. *SC Science and Technology, 1972–73*, Q. 443.

96. *SC Science and Technology, 1973–4*, Qs 337, 351. Leadbitter appeared to claim that not only he but local councillors and local government officials were 'capable of making decisions on highly technical matters in the nuclear field'.

97. *SC Science and Technology, 1972–3*, Q. 420 and above, p. 234.

98. *SC Science and Technology, 1973–4*, Q. 627.

99. See Burn, *Economic History of Steelmaking 1867–1939* (Cambridge: Cambridge University Press, 1940) p. 297.

100. *SC Science and Technology, 1973–4*, p. viii and *Hansard*, House of Commons, 2 May 1974, c. 1378.

101. *Financial Times*, 7 June 1974.

102. *Ibid.*, 26 June 1974.

103. *SC Science and Technology, 1972–3*, p. 23 (1973).

104. *Ibid.*, Q. 398.

105. During the evidence he heard Brown assert that in Germany 'everyone who has got them [LWRs] is scared out of his pants'. *SC Science and Technology, 1973–4*, Q. 429. This was news to him – and indeed to Williams; it was of course utter nonsense, and German utilities continued to order LWRs.

106. Report of the NPAB, 1974, p. 9.

107. *Ibid.*, p. 18.

108. *Ibid.*, p. 10.

109. *Ibid.*, p. 13.

110. *Ibid.*, p. 13.

111. Report of the House of Commons Select Committee on Science and Technology on the *SGHWR Programme*, Minutes of Evidence (London: HMSO, 1976) Q. 1 (2 August). Hereafter cited as *SC Science and Technology, 1975–6*.

112. Hansard, House of Commons, 2 May 1974, c. 383.

113. Ronald Brown did ask whether Palmer would care to live near a LWR in his constituency in Bristol, to which Palmer replied he was not speaking of safety. *Ibid.*

114. Above, pp. 15, 161–2, and below, pp. 269–70.

115. I have been told this privately: I do not think it was published.

116. *Hansard*, House of Commons, 10 July 1974, c. 1368.

117. *The Times*, 11 July 1974.
118. *Financial Times*, 20 July 1976.
119. *The Times*, 23 March 1976. The article was headed 'The New Generation of Nuclear Power Plants takes Shape on the Drawing Board'. The SGHWR was not a more advanced reactor than the LWR, only a later one.
120. *United Kingdom Atomic Energy Authority Annual Report, 1974–5*, pp. 20–3, 36–7. In February 1975 Dr Marshall, Vice-Chairman of the AEA and Director of Harwell, stated at a meeting of the Society of Chemical Industry that not only would new staff be recruited at Harwell for the SGHWR but the amount of non-nuclear work carried on there would need to be reduced.
121. There was no precise estimate of the cost of an SGHWR in spring 1976; there was no agreed design. Early in 1976 Benn gave a figure of £1000m as an estimate of the probable cost of the Sizewell plant which was to be a 2640 MW plant. This was a cost in March 1975 prices, and it included the first fuel. In discussions at the CEGB I was informed that by March 1976 it would be the cost for the plant without fuel, owing to further escalation. At both dates it was without IDC, and it was approximately £380 a KW. The SSEB had estimated £170 for March 1974; the CEGB figure for March 1973 adjusted as indicated in the Report of the NPAB on *Choice of Thermal Reactor Systems* (p. 26) was £200 to £205. Inflation would possibly account for a rise of 60 per cent, which would leave the March 1976 estimate over 15 per cent above the CEGB minimum and over 30 per cent above the SSEB figure. The most recent American LWR figure (above, pp. 31 and 83) gave a cost at September 1976 including IDC but excluding escalation of $695. At an exchange rate of 1.6 this was £435. The SGHWR price of £380, excluding IDC, in March would be approximately £500 with IDC (the 1973–74 estimates put IDC as 30 per cent of Construction Cost). Allowing for a further six months of inflation, this would be approximately £540 – to compare with the United States figure of £435. All these figures include a lot of speculation, but the margins are wide.
122. *SC Science and Technology, 1975–6*, Qs 11, 16.
123. *Ibid.*, Q. 20.
124. *Ibid.*, Qs 31–2.
125. *Nucleonics Week*, 27 August 1970.
126. *SC Science and Technology, 1975–6*, Q. 68.
127. *Ibid.*, Q. 55.
128. *Financial Times*, 16 July 1976; also, 20 July 1976.
129. *Ibid.*, 24 July 1976.
130. *Réponses de la Commission d'Evaluation en Matière d'Energie Nucléaire* (Brussels: Ministère des Affaires Economiques, February 1976) p. 34.
131. *SC Science and Technology, 1975–76*, Q 4.
132. *Ibid.*, Q. 58. Benn pointed out they had acted on ministerial orders.
133. *Ibid.*, Q. 26.
134. *Ibid.*, Q. 67.
135. *Ibid.*, Q. 57.
136. *Ibid.*, Qs 33, 48, 57, 75, 82, 84.
137. *The Times*, 19 October 1976.

138. The spread of working costs in 1974–75 is shown in the following table, for which I am indebted to the CEGB.

Range of Working Costs in Fossil Fuel Steam Plants, 1974–75

| Cost of fuel per KWh | Number of stations | Percentage supply[a] | Inclusive heat cost per KWh | Total working cost per KWh |
|---|---|---|---|---|
| less than 0.5 | 4 | 15.4 | 0.48 | 0.53 |
| 0.5 but under 0.6 | 20 | 39.3 | 0.54 | 0.62 |
| 0.6 but under 0.8 | 41 | 29.0 | 0.71 | 0.82 |
| 0.8 and over | 98 | 16.3 | 0.94 | 1.18 |

[a] The average price of coal was £10.97 per tonne in 1974–75, £15.41 in 1975–76 and £17.70 in spring 1976.

139. *SC Science and Technology, 1976–6*, Q. 82.
140. Above, p. 000.
141. *Financial Times*, 14 April 1976.
142. *SC Science and Technology, 1975–6*, Q. 30.
143. For example, Tombs in his evidence to the Select Committee in December 1973 said: 'There is something like a 50 per cent difference in capital costs between AGRs and LWRs.' *SC Science and Technology, 1973–4*, Q. 168.

# CHAPTER 14

# Retrospect

(i) MAJOR MISTAKES AT LENGTH ACKNOWLEDGED

Successive delays in completing and operating an AGR brought a growing awareness that something was amiss, either bad projects had been chosen or projects had been badly conducted, or both, at an enormous cost.[1] This was unmistakably the failure of a government enterprise. The initial choices of the principal managers and experts were ministerial, the whole organisational structure was designed by ministers, the main agencies involved were managed by men chosen by ministers and subject to ministerial direction, and the choice of projects required ministerial approval. Some of those involved and their friends attempted to shift all the responsibility on to the private enterprise group, the consortia, who were employed in a narrow subordinate role without scope for initiative, even on to British industry in general. But the question put increasingly was why did ministers make these ill-judged decisions, and why were they not quickly reversed, but instead reinforced and repeated?

A large number of ministers had been involved, first the Ministers of Supply before there was an AEA, then those who founded the AEA, and then the ministers in charge of the AEA, and the Ministers of Power or Energy. Lord Carrington, Lord Cherwell, Lord Eccles, Lord Hailsham, Lord Lee, Lord Mills, Lord Salisbury, Lord Sandys, Tom Boardman, Frank Cousins, John Davies, John Eden, Patrick Jenkin, Aubrey Jones, Geoffrey Lloyd, Harold Macmillan, Richard Marsh, Roy Mason, Reginald Maudling, Geoffrey Rippon, Eric Varley, Peter Walker, Anthony Wedgwood Benn were all at different times involved, not all in the most crucial decisions, but all at least in maintaining, and by implication defending, continuation of policies. Some were unmistakably distinguished, as were several of the senior civil servants who flanked them: Sir Richard Clarke, at the Ministry of Technology, for example, regarded by many as one of the most enlightened advisers on the relations between industry and government; Sir Richard Powell, at the Board of Trade, who presided over the Powell Committee of 1964; and Sir Dennis Proctor at the Ministry of Power, who had interrupted his Civil Service career by a spell of private enterprise – to name but three.

The position of ministers was already recognised as symbolic of British arrangements for nuclear power development and of their weakness when *The Political Economy of Nuclear Energy* was written. Subsequent developments in Britain, in Phase II, were more relentlessly disastrous than critics foresaw, bringing out the weaknesses with accumulating force. The analysis of these provides the main theme of this chapter. Development outside Britain confirmed the greater strength of the American arrangements, but also revealed political weaknesses in securing popular acceptance of the new technology, and economic weaknesses, imperfections of competition which probably contributed to the disturbing demand pressures of 1966–68.

It was universally accepted in the United Kingdom by 1973–74 that it had been a mistake to have a programme of AGR 'commercial' plants; that indeed, more generally, it had been incautious to proceed with Magnox as well as AGR 'from a first off prototype to a full scale programme of nuclear stations'.[2] The right procedure, according to the experts who gave evidence to the Select Committee, was to proceed in stages from a small to a larger protoype and then to a demonstration plant, and not to order in numbers until the economic advantages of doing so had been established.

It was widely, though not universally accepted that the AEA decision to concentrate for a long time on the $CO_2$/graphite type of reactor, first Magnox, then AGR, had been a mistake, because these reactors had inherent characteristics likely to prove a source of high costs, and an obstacle to quick development and to satisfactory long term operation. The AEA had agreed in 1969 that a water reactor, the SGHWR, their own design, would have lower costs than an AGR.

It was thus at length conceded that the two crucial decisions, on which Britain had based its development and ordering of nuclear plants since 1955–57, had been wrong. It was also widely accepted that the SGHWR would, when fully developed, not have enough extra to offer, which LWRs did not offer, to justify the cost of developing it, or to obtain export markets. It was recognised that Britain had fallen behind the United States and Germany in developing the HTR, whose further development in Britain was delayed because of the work on the SGHWR. Lastly, it was recognised that Britain was no longer first in the FBR field, and that there were formidable problems to solve before commercial operation. Dr Marshall thought we might have two FBRs operating by the year 2000.[3] He seemed more optimistic later. This judgement was passed less than a decade after Sir William Penney had fixed 1979 as the year in which the first commercial FBR would operate.[4]

It was thus now agreed that the two major decisions of 1955–57, which had dominated British policy for over fifteen years, had been wrong, and the time required for development of more advanced reactors grossly underestimated. The sources of the errors and of their persistence, however were not yet seriously probed, though this was of primary importance. Not, perhaps, for persons and institutions which had been in error.

## (ii) RESPONSIBILITY AT THE CENTRE

One source was referred to *ad nauseam* – the subdivision of the British engineering industry and the corresponding existence of three consortia designing and constructing AGRs, resulting in three designs with consequential inadequacies alike of design and manufacture. Subdivision was a problem in most European countries, and engineering standards had to be raised in all countries, including the United States, for nuclear work. The organisation and state of the engineering industry did not account for the British failure.[5]

To account for it one must explain the decision to have a ten-year Magnox programme, and a second programme in 1964, the choice of the AGR in 1957 against the judgement of several Harwell groups, the failure to run developments of different systems in parallel, the inflated estimate of the potential of the AGR even in 1967, the smallness of the AGR prototype, the waste of resources on the beryllium can, the slowness of early development, the belief expressed in the early sixties that AGR development was as solid as LWR development, the exaggerated significance given to early Magnox development as a basis for AGR development, the decision to plunge from a 30 MW prototype with a steel pressure vessel into design and construction of a 600 MW unit with a concrete vessel and with all major parameters much more severe, and with major new features incompletely developed and untested. Crowning it all was the most bizarre episode, the launching with AEA assistance of the APC Dungeness B tender and its selection by AEA-CEGB experts as better than any LWR. It was only a sketch design, not a design in depth, nevertheless it was endorsed by the AEA-CEGB experts after exhaustive examination by a vast pyramid of committees which the CEGB found 'essential if the operation, maintenance and safety' of a station is to be 'fully satisfactory'. The experts found that adequate technical information supported the design, the engineering was of the required quality, the stated performance would be obtained, and there were 'no technical problems which could not be solved on the Dungeness time scale'. Though the chairman of the CEGB said later they had been concerned about APC's ability to 'tackle and programme' the plant and 'implement the sketched design into working detail', work was started before a completed design had been produced. Soon it was clear the technical information was insufficient (some supplied by the AEA itself was inadequate), engineering of requisite quality was not established by the sketch, many technical problems were unsolved, prevening completion on the promised date. Nearly a decade later, before any AGR had even started to operate, Hill claimed the assessment had been a 'meticulously fair and proper analysis'. He appeared oblivious of the irony, because the assessment was so wrong and misleading that the more meticulously fair and proper it was, the more utterly incompetent were those who made it. They remain anonymous.

The troubles of the later AGRs, designed and constructed by the two stronger consortia, showed that many generic problems posed by the 600 MW AGR had not been solved by 1965, and that solutions ensuring completion within the five-year lead time had not been visualised. Nevertheless the AEA, CEGB and SSEB had thought it appropriate, on the basis of the Dungeness B decision and on the assumption that the AGR was the most economic choice, to go ahead with orders for ten reactors. The AEA had wanted an even bigger programme in 1964–65. All the experience ultimately gained could have come from two reactors.

The experience of Phase II thus reinforced the conclusions set out in 1967, that what went wrong in British nuclear power development was due primarily to misjudgements and poor performance at the centre, not to weaknesses on the perimeter. While there were weaknesses on the perimeter, especially in the relation of the perimeter to the centre, these arose mainly out of decisions at the centre, in the conception of the restricted role of consortia as groups to exploit AEA developments, not as strong firms to carry out the whole industrial job, supported financially in development work directly by the State and free to use what technology, British or foreign, they thought best.

Experience in Phase II, including the ultimate acceptance by the atomic establishment that the two major decisions of 1955–57 were wrong, followed by the choice of the SGHWR, established with still greater force and clarity the dangerous inadequacy and inappropriateness of the highly centralised and political British organisation for nuclear development, whose failure could be identified long before 1967 and was analysed in *The Political Economy of Nuclear Energy*.

(iii) THE UNITED STATES PATTERN – AT HOME AND ELSEWHERE

The two cardinal errors which were the genesis of most subsequent troubles were not, as was pointed out in 1967, possible under American arrangements. The United States policy aimed at bringing nuclear development 'within our normal economic and industrial framework' as soon as possible, and the development was intended to 'strengthen free competition in private enterprise'.[6] Manufacturers were encouraged to develop plants competitively – with AEC financial help, and full access to AEC data – and utilities to help in the development. The decision to buy a nuclear plant rested entirely with utilities, and a system deemed commercial received no further government assistance (except that important government safety research supplemented work by firms). A system unlikely to become competitive lost government support. The Government could not institute a programme of commercial plants or impose a reactor system. The outcome was that, while there were necessarily unsuccessful projects, strongly based development programmes having no parallel in the United Kingdom established the BWR and PWR systems competitively.[7] There were the appropriate sequences of prototypes and demonstration plants in ascending sizes and the major firms, Westinghouse and GE, organised the sequences from the research stage onwards, and carried out much of the manufacture including fuel. Members of the Select Committee were rightly impressed with these organisms in 1967.[8] The LWRs – the choice of the systems and their development – were the successful outcome of a deliberate policy of using quickly the existing industrial structure with the possibilities of comparison, of systems and development skills, and incentives to secure quality and speed, which the competitive structure provided.

The same course was followed on the Continent, except in France, and in Japan. The greater force of American development when compared with British was recognised early – this was manifest in both Germany and Sweden by 1956.[9] In many of these countries GE and Westinghouse were already strong forces. Most governments were quick to see that the right approach was to encourage development by major manufacturers – working on their own initiative in association with utilities and with licences usually from GE and Westinghouse – and to leave decisions on the purchase of nuclear plants to utilities, without any restraints other than for safety and environmental protection. Except in France, and in Italy after 1964 (before which the private utilities had ordered nuclear plants of different systems) there was more than one utility able and ready (with government financial encouragement) to participate in a nuclear prototype and ultimately buy nuclear plants, which in several instances proved important. Even in France the position was not wholly different from the common pattern. Their Magnox type reached a higher point of fuel efficiency than in Britain but the nationalised utility quickly saw the attractions of the LWR and secured experience by promoting joint Belgian-French plants. Once enriched uranium could be used in France, the CEA agreed that the LWR was to be preferred to an AGR and that a Westinghouse affiliate or a GE licensee, preferably both, should provide the plants.[10]

In general the manufacturer-utility axis, with government financial help in many forms and control to preserve safety and the environment, was impressively successful.

In Britain the assumption from the outset was that you could not leave nuclear plant development to the normal framework of the utility/plant-making industry, and should not try and push it in that direction. This remained so in Conservative policy decisions as late at 1972–73. The party's views may have changed since then: and Arthur Palmer showed signs of a change in May 1974, but this may have been an aberration.[11] The Conservatives opted at this (1972) stage for a special designated single purpose D and C company with a monopoly but subject to ministerial directives. They wanted to have GEC involved, but never seemed openly to contemplate leaving the job to GEC, with government support uninhibited but without a monopoly, as would have happened in practically any other Western country.

### (iv) BRITISH MISTAKES IDENTIFIED EARLY

Before the basis of this approach is explored there is one general point, unconnected with organisation, to establish. It is often implied that the policies now acknowledged to have been mistakes were natural and almost inevitable when they were made, and were rightly maintained by the AEA and Government, when challenged for instance by the consortia in 1959, and by advocates of the LWRs in 1964–65.[12] The plain fact is that it was easy at the beginning to see that the policies were probably unsound, and doubts were expressed in growing volume in public and private from 1955 onwards. The case argued against the twin errors, the concentration on $CO_2$/graphite (the 'narrow front') and the adoption of power plant programmes in the period 1955–59, was five-fold. It was progressively established, and acknowledged, that the 1955–57 forecasts of low 'competitive' costs were based on wrong economic, technical and commercial assumptions, and it was quickly recognised that a programme could not be justified as an answer to Suez and that nuclear power was not urgently needed to fill an approaching gap.[13] More crucial, it became known by 1956 that American experts had decided the $CO_2$/graphite system was one that should not be developed in the United States on account of its 'low material economy', and that, although there were a few advocates of the Magnox system on the Continent, utilities and manufacturers were tending quickly to favour the LWR system as the more promising based more securely on solid development.[14] This established a strong argument against the narrow front and in favour of the parallel development of prototypes of different reactor systems for the first generation:[15] this would have involved a different disposition of resources, probably of fewer resources.

Next, the system whereby industrial groups were used only to construct plants based upon systems developed by the AEA, and did not make fuel but used fuel made by the AEA (later BNFL), proved extremely frustrating to some at least of the consortia by early 1959, and in fact earlier, but they had a platform in 1959.[16] Finally it was evident that there was no logic in the argument that a programme of broadly uniform plants was needed to build up a nuclear industry which would be available when nuclear power plants were required in great numbers. If prototypes of different systems, some under licence from American companies, had been built by autonomous firms, and companies had been encouraged to increase their development capacities, this would have given a stronger industrial base – as Benn recognised in 1969.

So in explaining policy decisions in the United Kingdom and the working of the British system it must be recognised that the strong possibility that other reactor systems might be better than the AEA's choice could have been known by ministers, civil servants and MPs in the later fifties, as it no doubt was known by people in the AEA. That ministers, civil servants and MPs were probably not aware of it is a factor that has to be explained and taken into account in the reckoning.

### (v) MINISTERS' OPTIONS

Ministerial decisions are rightly seen as the mainspring of the costly misjudgements in British nuclear plant development, although the policies adopted and the performance in implementing them are to be attributed to the AEA. But when the question put is the one with which this chapter started why did ministers make so many mistakes for so long? – this gets the emphasis wrong. The right question is – why did ministers, and civil servants, and MPs, and most advocates of the mixed economy ever think, and despite all that happened go on thinking, that ministers could get the answers right?

This is an unfashionable question in Britain, where it is currently assumed that if ministers decide to do something they are competent to do it. They are even thought competent to do it, whatever it is, at a moment's notice, on taking over a new ministry, which emphasises a second assumption, that just as ministers are competent to do anything they choose so too the civil servants who advise them are competent to give the right advice on any subject on which a minister asks advice or intends to act.

At the outset, civil application of atomic energy was a matter on which the Government had to act because governments had developed the atomic technology for war and owned it and all the facilities for dealing with the weapons applications. Government would also need to remain associated with civilian applications to ensure safety. It was the same in the United States.

But the United States Executive and Congress were determined that development of nuclear power plants should, as soon as possible, be managed and financed by firms, who should choose the system and have free access to government sources of information, and financial help in early stages, and that utilities should decide freely when the time came to buy a nuclear plant as commercial and competitive. They were encouraged to help finance development of promising systems, but were under no compulsion and not (at least not heavily) subsidised.[17] The same prescription as we have seen was adopted universally, except in Britain, France and Canada. Manufacturers were regarded as experts in this kind of development, utilities as experts in assessing development achievements. The British ministers decided they could do successfully what the United States Government – and almost all other governments – decided they should not do, because they could not do it efficiently. When the AEC experts were saying, by way of analogy, 'You don't know how many to breed to get a Kentucky Derby winner... You have to raise several of them and take good care of them to get a winner', British ministers had decided they could foresee which reactor type would be the winner and would soon be adopted as the basis of a ten- to twenty-year programme of commercial plants, to be followed in year twenty by a breeder reactor which it would be possible to make 'in numbers' and which would be able to generate all Britain's electricity. A minister was telling Parliament of the prospects by January 1953. In April 1953 a committee to recommend an appropriate organisation for the new phase of atomic energy development was appointed. Soon the building of the Calder Hall reactor, whose assumed economic success was to be the basis of the scheme, was started. In April 1954 the proposed new organisation – the AEA – was unveiled. Early in 1955, nearly eighteen months before the first prototype was completed, let alone tested, a ten-year programme based on it was announced, with a possibility that a water-cooled reactor might be introduced in the later part of the programme; but that had faded out of sight by the end of 1956.

(vi) THE REASON WHY

On the evidence published so far one can only surmise why ministers took the amazing decision to base a ten-year nuclear power plant programme on a reactor system of which not even a small prototype was completed.

For a minister who wanted to see a rapid start and rapid growth in supply of power from nuclear energy, the only available reactor system in the early fifties was the $CO^2$/graphite natural uranium type. There were a few ministers who did think speed of development important: Cherwell was outstanding in this. A participant in government organisation of atomic energy from the wartime beginnings, he passionately believed that Britain's whole future prosperity depended on leadership in its use for civilian purposes; unless we adopted the right organisation and policy and moved fast 'the Military and Industrial Revolution will pass us by'. While Cherwell was mainly concerned that the United Kingdom should be in the van of industrial progress, others saw speed as necessary to supplement coal supplies, which could not rise as fast as it was assumed energy consumption would rise. Oil appeared to be overlooked.

Within the atomic organisation there were two approaches to the method of developing nuclear power plants. At Harwell, which Cockcroft believed until 1953 had the prime responsibility for developing power reactors, several systems, including gas/graphite/natural uranium were explored, but the general view was that natural uranium reactors had no future. Fast breeders were seen as the ultimate object by all the British groups. This was so everywhere; in the United States fast breeder development was started in 1945 and some electric power based on a small experimental fast reactor was used in Idaho in 1951. The study of enriched uranium systems in Britain may have lacked firm direction, but one major misjudgement occurred. The PWR system was studied in 1951–52 because it was being developed in America for submarines and the Harwell group failed to see what the Americans saw, its scope for large-power reactors.[18] Harwell organised in 1952 a feasibility study of the $CO_2$/graphite system ('Pippa'), regarded as a simple system industrial firms could largely design and construct; it was not undertaken as the system for the future. Cockcroft thought 'full-scale atomic power stations would be a collaborative affair between their establishment [Harwell] and private industry, which must be brought into the business.'[19] This had affinities with the American approach, lacking some of its dynamism. Industry, learning the simple system, could go on to things more advanced. Harwell's Pippa feasibility study was completed at the end of 1952.

The second approach may be called the Hinton approach. Hinton advocated exclusive concentration on gas/graphite reactors until the fast breeder was ready. He wanted this adopted in 1948 for the reactor to be built at Windscale to produce plutonium. A gas/graphite reactor would have also produced power. An air-cooled, graphite-moderated reactor was chosen instead, to speed construction. Hinton argued the gas/graphite system woud be the best route to power production because experience must be gained in simple schemes before going to more complex ones. The 'down to earth engineers' philosophy', as Gowing describes it, was that 'for immediate development there should be a bias towards projects which were extensions of existing systems', and gas/graphite fell into this gategory.[20] The United States had a more radical 'down to earth engineers' philosophy' about these reactors: 'we are not charmed by the gas/graphite reactors because of their low material economy'.[21] They had to be competitive-cost-conscious, which had never been Hinton's problem since 1939. He was not alone in advocating the $CO_2$/graphite natural uranium system for a programme, but he was early on the track. R. V. Moore suggested a programme with striking similarities to the Calder Hall–Chapel Cross complexes in 1950.[22] The system found more support outside than inside Harwell, among heavy plant makers in the electricity supply industry, and civil servants in the Ministry of Fuel and Power.[23] Cherwell agreed in Canada in 1952 that Britain would concentrate on the gas-cooled system in the middle term, leading up to the FBR. Canada would concentrate on the heavy water system: the two countries would thus have complementary, not cooperative, programmes.[24]

Hinton was wholly absorbed from 1947 to 1952 in managing the design and construction of the network of plants needed for producing plutonium. Though this work took years longer than the remarkable American wartime creative effort, and longer than the Russian post-war job, it was a substantial achievement and Hinton earned a great reputation. With completion he could contemplate more work on civil power reactors, and was looking for new fields to conquer. He claimed in a memorandum to the Atomic Energy Board – which under Cherwell was put in control of atomic energy affairs in April 1952 – that it had been agreed verbally in 1947 that Risley should be responsible for all atomic energy projects on an industrial scale. He claimed that if it was decided to build Pippa it should not be built at Harwell by industry, working with Cockcroft's organisation, but that Risley should take over the design and construction. The Board agreed.[25] Lord Cherwell was in fact anxious that industry should be attracted to nuclear work because he wanted large-scale activity; but it would now be on Hinton's terms. It was three months before the decision was taken to go ahead with Pippa and then because more plutonium was urgently required. The enormous value placed on plutonium was the economic justification.[26] It was not immediately supposed that a programme of many gas/graphite plants would follow.[27]

Nevertheless, the decisions over Pippa were a turning-point, and contained the germ of all major subsequent bad ministerial decisions. Hinton after initial opposition had by 1952 welcomed Cherwell's initiative to take atomic energy out of the Ministry of Supply and put it into a public corporaion – the future AEA. He was delighted to discover an able minister who knew something about atomic energy and who had actually visited Risley.[28] Cherwell probably welcomed Hinton's driving force as likely to bring that speed into nuclear development which was to keep Britain in the industrial race. Building the network of atomic factories which Hinton had directed with great energy and skill (enjoying priorities fully exploited, great help, goodwill and know-how from

ICI and other firms) was described in the first official account as bearing comparison with any achievement in British industrial history.[29] Even if this were true (I regard it as hagiography and as a comparison quite undefined[30]) the work had been entirely within the realm of Hinton's pre-war and wartime experience – quite different in character from that of sorting out, evaluating and developing reactor systems, choosing which horses to back, when to abandon them, when to choose new ones, and riding them well. At the time this distinction of task was probably not observed.

After the decision to go ahead with Pippa for plutonium and by-product power ministers gravitated steadily towards decisions to have a ten-year plant programme (to be trebled within two years), to concentrate on the $CO_2$/graphite family, to have a 'narrow front', to discourage broad exploration at Harwell, and to give Harwell no outlet for development projects except Risley, to bring in industry in a permanently subordinate role. My guess is that the driving force in this was Hinton, with Cherwell perhaps a willing victim.

In retrospect we know the basis for rational decision was lacking. The reactor types out of which a choice should be made were not known even by name, let alone by quality. The PWR had been ruled out. There were grandiose pretensions about making the best use of scarce resources when there were no adequate criteria to justify the choices made. The decision to go ahead with a Magnox programme based on Risley's work, expected to cost £1000m, was treated as manifestly a good use of resources. A smaller programme using fewer of the same resources but designed to test alternative systems, developed by individual firms, in one or two instances based possibly on foreign licences, with state aid and some staff from the state atomic activities, would have been described as waste. Yet the first brought a disastrous waste of resources, and left Britain without a substantial nuclear engineering industry, and the second could have been fruitful.

All the ministers closely concerned (Cherwell, Sandys, Lloyd, Eccles) would presumably have said they would not and could not assess relative or absolute merits of reactor systems, or identify good and bad development practice.[31] They would have claimed they could choose the persons who *could* make such judgements, or could appoint persons who in turn would appoint the technical decision makers. With the AEA they did both; they appointed as full-time members of the Authority the technical heads of the three atomic energy activities of the Ministry of Supply, and three other senior civil servants. Cherwell was a part-time member (this was not possible for a minister from the Commons) and three industrialists and a trade unionist were appointed – but none of the industrialists came from firms which had close links with the United States companies who had already great nuclear experience.

Eccles' second reading speech was a commitment to a *policy*, but the rationale of the appointments was not explained. The creation of the Authority was no doubt meant to insulate the ministers from technical and management decisions (and *vice versa*) in this activity which was expanding phenomenally: there were 11,000 atomic energy staff in the Ministry of Supply early in 1952, 24,000 in the AEA in March 1955, 28,000 in 1958, over 40,000 in 1961. Eccles spoke of good results emerging from the leading scientists and engineers meeting round the boardroom table, but how do you know your selection are the leaders in the implied sense, not just those at the top, and that their choice will be good if *you* do not know what is good? If Harwell wanted to try a light-water system at once and Risley said it could not take an additional load – and Risley won (which with Hinton's strong personality was likely) – would this necessarily have been a good result because the leaders were round a table? And if no one wanted to get on with light water – was this a good result? The problem could not have been put in those specific terms in 1954 because of the ignorance of the participants, but it could have been put in general terms. What criteria could members apply to judge the performance of this enormous mass of resources under narrow monopoly control? It was clear when the Chairman of the AEA gave evidence to the Select Committee on Estimates in 1959 that he did not realise the significance of the BWR, and when Penney gave evidence to the new Select Committee in 1967 his optimistic assessment of the AGR argued a disturbing remoteness from what was happening. The AEA experience does not sustain the view that ministers could make good appointments and secure the best use of resources.

It is highly unlikely that ministers asked whether they were competent to appoint and control the monopolist organisation. The form of organisation was recommended by a distinguished committee, of whose three members two had had exalted administrative experience.[32] Defence research chiefs were departmentally appointed (there was a big defence element in the AEA). Ministers were appointing a growing number of boards of nationalised industries – it was part of their job. Private industrial firms could not afford atomic development alone: the technology had to be transferred to them – it would need government money, staff and possibly facilities. Ministers would have to appoint people to allocate these: they could not avoid making choices. Ministers, moreover, appointed utility chiefs and had to approve, if only as a routine, their investment programmes: there was no question of utilities being autonomous as purchasers, as in most countries.

It is at times argued by way of apologia that Ministers were as competent to make this kind of choice – of persons and policies – as the chairman and director of a company large enough to handle nuclear power development. Such a company would inevitably – the arguments runs – be a bureaucratic institution. It is a misleading analogy. Companies vary in character and quality and over time. But normally persons making major decisions and appointments in them would have been dealing over a long time with the same kind of situation, and with the type of technology and development processes involved, and would have come to the top through some success in this kind of work. Since they would usually expect to remain in the job for a reasonably long time it would be important that policies and appointments should succeed, and they had to succeed in competition, where people would be making comparisons on an international basis. The same was true for those in the firm who recommended the policies to top management and those chosen by top management to administer them. Long-term success was their object, not short-term acclaim.

A minister's situation and background were quite different. Usually he had no experience of making this kind of decision on policy or persons in *any* activity, let alone in relation to the type of problem presented by nuclear power development. Though Cherwell was a partial exception and knew the scientific background, ministers could not be expected to know the technology. Unlike company heads they would not be personally familiar with the principal technical and managerial staff involved. They did not normally expect to stay long in the job, since ministers look for promotion and broad experience and Prime Ministers move them around. They were secure in a domestic monopoly against a quick competitive challenge. Hence, though no doubt they would wish to believe their decisions in the long-term national interest, whatever that might be, immediate acclaim mattered more to them personally: if things went astray their part in the affair would be forgotten, and they would lose nothing, none of those things – Riches, Knowledge, Honour, which Hobbes identifies as forms of Power.[33] If a decision would give an immediate boost to public morale or national prestige or (for a socialist) national ownership, or would offer tribute to environmentalists, this would recommend it.

The normal first step towards becoming a minister, to be elected as a member of Parliament, is not a process one would choose deliberately as a way of selecting persons to make the most crucial decisions for a novel and complex industry. The candidate is chosen, usually on a party ticket, to vote – nominally for a constituency – on a whole range of activities which governments or MPs may decide or feel compelled to deal with, and while good public relations techniques – advertising skill – will be an essential qualification, skills in judging difficult industrial situations will not count as a qualification at all. Edmund Dell, a practitioner (he was Secretary for Trade at the time of writing) has written as though ministers are likely to have a sort of general-purpose judgement which will be appropriate for all industrial problems. 'The searching eye of judgement selects a few acceptable courses . . . from a multitude of possible courses which calculation may suggest.' (What is an 'acceptable' course – acceptable to the Party? And what can calculation suggest about choice of persons?) 'It then decides on one. How it does it we do not know, how to train it we do not know; but it is the great factor and it cannot be replaced.' He goes on to say later, comfortingly: 'It has been observed that good judgement [nowhere defined] does not *necessarily* [my italics] require high intellectual capacity.' Does this mean it sometimes does and sometimes does not? 'It does however', he says, and this means *always*, 'need to be informed.'[34]

'Informed' in what sense? In 1952–55 ministers and civil servants based sweeping decisions on totally inadequate information. What they needed was to be informed of the extent to which they were uninformed, and of the ways in which they would necessarily remain so, and that their 'decision' should be an outcome of this recognition – a recognition of what they could not do on a rational basis. Governments in most other Western countries recognised the limits and reacted appropriately.

## (vii) RESPONSE TO INFORMATION

How would ministers, Opposition leaders, Select Committee members, the AEA, trade union leaders, and the media, respond to information as it became available? – when for example the range of reactor systems from which a choice could be made and foreign methods of making the choice and their relative success, were known. As we have seen, the response to information can equally well be called response to failure – and to a surprising and alarming extent also failure to respond.

The weakness of the foundations of the decisions of 1953–56 (the narrow front, concentration on gas/graphite, ten-year programme and AEA development monopoly) began to be seen quite quickly, and the process of exposure continued relentlessly for two decades. By 1958 it was known that the cost forecasts on which the Magnox programme was launched were wrong in most respects, had been assembled without adequate care, and all the errors were pro-Magnox.[35] It was recognised that there was no urgent need for nuclear power and that nuclear power could not satisfy an urgent need if it existed. Most important of all it had become clear (and was reported in serious newspapers so that ministers could have known it without the help of their civil servants) that continental preferences increasingly favoured the American LWR systems. Important groups of Harwell scientists agreed with them. There was, by 1958, an account from the inside of fundamental weakness in AEA development techniques.[36] By the end of 1958 some of the consortia, who had built up strong R and D organisations were seeking freedom from AEA domination, wanted to work on licence from American firms, and pointed to the nonsense of a situation where five consortia were all confined to developing one AEA system.[37]

The climax of the process of exposure occurred in 1964–65 when the CEGB, having first shown great interest in the Candu system and then in the BWR, as probably offering lower costs than the AGR, was allowed to invite competitive tenders for both AGRs and LWRs for the second plant it proposed to build at Dungeness. Britain now had a Labour Government again. An evaluation of the tenders by AEA and CEGB experts established, so it was said, that the AGR offered the cheaper power. Britain had made the greatest breakthrough of all time. The subsequent unfolding was a ruthless demonstration of how abysmally incorrect the judgement was – wrong in its generic comparison of the systems, wrong in its assessment of the specific design tendered (which was only a sketch!), wrong in its view of the stage of development which the AGR had reached.

Towards the end of Phase II the AGR disaster had led the atomic establishment to agree that choice of the gas/graphite system had been a mistake, that the decision to base a programme on one small primitive prototype was a mistake ('we are proceeding much more cautiously' was the way Hill put it), that the development of the AGR had been inadequate, that it had been wrong to step from a 30 MW prototype to a 600 MW commercial plant, that the procedure followed by the Americans in developing the LWR (it was not put this way) had been right – moving from small experimental reactors to large ones of about 100 MW(t), then to prototypes of from 80 to 200 MW(e) and from 200 MW through 350 MW to 600 MW,[38] and organising component development for sensitive parts.

It was thus recognised that ministerial choices and the whole strategy of development undertaken, as advised and organised by the AEA, had been wrong. It was also obvious that the collapse of the AGR programme and the absence of new orders for nuclear plants since 1970 had undermined the nuclear engineering industry; the professed intention of building up by programmes an industry capable of constructing nuclear plants in numbers when they were needed was not fulfilled.

One might have supposed it impossible for ministers – as it certainly was for the civil servants and the atomic 'establishment' in general – not to recognise that mistaken policies, development strategy and performance at the centre betokened the failure of the British compared with the American type of organisation. Mr Callaghan was to show in 1976 that it was still possible for the Prime Minister not to acknowledge it.[39] Recognition of the failures of central policies went hand in hand with a remarkable adherence to the old policies and structures. There was a continuous discussion of the appropriate organisation for atomic affairs from the early sixties, dominated by a drive to establish a single design and construction company under the control of the AEA. This discussion was insulated from the process of development, which by 1973–74 was recognised to have been misconceived. The single D and C company formula, advocated by the AEA from 1964 onwards, was supported on the assertion that export orders could be obtained for British reactors. Failure to get orders betokened ineffective selling or excessive costs. Selling was ineffective, it was said, because it was competitive, and costs were too high because production was fragmented. It could be seen from the outset that this was based on a completely incorrect diagnosis, because there was nothing exportable to sell. Nevertheless it was sustained, its proponents – MPs, trade unionists, the AEA – were undeterred by the growing weight of 'information'. Benn resisted; and for several years some Conservatives resisted. In the end the Conservatives fell for it in their 1972–73 reorganisation.

Which was a token of the fact that the Conservatives in reorganising failed to ask the most pertinent questions. The nuclear edifice built on the Conservative decisions of 1953–55, reinforced by the Dungeness B decision of the Labour Government in 1964–65, had fallen like a house of cards. Reorganisation and new direction of activity was needed because the initial decisions had been blunders and bred more blunders. Ministers and their civil servants, on the advice of the AEA and its precursors in the Ministry of Supply, had made a series of premature choices: in effect they had made technical choices in a technology about which they were wholly ignorant, had chosen people to manage the development of the technology, and had given monopolies to the AEA (hence to a dominant group within it) as developer and in effect also to the consortia collectively as architect-engineers for nuclear plants. Failure sprang from bad ministerial decisions, based on bad advice from the technical experts they knew and hence consulted, and from bad performance by the AEA in its development work. At times there was connivance from the CEGB.

But these points were not brought out in the reorganisation discussions. Peter Vinter once advanced the unimpeachable doctrine that for the reorganisers history was only of interest if it could help solve their problem. For him the sheer failure of ministers and their civil service advisers and the AEA appeared to have no message, nor, it would seem, had the success of other countries where major reliance was placed on the judgements of utilities and manufacturers.

In public there was no ruthless probing into why things had gone wrong, or why they had gone better elsewhere. Nor was there any obvious attempt to find out who had been right in Britain and why their advice was neglected. It is easy – it was easy at the time – to discover that in 1964–65 some of the consortia, and leading engineers in them and in associated firms such as Rolls Royce, recognised that LWRs were to be preferred to the AGR, having lower costs and good prospects. They were right. What are they doing now? Not playing a major role! The significance of dissident Harwell views and of the claims of the consortia who wished in 1958–59 to spread their wings and have more autonomy in development, appeared to pass unnoticed.

The extraordinary episode of the Dungeness B contract and the laudatory CEGB *Appraisal* was left without thorough study, despite disturbing information in evidence to the Select Committee. In retrospect the propaganda for the AGR, the claim that its development was as solid as that behind the LWR, looks ruthless. There are still many serious gaps in our knowledge of facts. Why was a firm which was regarded as the weakest consortium, whose staff was not adequate to make a detailed design, given the contract on the basis of a sketch design? Simply because the firm had done some costings of the most advanced AGR design for the AEA and was not, like the other consortia, tendering both for an AGR and an LWR and so comparing them? How was it that the CEGB–AEA group of experts, whose immensely detailed and probing methods of analysing tenders were described pretentiously to the Select Committee, concluded that the design tendered was such that the plant offered would do what was asked of it and promised by the tenderer, and that no engineering problems were presented which could not be solved within the contract period? Whereas within a short time it was discovered that many important engineering difficulties were encountered which had not been foreseen in the designing, that basic data on corrosion risks had not been available from the AEA, and that drastic redesign of important parts was quickly needed, involving the scrapping of much work already done. How was it that building was started before detailed designs were ready? Much of this looks in retrospect extraordinarily irresponsible behaviour by two public corporations. Who were the people who made these collossal misjudgements, the nameless experts of the public corporations, and where are they now? Did the AEA experts include persons responsible for developing the AGR – which might be expected to give them a bias? Did the specific developers, not merely distant colleagues in the vast organisations, sit as judge and jury on their rivals' products? There is the nagging question, was there *political* pressure on the CEGB, or was it merely pressure from the AEA to which Hawkins referred? If there was political pressure how did this impinge on technical assessments? It was being said by senior civil servants in science-based ministeries in 1964 that a decision not to go ahead with the AGR would be taken as something like ending the work of the AEA on civil reactors, which introduced a strong emotive element into the decision. Did this affect ministerial attitudes or 'expert' appraisal?

Not only was there no probing of the past in this period of decision in 1972–73 but the investigation of the present had one crucial gap – there was no attempt to obtain an assessment of the safety of the LWR by the British Nuclear Installations Inspectorate, although it was known that the LWR was again the system most likely to provide the cheapest power. The Chief Inspector was intimately involved in the working party discussions on the choice of the next reactor, he knew the key position of the LWR, but ostensibly he did not seek powers to investigate and was not asked to do so.

Hence, as we have seen, the reorganisation of 1972–73 (although a consequence of the collapse of policies based on ministerial decisions and AEA monopoly) largely retained the features which were the source of the collapse, and actually increased the degree of centralisation. Ministers retained power to decide the choice of reactor system 'in the national interest', to decide who should design and construct nuclear plants, to allocate R and D functions between the AEA and the single D and C company which was designated (in which the AEA was a shareholder) to decide what links the D and C company might make with foreign firms, and in a measure what sub-contractors it should use. The utility could not automatically have the reactor of its choice and could not negotiate directly on an autonomous basis with the manufacturer. The Government still thought in terms of fixing a programme of nuclear plants, as well as having the last word in choosing the type of reactor. Ministers indicated the significance they attached to their role in decision-making by setting up a new advisory body (NPAB) to help them.

Varley's announcement in July 1974 that the utilities must buy the SGHWR, which the CEGB did not want, and that there should be an SGHWR programme, modest, but still 4000 MW in four years, not just a prototype or demonstration unit on the American development pattern, set the seal on the Conservatives' new structure, which was rather a rejection of information than a response to it. Conservatives might have said they did not envisage acting in this way – but they endowed the Minister with the function and the power. Varley's act (based on a Cabinet decision) was a more categorical assertion of the ministerial power of making technical decisions than any made hitherto. It was made against the judgement of the CEGB, against the judgement of the heads chosen for the one designated D and C company and the majority of component manufacturers, possibly against the judgement of the AEA – though the presentation of this judgement to the Select Committee had been finely balanced. It was a remarkable assertion of personal or political preference.

The timing was interesting. The CEGB choice of the LWR came just after LWR plants in the United States had been experiencing a fierce attack by environmentalists as a potential source of catastrophic disaster. The attack was weakly based technically though promoted with exceptional propaganda skill. It was rejected by the Regulatory Authority in the United States which asserted, with good supporting evidence, that its conservatism in licensing – that is, its insistence on exceptionally large safety margins – ensured safety. By the time Varley made his announcement the conclusions of the Rasmussen Report were known to the effect that the LWR involved negligible risks,[40] and a high-level British investigation of the safety of PWR pressure vessels had concluded in a preliminary report that, provided the specified conditions were observed in manufacture, the vessels were safe.[41]

Varley – and the Government – were deciding politically between two groups of engineering experts, of whom the larger and more impressive was on the side of the LWR. Safety was never in jeopardy, unless the Minister distrusted the NII, and if so he should have strengthened or replaced it. Time and cost were at stake. The CEGB and NNC said it was possible to provide nuclear power faster and more cheaply by the LWR route. The SGHWR camp denied this.

Many claims made for the SGHWR had been disproved by summer 1976: the time required for design, especially for safety (the proponents of which had said presented no problems), the cost of development, and probably the ultimate cost of the plant, had proved longer and higher than the proponents claimed. Varley's intervention, based partly on misjudgement, was not only a remarkable assertion of power but symptomatic of the British approach in that the Minister, in carrying out, no doubt, a government decision, thought himself competent to back a minority group on a technical issue.

## (viii) THE ALTERNATIVE COURSE

The alternative to the 1972–73 organisation was to encourage the utility-plant maker axis which succeeded in most other countries. Under these arrangements utilities would have been left free to choose the reactor system and the supplier: manufacturing firms would have been free to offer (and develop) systems they thought commercially most attractive. The Government's sole function in regard to sales of *really* commercial plants would have been to be satisfied on safety.

The immediate reaction to this proposal would doubtless have been that it was wholly impracticable, that no firm could take on the job and carry the risk, no firm had a reactor to offer which was commercially attractive. What had been done in Germany, Japan, Switzerland, Belgium, Italy and Sweden could not be done in Britain. The D and C companies would have been at one with the Government in taking this view. Given the greater rate of growth normal in these other countries it was a good pragmatic conclusion that the United Kingdom could not in 1973 do as they did.[42]

It was not, however, necessary to decide about the new D and C company at that time. No doubt some cracks had to be plastered over. But it was absurd to decide on a form of organization before it was known what the organisation was intended to do, in short, what reactor system was to be built. The difficulty was greatly increased by the fact that the rate of growth under the Labour Government from 1965 to 1970 had been half the rate in the previous quinquennium, which checked the increase in total demand for energy; while consumption of electricity was also reduced by the increasing use of North Sea gas from 1966, which for some purposes was a substitute. Electricity consumption rose by almost 50 per cent between 1960 and 1965, but by less than 30 per cent between 1965 and 1970. The additional generating capacity ordered in the first half of the sixties greatly overprovided for the end of the decade. For this reason delays in completing conventional and nuclear plants were not an embarrassment, except in forcing the longer use of some high-cost plant. In the early seventies there was no need to place many new orders. It was natural that, if the choice lay between an oil-fired plant and another AGR in 1971–72, the choice was inevitably oil. So when the 1972–73 reorganisation was being planned no nuclear plant orders were in sight, nor was it known what specifically would be ordered. No firm would go into such a business with such a history without a guarantee. There was no prospect.

But if it had been known that the LWR would be acceptable, provided the NII found it was safe, that the Government was leaving the question of choice to utility and manufacturer, as it could have in 1971–72, this would have changed the whole situation. The LWR was a competitive reactor, recognised by virtually all major Continental utilities as capable of providing electricity more cheaply than any fossil-fuel plant – including oil-fired plants with oil at 1970 prices. In 1955 nuclear plants were *going* to be quickly competitive in the United Kingdom – they were not. One felt that programme II had been decided upon on the assumption that one must give the nuclear engineers something to do. But if LWRs had been chosen in 1972–73 it would have been in the knowledge that in the future possibly almost all large power stations would be of this type for twenty years and more. It was no longer a matter of speculation and hope. The commercial attraction to be in on this business would have been substantial. It was neither necessary nor desirable to give anyone a monopoly. But if a firm, even GEC, could secure monopoly, it would welcome it.

By the end of 1973 the LWR was made still more attractive by the enormous increase in prices of oil and coal, so that the total costs of LWRs were likely to be lower than the prime costs of many coal- and oil-fired plants.

Hence, although there was no immediate prospect of orders being placed for nuclear plants in 1972, had it been known that the LWR could be used there is good reason to suppose that firms would have been ready to do the job without monopoly and with a good prospect of growing business, provided that, as expected, the British economy would continue to pick up, which it did from 1970–73, so that by the end of 1973 the CEGB was envisaging the substantial programme it put, to their horror, to the Select Committee.

The adoption of the LWR involved a transfer of another technology from American to British firms, GE and Westinghouse, and the other American reactor manufacturers, Babcock & Wilcox and CE, would inevitably be the sources of the technology; willing sources, naturally; and British engineering stood to gain. GE were prepared in 1972 to organise the supply of BWRs in the United Kingdom using mainly British subcontractors; they had it all mapped out, and contemplated possibly making Britain the main centre of their nuclear activity in Europe. They offered delivery in four and a half years.[43] Westinghouse were not less active though they took a different route, and were already strongly centred in Belgium and France. They studied and cultivated the British market intensively.

It was highly probable that GEC would have been prepared to participate, without monopoly, if there had been a clear prospect of orders, and freedom for American and other overseas firms to look for business. It was a sign of weakness and unwise for a government to put a high priority on attracting one specific company and one specific manager. The bulk of manufacture would inevitably have been by British makers, who would initially have relied on American mentors. Experience in Germany, Japan, Sweden, Switzerland, Italy and Spain, to name but six countries, showed association with American firms did not destroy national industry: quite the contrary. When the United States has the best technology other countries thrive best by using it and building on it. It is a waste of resources to develop either a second-best, or something which will be no better than existing technology, in an industry where development costs are phenomenally high and there are more advanced systems in prospect. Britain's policy of exclusion contributed to the persistence of her slower relative national growth: more resources were used unproductively.

Foreign experience showed that while design, construction and component manufacture should be in some form international, expressed in organic links between firms, and offering competitive supply in all industrial countries (where the market is normally too small for one D and C company or supplier of major components), it was also desirable to have several utilities which made it possible to get greater advantage from competitive supply. Based on this, the 'alternative' policy in Britain would have included subdivision of the CEGB. This had been widely advocated. The appointment of the Plowden Committee by Varley in 1974 to examine the structure of the electricity supply industry implied that there were doubts about its efficacy; perhaps the Board's open advocacy of the LWR when they saw the light had raised the Minister's doubts.

The Plowden Committee shows the record of the CEGB as disturbing. But the Committee was impressed with the Board's efforts to set things right, and doubted whether 'further reorganisation...would bring greater benefits than would a period of consolidation'. They also said, however, that neither the Department of Industry nor the Electricity Council nor the CEGB 'presented a reasoned appraisal of the Industry's effiicency or showed that they made systematic and regular comparisons of its performance compared with that of other undertakings'. They urged all concerned, with the CEGB taking the lead, to fill the gap. The gap did not deter the Committee from concluding that 'the single organisation for power station design and construction must be preserved. Arguments for intellectual rivalry between different purchasers of power stations have been put to us. We find them unconvincing. The country can no longer afford to try two different approaches at once to major strategic choices such as that of a new reactor type' [as if the LWRs were new!]. And 'if there is a single approach it must be worked out by the electricity supply industry and its suppliers together...the dialogue between them is most likely to be balanced and fruitful if the electricity industry continues to speak with one voice to match the increased concentration in the nuclear design and construction and heavy electrical plant industries.' We are still in the world of getting the best brains together, as happened in the AEA when Plowden went there, and with the CEGB under Hinton, still in the make-believe world where foreign success and its institutional basis counts for nothing. No measure of your institution's efficiency has been given but none the less it is the best.[44]

(ix) MINISTERS AND MONOPOLY

We turn finally to the question how far experience in Phase II – and the greater knowledge of Phase I gained in Phase II – has confirmed or disturbed the criticism of the British organisation for nuclear power development presented in *The Political Economy of Nuclear Energy* in 1967. The summary of this criticism has been quoted above (p. 13). It is enough at this point to recall three conclusions:

(*a*) to give a monopoly of research and development in an advanced technology to a strongly centralised authority owned, financed, appointed and supervised by the Government is not a means of promoting rapid growth;

(*b*) if the object is rapid technical advance...it is dangerous to dispense with variety, the possibility of comparison, the stimulus of competition between autonomous groups, and the measure of profit, crude and blunt though these may be;

(c) a Minister in charge of a monopoly cannot make useful judgements or exercise control to protect the public when he has no basis of comparison, and the public is particularly unprotected when the Minister is nominally responsible for the monopoly's programme.

The greater experience and knowledge acquired by 1976 reinforced the earlier criticism. The concluding section of this chapter analyses first the ways in which understanding of the operation of the British system has been amplified. It is in almost all respects disturbing. Next the suggestion of the Select Committee in 1967 that the situation might be put right if the Minister had an expert 'technical assessment unit' is examined. Finally the argument that Britain could not have adopted the American system – under which the mistakes which led to the AGR disaster would have been impossible – is looked at again.

### (a) The Working of the System

The greater knowledge of the Dungeness B episode culled from evidence given by participants to the Select Committee and from the contrast between promise and performance is particularly enlightening. It was recognised immediately in 1965 that the extrapolation from 30 MW to 600 MW was excessive, the long-term economy of the AGR overvalued and the pro-AGR verdict biased.[45] In retrospect, with later evidence, the degree of exaggeration involved, alike in the assessment of Magnox experience and of AGR development itself and its development potential, and in the valuation of the APC tender, far exceeded initial fears. At the outset the exaggeration was all by the AEA; in the *Appraisal* it was by the Authority and the CEGB. They were jointly responsible for stating that there was no engineering problem involved in the APC tender not soluble by 1970. The Select Committee was assured in 1974 that this assessment had been scrupulously fair, and the CEGB described in great detail the thoroughness of the comparison between the LWR and AGR. Nevertheless in retrospect it is inescapable that the result of the work was grotesquely misleading both on the BWR and the AGR, the AEA's own product. The report gave no hint that there was no detailed design of the AGR tendered and that the staff of the APC was weak. The public were seriously misled by the report issued by this combination of two public corporations, and ministers provided no protection.

If the assessment of one participant that the *Appraisal* was 'honest but wrong'[46] is accepted (Hill's claim that it was 'meticulously fair' implies it was fair in result – it can at the most have been fair in intention), then those who made it lacked the necessary engineering knowledge and judgement or they suffered from exceptional and dazzling bias. There may have been a mixture of both.

Bias could certainly be detected in the presentation of the case for the AGR by the AEA in 1964[47] though the extent could not be judged precisely by an external observer. It would not be surprising if bias entered into the AEA contribution to the joint *Appraisal*. This under-lined a fundamental fault in the administrative structure to which the Select Committee referred lightly in 1967[48] whereby the AEA com-bined the function of being principal and almost exclusive adviser to the Government on atomic development with that of monopoly developer. There was a conflict of interest within the AEA once a com-parison was to be made between an AEA system and any other – necessarily foreign because the AEA had the domestic monopoly.

It was a conflict for the AEA as a whole. But the pro-AGR bias would almost certainly be greater for those engaged in its development, for whom the system's advantages were likely to be exceptionally clear, who might be less sensitive to the risks it involved, and whose future would be immediately brighter if it was adopted. It was therefore a further important lack of information for the public that the names and functions of those AEA staff members who participated in the Dungeness B *Appraisal* were not published. Gordon Brown told the Select Committee in 1969 that he had been one of the team, though not one of the front runners; the full extent of participation by leaders of AEA reactor development was not, I believe, published. The Select Committee did not press for this.

There was a further disturbing aspect of the Dungeness B episode which was less conspicuous. The best guides to the relative value of the LWR and the AGR before the *Appraisal*, were in the consortia. There were strong supporters of the LWR in the CEGB early in 1964; but if this support survived it went underground. Support for the LWR from the consortia became muted after the government decision in favour of the AGR, and further work on the systems ceased. Britain was cut off again from the most promising system. There was in this silence from the consortia on the LWR some deliberate dissimulation. Leading members of partner companies concerned – and the consortia – were persuaded that they could now only obtain orders for AGRs. How, one can only surmise. In 1963 Lord Coleraine complained in a House of Lords debate of the methods used by Hinton ('a man who thoroughly and completely knows his own mind, until he changes it') in persuading four consortia to arrange mergers to form two groups only. There would be no compulsion, it was not for Hinton to tell companies what to do – but there would be no orders unless![49] How-ever rough the method the object was a good one; though Hinton, who had exerted pressure on firms to form consortia[50] was hardly the best person to treat them roughly. He was of course no longer in the act in

1965. Suppression of discussion or further development of the LWR cannot be regarded as a defensible cause. A more recent parallel followed Varley's decision that the SGHWR should be ordered; some of the major supporters of the LWR were asked, no doubt verbally, to abstain from speaking in favour of the systems they knew the best.[51] The abstinence was conspicuous, but views had not changed.

This situation could not arise without a dual monopoly (or near monopoly) – of developer and buyer – with the additional feature in this case that both buyer and developer were supervised and controlled by ministers. It is a symptom or symbol of the British arrangements, designed to serve only the public interest, which have effectively concealed more from the public than can be concealed in a competitive situation.

Moreover, as the Select Committee became aware in 1967 (it did not follow the trail further in later reports), ministers were more likely to be controlled by, than to control, the AEA.[52] This was largely due to the fact that at an early stage the AEA, and the atomic 'establishment' before it, had acquired an exceptional status, as though they could claim an intellectual ascendancy. In a narrow sense they could make this claim, since for years they had a virtual monopoly in the new science-based technology and grew rapidly into an enormous group, of which an exceptionally large proportion were highly qualified scientists and engineers. At the peak, when they numbered 40,000, many were able, attractive, distinguished: it is one of the grim aspects of the story that valuable resources have been wasted. The group retained something of the aura of the heroic days of international scientific discovery, though those days were in the past; the science had become fairly routine, high grade 'master-craftsman' science, and the predominant work of the Authority was high-grade engineering. The world at large did not observe this. Harwell remained the symbol of the AEA in the popular mind, Risley had not registered.

The Authority had a number of scientists and engineers who were very articulate – they always made a splendid case to the Select Committee – and with its planning leadership it presented the development programme which it wished, and intended, to pursue in an appropriate setting of long-term energy forecasts (the Authority had to look further ahead, it was said, than the electricity supply boards) and cost-benefit analyses. It was powerfully equipped to 'put itself across' – especially to civil servants and economists (above all the macro-economists who were increasingly popular though inappropriate as departmental advisers) and to MPs and Ministers, and the media.

Vinter showed unwittingly how misleading apparent ease of communication can be when he gave evidence to the Select Committee on the Nationalised Industries in 1963. The Treasury discussed with the

CEGB the criteria adopted by them in making investment decisions. He remarked: 'Our relations with the Generating Board are probably more intimate than with other nationalised industries. We tend to talk, subject to the difference of our jobs, rather the same language.' The Treasury did not, he said, try to take on the Board on 'generating engineering' – not many people could take them on 'on their own ground', just a few consulting engineers. Plant manufacturers he did not even mention. Despite this limit on understanding, his 'impression as a layman' was that 'they do a very good job on this.' How could he tell? Their policy in ordering was privately strongly criticised at the time; in retrospect it seems generally agreed that this and the organisation of their field services were highly unsatisfactory.[53]

Departments probably spoke more completely in the same language with the Authority than with the Board. Compared with the Authority the consortia (who would not come into the Treasury picture but *were* in the purview of the sponsoring Ministries of Technology, Industry, Power, *et cetera*) were small and undistinguished, relying until 1964 for the starting point of their design and development on AEA work (and then for a short spell upon work by GE or Westinghouse). The consortia lacked the breadth of the Authority, the expertise that comes from numbers, the scientific heritage: they were only engineers, private enterprise engineers. As late as 1973–74 the Select Committee thought it inappropriate that the single D and C company should be represented on the Minister's new advisory board because its view could only be wanted on technical matters, which the CEGB and officials could deal with, whereas if it were on the Board its view could be put on all other problems raised.[54] 'Is it not undesirable', Palmer asked Boardman 'that the company with its commercial interests should serve on a body ... which is intended to give the Government the best and most objective advice?[55] The AEA, the BNFL and the CEGB would qualify for the Advisory Board in Palmer's view – and he probably spoke for the whole Committee – as capable of giving an objective view untinged with commercial interests.

No clearer expression could be asked for of the blinding effect of what may be called the 'Upstairs–Downstairs' view of the British industrial economy which confused the debate. The publicly-owned, highly centralised bodies were upstairs (the CEGB sometimes only half-way up), along with civil servants and academic advisers and the ministers and MPs who happened at the time to be in the ascendant. These sought no profit out of decisions (and took no risk): for them there could only be the 'public interest'. The commercially-minded firms were downstairs, motivated one might suppose only by profit, and clearly belonging to a lower order.

Yet there was now almost two decades of experience in which the combination of publicly-owned monopoly bodies, appointed and supervised by ministers who were aided by civil servants and who were responsible to Parliament whose members were believed to be zealously watching to see that public money was well spent, had resulted in a series of disastrous misjudgements of the most fundamental kind. These mistakes were admitted; wrong systems, premature programmes, wrong methods of developing, grossly excessive extrapolation, bad choice of contractor. Their cost ran into several billions, which weakened the British energy position and destroyed the nuclear plant industry it was intended to create, secured none of the substantial export business in nuclear plants, components and systems which developed, and slowed down general industrial growth. The groups responsible nevertheless sought understandably to hold on to their functions – if in modified forms – 'in the public interest', and thereby to hold on (though they may not have seen it in this way) to the power, position, status, honours and prestige which they conferred.

It was not surprising that by the early sixties the best guides to a competitive reactor were to be found among the consortia (though for a period the CEGB was in contention). They sought autonomous individual development. Their ultimate success in (as they hoped) international markets must rest on a competitive reactor. They responded – as the CEGB did initially – to the Oyster Creek landmark. Profit, if they secured any (no consortium did) would be a measure of success in providing plants which reduced generating costs. But this was effectively disguised in the distorting mirrors of contemporary controversy.

### (b)  *A Technical Assessment Unit?*
In 1967 the Select Committee recommended that since ministers appeared to lack an effective technical check on the activities of the AEA and could not therefore judge whether AEA policies were necessarily the 'best', a technical assessment unit should be set up to advise the Government on 'the merits and prospects of particular projects proposed ... by the Authority'.[56]

They wanted to improve the system of 'minister plus monopoly', sensing that major errors might have been made – possibly the choice of the gas-cooled technology. It was enlightened of the Committee to recognise that something fundamental in the core of the organisation, the relation of ministers and the AEA, had gone amiss, and regrettable that they did not return to it. The way in which a technical advisory unit could help was not explored, and obviously it had to be. This leads to a more general question. Most people agree that the arrangements set up in 1953–55 led to major errors of policy, and that these errors were perpetuated and new ones added by subsequent ministerial, AEA and CEGB decisions. Does this experience lead one to suppose that such a series of errors could have been avoided if there had been additional administrative devices to make the 'Minister plus Monopoly' situation work better? The errors could hardly have occurred under the arrangements in the United States, Germany, Japan or Sweden. Could they equally well have been avoided under the United Kingdom's 'Minister plus Monopoly' arrangements, if these had been fortified in some way? Could they be made fool-proof for the future?

Ministers could, of course, in 1952–55 have supported Harwell plus industry against the Risley takeover of all industrial-scale development. They could have abstained from the programme in 1955, turned down the narrow front, encouraged work on the LWR from 1956–57 onwards, brought firms in to develop and design plants with direct state grants, given more freedom to the CEGB in its dealings with the AEA, and to other potential developers and suppliers, and so on. All these steps – assuming ministers decided to do something – would have been steps towards the American system. The ministers would have been withdrawing from making technical decisions and would have been avoiding monopoly and lessening the role of ministerial appointment. Ministers would not have needed expert technical advice to have taken such actions – shrewd lay advice would have been enough.

But this was probably not the kind of advice which the Select Committee expected its technical assessment unit to provide. Most if not all members probably still adhered in varying degrees to the idea that ministers, and even MPs in general, knew what the national interest was in any matter provided they had the necessary information; and this they could ask for. Ministers could ask for it in regard to the AEA; and they could choose an appropriate group of experts, as they chose their scientific advisers.

*The Times* remarked in 1946 that 'Members of the present government appear to regard the public interest as their private property, or even prerogative; yet Ministers tend to be tongue-tied when they are pressed to explain how they proposed to measure it'.[57] Though Oliver Lyttelton remarked wisely in 1949 that the national interest was not something that ministers instinctively recognised and protected, the Conservative Government in the early fifties was ready like its predecessors, if not quite so frequently, to give powers to ministers to decide the national interest without giving it any definition: they did it over steel in 1953 ('The Government of the day and Parliament alone are competent to judge which is and which is not in the national interest', said Duncan Sandys[58]) and as we have seen the constitution of the AEA fell into the same pattern. Politicians of both parties and heads of consortia continued to agree in the 1960s that 'nuclear policy must be determined by the Government', and this covered decisions on how big a programme there should be, what reactor should be chosen, and so on.[59] Hailsham in 1963, as Minister for Science, told the House of Lords that he had 'made it his business to advance as far as he could the progress of the AGR', in which he personally had 'great confidence' – though he thought the CEGB reasonable in waiting for it to operate a little before deciding to buy it: they 'should see an ounce of practice as well as a pound of theory' – an illuminating sense of disproportion.[60] The *Second Nuclear Power Programme* of 1964 stated that the Government would decide 'the type of reactor to be built'. The next – Labour – Government did decide. The Select Committee in 1967 was, I think, envisaging that an expert assessment unit would contribute to this kind of process, advising the Minister in evaluating AEA development work and R and D projects advising him how to intervene, but not advising him to abstain from intervening in programmes, reactor selection and so on.

It was not as simple a concept as it sounded. Who, and how numerous, would the assessors have been? Who would have chosen them? Would they have been drawn from the AEA and the consortia? How senior in status, and how authoritative, would they be? Would they have merely vetted the AEA proposals or would they have submitted alternatives or proposed alterations? Would they, had they existed, have provided an alternative Dungeness B *Appraisal*? Would they have contacts with foreign experts and firms? Could they commend an American system to the Government? Were the ministers expected to stage confrontations between the presumably small body of assessors and the vast Authority – and then decide? Would ministers in effect always accept the unit's advice, so that it would in effect become a new technical monopoly? With whom would responsibility have rested? Edmund Dell referred in *Political Responsibility and Industry* to the political as well as the technical difficulties which would face a minister who has two independent sources of information – 'the full fury of the lobbies will be let loose'.[61] And, he added for full measure, to have advice from two sources 'provides no assurance that he will choose wisely'. It provides no assurance moreover that he will know all there is to choose from.

This leads back to the central question in the context of this chapter – why should the Minister make decisions on utility programmes, and the choice of reactor for utilities, and on the number, constitution, research programmes and foreign connections of D and C companies who were to construct plants for utilities, and why should he appoint the top managers of the atomic industry? In these respects the British arrangements differed conspicuously from those in the United States and most other Western industrial countries. Obviously ministers could have been more informed than they were – the Select Committee was right in this. But what would be the gain?

In the beginning, in 1953–55, ministers alone could decide whether the development of nuclear power plants should be promoted, and when they decided yes, they had then to decide how the promotion should be arranged. They chose the highly centralised methods and policies which gave ministers a continuing role in making major decisions and appointments and created a monopoly which excluded the advantages of competing autonomous initiatives. By 1973, as Weinstock said, there had been a great expenditure of resources, with little to show for it.

When the Select Committee recommended its technical assessment unit, and still more when the Conservatives later made their reorganisation and Varley made the most assertive of ministerial technical choices, the situation bore little resemblance to that faced – with disastrous results – in 1953–55. There was no longer the question of how to organise development of a competitive reactor – the LWR was competitive and developed. The AGR would probably be competitive with coal but not with the LWR when its development difficulties had been overcome. Ministers would still have to decide whether, and by how much, and in what ways, to support development of more advanced reactors and of ancillary fuel industries: problems faced by governments in other major industrial countries. But utilities in Britain as in the rest of the industrial West (and in the industrial East too) could assess the LWR from considerable and rapidly growing experience, and could assess, though with less certainty, the potential economics of a few rival but probably higher-cost systems – Candu HTR, SGHWR. Because of the scale of use of the LWR many of its components were standardised and production techniques were well advanced and organised – and transferable.

Nevertheless, in this situation the Conservative Government in its reorganisation retained the power of the ministers to decide which reactor utilities should buy, and retained ministerial power also to create a single D and C company which would have a monopoly of the business, subject to ministerial control of its reactor choice, the scope of its research and its foreign relations. Ministers clearly intended to institute another four- or five-year plan for commercial nuclear power plants. Varley, in the succeeding Labour Government, introduced a four-year plan and decided that the utilities should use the SGHWR.

These actions, particularly Varley's, implied a refusal to accept the view that the LWR was established as competitive for British conditions, even if it was competitive everywhere else. Still less was it accepted that it was established as the most competitive reactor with the lowest costs. On the contrary the SGHWR was accepted as equally or even more competitive. One may ask whether, if ministers had had a technical assessment unit, their judgement on this point would have been different. Varley was able to consult the newly created NPAB, but even without this he knew that the CEGB and NNC were convinced that the PWR was not only competitive but offered by far the lowest available costs, and that the SGHWR would take much longer to bring into quantity use, was likely to have costs above the PWR and involved much more uncertainty. He no doubt also knew (it was not published at the time) that Franklin, the Managing Director of BNFL, also supported the LWR, and Hill had said publicly that the PWR could be used as a stop gap while the development of the SGHWR or the HTR was being completed. Varley said in Parliament that he thought the majority of the NPAB members supported the Government's policy; unless some of the members expressed different views in private than they had in public the majority who supported him must have been made up mainly of the pure scientists and the lay members. Varley knew that the majority of component manufacturers wanted to see the LWR adopted. Tombs alone supported the SGHWR publicly as likely to be available in large numbers faster than the PWR.

Varley, expressing government policy pitted himself against the majority of nuclear engineers and decided that he knew what was technically good for the CEGB better than the Board knew itself. Is it likely that it would have made any difference if there had been a technical advisory unit? In the circumstances it seems highly doubtful. Varley had only recently come to the job. He could not have known the unit for long and it would probably not have had time to establish any kind of ascendancy over him. Since he was prepared to dismiss the views of the majority of senior nuclear engineers, he would presumably have been prepared to dismiss those of the unit.

Varley in explaining the choice – which he did not do in a systematic or sophisticated form – defended the SGHWR because it was a British technology, and because it would be publicly acceptable since no one spoke of it as unsafe. He pointed also to the support he had from the IPCS, as though this was support from an independent body of technical experts, whereas it was from the pressure group representing the AEA staff who had developed the SGHWR. There was no pressure group for him to worry about representing those who would have had jobs on the LWR had it been ordered. His decision to disagree with, and disregard, the majority of nuclear engineers – though his own view could have no objective value – thus fitted a convenient political course, offering 'short term political dividends', something for the next election which clearly interests ministers (not only of one party). Like the aluminium smelter policy of 1968 it 'held out a promise to excitable nationalist opinion'[62] – which Dell saw as 'imperialist opinion' in the nuclear context. It avoided a confrontation with environmentalists and may have brought in some environmental votes; it avoided confrontation with the IPCS, it fell in with coal-mining views and it went along with the Select Committee. There is no reason to suppose that a technical advisory unit would have diverted Varley or the Government from such temptations. *The Times* described the decision as 'one of the worst political decisions in the industrial field for many a decade'.[63]

Dell, in a more general context, remarked: 'We are terribly dependent on the judgement of Ministers.'[64]

## (c) The Scope for Competition

Edmund Dell also surprisingly argued that 'there is no rational reason for refraining from intervention in the fact that intervention is often badly handled'.[65] He could say this merely meant there was no reason against intervening in some form provided your particular intervention was not badly handled: he seemed to imply *his* interventions had been well-handled.[66] But if certain forms of intervention are almost always 'badly handled' there is a strong 'rational reason' for refraining from intervention in these forms. If there are plain reasons why such forms are likely to be 'badly handled', and there are alternative ways of handling the situations involved which experience shows are likely to work well, the 'rational reasons' against government – ministerial – intervention in such forms are overwhelming. The 'terrible' dependence on ministerial judgement could be avoided; and if 'net positive gains to the community' (Dell's definition)[67] is taken as the object, the avoidance would be 'in the public interest.'

Nuclear power policy is an area where these three conditions were – and are – satisfied. First, the forms of intervention chosen have undeniably been 'badly handled'. The choice of policies and of persons to implement them, the grant of monopoly, the exclusion of competitive indigenous initiative and the prejudiced rejection of foreign initiative, all resulted from a flow of ministerial interventions. It was not a case of one initial misjudgement; the misjudgements have been by both parties – the initial ones Conservative, the more recent major ones by Labour governments.

Second, there is good reason to expect British arrangements to work 'badly'. The ministers who exercise the powers – to repeat the earlier analysis – unlike the chairmen of big companies, are temporary, they come to the job with no training and no knowledge of the specific problems involved, or of the general type of problem, they have not had to rise in competition with other budding top executives, they do not know the chief members of their civil service staff as a company chairman will know his staff, they do not remain on the job for long, they may have distracting interests, they will not be in the same office when the results of major decisions become manifest. In brief, they have no knowledge on which to base their decisions, no experience to guide them in this kind of decision-making, no incentive in possible gain or loss of reputation, position or wealth to make good decisions since they will not be accountable for and will get no significant credit or discredit from remote ultimate results; and finally, they have no temperamental inclination to concern themselves with long term prospects: in politics a week is a long time, and they are adapted to this, and are likely to be most alert to 'short-term political dividends', all the sorts of things that Varley called in aid in support of his decision.

Dell (I make no apology for quoting him again because a committed contemporary practitioner rarely sets out as he did to make even a half-detached analysis of these problems) after a bleak survey of ministerial and civil service capacities and attitudes concluded: 'ministerial responsibility does not in the case of industrial policy provide the protection that constitutionally it should provide.'[68] This was plainly right, yet by focusing on the ambiguous concept of responsibility it concealed – unintentionally – what was most important, and suggested a way out which is not available. Lack of knowledge and experience made – and makes – ministers and their senior civil servants incapable of choosing reactors, managements and decision-makers in the nuclear power field on the basis of objective judgement. At its most austere level, a sense of moral obligation, responsibility might be expected to have led civil servants to advise abstinence from doing things which neither they nor ministers were qualified to do, and ministers and governments might have been expected to refrain from intervention. There are few signs that this austere level is ever attained: Dell does not entertain it. 'Big government is here to stay', which seems to mean that there are pressures from sections in the electorate for intervention, and politicians seeking power will look for scope for popular intervention, and ministers will intervene more and more, irrespective of their capacities.

The way out, which is sometimes now suggested, is that responsibility shall be pinned on particular ministers or civil servants: they shall, for example, be called on to explain in detail with all relevant statistics or records (and possibly with dissentient views stated) possibly to a Select Committee, why a technical or technically-based decision has been taken or some appointment made or an institution created. Having done this they shall be counted as responsible and answerable for it. The mere fact of being forced to explain (and to present dissentient views) will, it is suggested, induce greater care and responsibility. I believe this is illusory. No minister or civil servant is likely to find difficulty in explaining a decision or explaining away dissents – they do this as things are. They are all experts in 'justification by words'. A minister is unlikely to be made more conscious of his lack of qualification to make a decision or appointment by the fact that he has to defend it; a senior civil servant would be covered in respect of major decisions by the fact that he was doing the best for his minister. This device of requiring an explanation on the grand scale and placing responsibility can only, at the most, discourage bad decisions (and it is improbable that it will do this), it does nothing to make good decisions likely. And if a bad decision or bad appointment or bad institutional change occurs the fact that responsibility has been fixed will be of relatively minor importance; there will be an important loss of time and money in development with presumably nothing to offset it. The 'minister plus monopoly' structure works very differently, in this respect, from competition – as will be seen.

Third, the alternative solution, adopted by the Americans, of encouraging development of nuclear power plant as quickly as possible 'within our normal economic and industrial framework' had worked well, both in the United States initially and later in modified forms in most of the Western industrial countries. It was not surprising that the American type of arrangements succeeded, because under them the leading manufacturing companies who were experts in this type of advanced engineering development were encouraged – with initial subsidies – to undertake competitive autonomous development of systems selected by them from AEC initial experiments. The competition allowed utilities to compare systems and developers. It was open to them to participate in development, to back their fancy, and to contribute finance; the decision to buy plants as 'commercial' (and to select a system) rested exclusively with them and, like their participation in development, was based on commercial considerations only. The conditions which made it – and make it – impossible for ministers and civil servants to make good decisions and appointments except by accident were reversed. Those who in these large concerns make major decisions and appointments and organise research, development, manufacture and sales are permanent executives, are trained for the job, have risen to the top in competition with other budding top executives, do have (between them) knowledge both of the specific and general problems involved, do know their staffs well and have straightforward relations with them, do expect to remain on the job for a long time and to be in office when results become manifest and expect, in the interim, to be guiding development with a perceptive eye on internal and external progress. They had in short the knowledge, experience, incentive and temperament for long-term development which was (and is) lacking in the British 'minister plus monoply' arrangements.

The principal executives involved in the company and utility decisions and management were unambiguously 'responsible'; they would be accountable for such failure or success as occurred, and would lose or gain, as the company or utility as a whole would. This was an important element in the stimulus and encouragement which the system gave to speed and quality in development and relative low cost in the product.

While encouragement of parallel developments would tend to stimulate speed, it also meant that, if one line of development proved slower than was hoped, the results of an alternative were likely to be available.

The delays which have characterised British projects – the AGR has been outstanding though not isolated and the SGHWR is following the pattern – would have had much less impact. If the LWR had been taken up by an autonomous firm in the late fifties in competition with the development of the AGR and SGHWR and entirely separate from them, the United Kingdom need not have fallen behind Germany France, Italy, Spain, Japan and Sweden in the nuclear power plant industry. She might have done so even with the LWR; but with the concentration on the indigenous reactors it was inescapable.

Finally, far more information was readily available about the American industry than about the British, where it was possible to conceal what was happening, to claim successes which had not been achieved, and to promise early successes which were decades out of reach. Public-owned corporations in Britain published – possibly in all good faith – misleading, incomplete and incorrect information, conspicuously over Dungeness B. In a competitive system the existence of rivals would have made this impossible. The myth of British leadership has continued to be sustained. Even in June 1976 Mr Callaghan, as Prime Minister, asserted that 'Britain had pioneered the use of nuclear energy for electricity generation', and then went on to say that there was no country in the world 'which could claim such a long experience of tried and tested systems as those now in operation'.[69] All those in operation had indeed been tried, tested and found wanting; all the systems needed radical redesign (which was said to be nearing completion for the SGHWR) if they were to be used again, all were regarded as having inherently higher costs than the LWR, and this the world's one successful tested system had not yet been tried in Britain at all. The assertions were made at a conference arranged by Benn, it was stated, 'as part of his "open government" philosophy'. No one, not even Benn, who as Minister of Energy presumably knew these facts, troubled in the interest of 'open government' to correct the Prime Minister.

This underlines what may seem a paradox to those accustomed to suppose that ministers and others who secure industrial power by the political route, or as political appointees, only have the public interest as an object and can infallibly determine what it is. In practice the social responsibility imposed on decision makers and executives in the companies and utilities under the American arrangements has been conspicuously greater than the social responsibility of the decision makers from the ministers downwards and the executive under British arrangements.

This was inherent in the nature of the systems. First, all the people involved in the United States were working within the field in which they were competent and expert; this was not so in Britain. It does not even seem to be regarded as important in Britain – not a matter of social responsibility! Next the line of decision-making was shorter and unpolitical in the United States; the possibility of delays in making key decisions – except over licensing – therefore less. In the United Kingdom there have been notable delays over decisions – conspicuously over the LWR both in 1964–65 and 1971–74.[70] Time is of immense importance in this development, but rarely mentioned in Britain. Third, under the American arrangements there is practically no possibility for drawing a veil over what goes wrong, or misrepresenting for years on end the whole trend, as has happened in Britain, where the State as single buyer could even discourage the firms who knew the facts from revealing them to the public. In the United States many autonomous buyers, advised by several competing architect-engineers, are checking and evaluating developments all the time, critically, and applying vigorous commercial standards. Fourth, and underlying the other factors, was the competitive process itself. Success would be measured by profit, but it would be achieved only if a firm was able in competition with rivals to achieve speed, quality and low cost in developing a competitive system which utilities would buy. To do this was everywhere regarded as the 'public interest', in the United Kingdom as well as in the United States. There was no mechanism in the British system to encourage speed, quality and low cost in development from domestic sources, and external forces were not (and are not) allowed to operate. In the United States competition made possible open continuous comparison by utilities and consultants – and the AEC – of the achievements of the major contenders, and this provided stimulation and incentive. There was no possibility in America of getting a stay of competition from a more successful rival. This occurred in Britain – by the Dungeness B decision in 1965 and by Varley's 1974 decision for the SGHWR. Responsibility in responding to the stimulus in the United States was in fact social responsibility, an acceptance of what under the system is inescapable, 'public accountability', which under British arrangements could be avoided or deferred by the principal architects of policy and organisation.

We return to the question, why in the face of the experience in almost all other Western countries did the United Kingdom move, even under the Conservatives in the Heath regime in 1972–73, to greater centralisation with the creation of a single D and C company whose Board must be acceptable to the Government, and more specific far-reaching ministerial powers such as Varley exercised in the SGHWR decision?

It has often been argued that the American type of organisation was impossible in Britain because there were no firms on the United States scale and because no British firms had the experience that many of the American firms had of acting as contractors to the AEC in building and managing their facilities, and because British firms refused.

Certainly two firms in Britain refused after the war to manage the building of the diffusion plant though one, ICI, contributed a great deal to it in technology and design. Carrington's remark in 1963 was representative: 'basic research and the development of nuclear power systems were so costly and such a long-term operation that it would not have been undertaken except at public expense by a body such as the Atomic Energy Authority'.[71] As I commented in 1967, the need for public expense did not require a body such as the AEA – why not one like the AEC? Or as in Germany no such body at all?[72]

British electrical engineering firms wanted by 1948 to be involved like the American companies in reactor development, and Harwell clearly recognised their participation as a good basis for development in 1950–53. Cockcroft had himself been an apprentice in Metro-Vick, who started him on his university career: he knew the strength of these companies from the inside.[73] (It is indicative that the Director of Research and Education of Metro-Vick from 1953 to 1961 – by then it was part of AEI – was Willis (later Lord) Jackson, who had been Professor of Electrotechnics at Manchester University and became Professor of Electrical Engineering at Imperial College in 1961.) Hinton's takeover at the end of 1952 was a major change of course; possibly it was this that P. M. S. Blackett had in mind when he said the United Kingdom had 'inadvertently' taken the wrong course in concentrating its atomic research and development in government establishments instead of in companies.[74] He like Cockcroft, recognised the companies could have done it – although, being much smaller, they could not have put in the mass of resources which the companies in the United States did. (But nor could ASEA, AEG, Siemens or the Japanese firms.) That they were willing to be involved was made abundantly clear by the readiness to form consortia, which impressed the AEA in 1955–56. They were willing because business was available and looked like expanding.[75] By 1958–59 the consortia wanted to spread their wings and have more autonomy, possibly as holders of American licences; they were repulsed. By 1964 two at least were anxious to work under licence with American companies – the natural course of development now in view of American leadership – and they were given encouragement by the CEGB. But they were successfully frustrated in 1965. By 1968–69 Benn recognised the desirability of having competitive companies with the development scope of the American companies but neither he nor his IRC advisers recognised the radical adaptation needed in view of the failure of British technology. By now companies in this field in the United Kingdom, if they were going to be competitive, needed to be international for two reasons: first, there was no satisfactory indigenous reactor; and, second, the British market for electric power, and hence for new nuclear plants, was growing more slowly than markets in more flourishing economies free from the disturbance of North Sea gas. These difficulties were not fully appreciated either by Benn or by the Conservatives. By this time the only practical technological base for a world competitive company in Britain was the LWR, as this was the only exportable reactor. This was not acknowledged, so as no export or home orders came the illusion that a competitive company solution was not possible in Britain gained still more support, though Palmer for a brief interlude recognised it was an illusion, when he suggested that an American company might provide the CEGB with LWRs if it insisted on buying some, as long as a major part of the actual work was done by British sub-contractors.

The increase in centralisation did not occur in the United Kingdom because the alternative, which was adopted with success in almost all other major Western economies, could not have succeeded in Britain. It resulted from the domination of policy making by strongly entrenched groups of persons within the industry (including all stages from ministers downwards) in the highly centralised form in which it had been developed, and by political and economic concepts and ambitions which were unrelated to industrial experience.

Those who managed the AEA naturally resisted the granting of development autonomy to consortia and the use of American technology, which would have exposed the Authority's fallibility and failure and lessened its size and influence. They aimed instead, successfully, at spreading their tentacles more widely. Even the Select Committee in 1967 recognised that the heads of the Authority would be biased in favour of their own activities.[76] What was important was that they found so many allies for so long.

They started with great advantages. The AEA was, in the power struggle, 'Upstairs'. Leaders of both major political parties had succumbed to its glamour, its promises, its programmes, its personalities. It had for long a near monopoly as a source of information: its internal critics had no publicity, external British critics no status, early foreign critics no platform in the United Kingdom. Its long-term forecasts and cost-benefit analyses made it look to politicians, the media and some economists a planning oasis in the industrial desert.

This was a mirage, and gradually seen as such,[77] and in the early seventies the errors in forecasts, reactor building programmes and development methods were acknowledged, but the organisation weakness of which they were the symptoms were hardly mentioned – except the position of the consortia which was, illogically, used to boost more centralisation and more power for the AEA. But there was no rigorous examination or public discussion of the sources of the AGR troubles or of their antecedents or their cost, nor was the fact that under the American-type organisation the troubles could not have occurred, mentioned.

This was symptomatic of the state of Parliament's and the Government's handling of industrial problems, and could easily be explained in political terms. Both Conservative and Labour Governments had been deeply implicated in the development. No politicians would welcome a probing which would underline ministers' fallibility. Since the thrust of both main parties in industrial policy in the early seventies was still to assert ministerial ability to intervene 'in the public interest', it would not be helpful to them in this cause to emphasise that in the most glamorous instance of government industrial promotion the objective counterpart of ministers' assessments of the public interest was a monumental loss, the largest waste of resources from government action so far recorded, resulting from policies and developments recommended and carried out by a nationally-owned, government-appointed monopoly. Many MPs, notably Palmer, had gone along with the various governments in their choices, so that they in turn had nothing to gain from a deep probe. The civil servants and economic advisers who had been involved were in the same boat.

The episode underlined the lack of accountability in ministerial intervention which Dell showed forcibly: the immense scope for concealment, the irresponsibility of ministers and civil servants in designing the new organisation without understanding the operation of the old one and the successful operation of alternatives elsewhere, indeed the whole 'unconscious levity' of government intervention in industry to which we have referred earlier.

It is important for perspective that the AGR episode was not an isolated instance of 'levity' and loss, though it would be inappropriate to analyse the others here. Dell said clearly that government intervention had often been 'badly handled'. Concorde provides the largest and most familiar parallel to the AGR,[78] but post-war industrial history is studded with large investments blessed, sponsored, and where necessary – indeed usually – subsidised by ministers; in nationalised industries whose managers were ministerial appointees, and in private firms persuaded or cajoled into conforming with government plans and strategies rather than their own. These have commonly resulted in waste through misjudgement and inefficiency in technical conception and management and through the creation of excess capacity, expensive subdivision and dispersion of production, encouragement of high cost and inappropriate locations. This in turn has contributed largely to the relative decline of British industry internationally in terms of production costs, productivity of capital and labour, and in some areas innovation, in others quality.[79] The decisions have normally, often pretentiously, been justified by some possible, often ambiguous and always unmeasured, short-term gains. But no careful assessment of the results individually or collectively has been made: no pretence at *ex post facto* cost-benefit analysis, no study of the organisational background, and the qualifications of the decision makers, thrown up by the organisation, has been embarked on. One gets the impression that those responsible for policies (ministers and their civil service and economic advisers) regard the adoption of their policy as the 'end of the affair'; its outcome, which is the important thing socially, is not of interest to them. In political terms this is understandable: to secure the acceptance of a policy is an exercise of power for the minister and his advisers, it often results in a transfer of power to friends and supporters, and for the government it is expected to gain some immediate additional voting support. The myth of accountability nevertheless survives.[80]

The counterpart of the failure to analyse the institutional sources of the AGR disaster and its precursors was the absence of interest in the scope for competition. The decline of such interest as there had been in 1967–69 could be presented as pragmatic: all – or almost all – were now agreed that the domestic market was too small for competition. But the way out of this difficulty had been familiar for years: to encourage international competition and international companies.[81] This would have had the added advantage, as Aldington pointed out in regard to the NNC plans for the PWR, of bringing some American and possibly other advanced engineering practices into Britain.[82]

The absence of interest betokened a widespread failure to recognise what should have been plain enough – that, unlike the AEA, a competitive company could only survive in the business if it could do what it promised the utilities: produce a competitive plant. Its judgements of competitiveness, of its own and alternative products, had to be objective; when its judgements were allied with those of utilities they were the most authoritative assessments. If it was wrong it suffered, financially and in loss of business; and this would involve a setback for the management, by and large a professional management, the policy makers. This was inherent in the system – an inbuilt safeguard sometimes called automatic, but only automatic in the sense that under these arrangements firms must keep promises if they are to survive, and could only succeed if they are responsive to the market, to potential customers whom they set out to satisfy. Under British arrangements the situation was entirely different. The top management, who made or contributed to major decisions, made promises and assessments and forecasts, recommended programmes and organised development, did not suffer from the long series of misjudgements, mishandled development, unfulfilled promises. Ministers with civil service aides continued to make major decisions, the heads of the AEA extended their powers and responsibilities. The chairman was also, incongruously, the Government's chief adviser on nuclear matters; the vice-chairman was also Chief Scientist of the Energy Department, but not wearing his nuclear hat[83] – a modern pluralist. The Authority,

owned the fuel company, was the British partner in various international companies, and was 35 per cent shareholder in the single
monopoly D and C company in which former AEA engineers held
most of the principal positions. It was natural for those at the top of
the nuclear industry – ministers, civil servants and members of the
AEA – to seek (in varying degrees according to their position in the
hierarchy) not only to retain but to extend their authority and influence,
obtain greater power, wider functions, promotion, continued job
security, and comfort (Benn had remarked on the comfortable position
of AEA R and D workers[84]), to win greater respect, and distinction, to
improve their status; and for the politician to continue to get political
advantage out of the industry's woes. These, excluding the politicians'
aim, were normal objects of professional managements who controlled
the policies of competitive private companies, and in this context were
recognised as self-interested, which was often presented as the anti-
thesis of public interest. Economists have never tired of quoting Adam
Smith's dictum that 'people of the same trade seldom get together even
for merriment and diversion but the conversation ends in a conspiracy
against the public';[85] even in the bicentennial year they had not got
round to recognising that all other groups with common interests have
similar habits. What distinguished the private from the public sector in
this respect – in the nuclear industry, but also of course more generally
– was that in the private sector the desired objects could only be
secured if the institution was successful.

The postponement of the SGHWR programme after the AEA had
advised it should be abandoned provoked a fresh round of discussion
of the way in which the British nuclear power development was
organised. This tended to be overshadowed by the controversies over a
plutonium economy and on the political level it did not lead to a
radical perceptive probing of the 'minister plus monopoly' system,
though on the academic level there was an encouraging beginning.[86]

When the Select Committee started to examine the status of the
programme in August 1976, with Benn as first witness, they looked
outside the political world to see where blame should rest. The
favourite ploy was tried – that industry, meaning private industry, was
responsible – but Benn rejected this; industry had not 'dragged its
feet': and the consortia myth could not be invoked – there had been no
order, the AEA development team had been incorporated in TNPG,
there had been no costly time-wasting transfer of know-how, no competitive designing, no parallel manufacture without 'replication'. It was
complained that the CEGB had not placed an order in 1968–69. But
when the AEA said the SGHWR had 'no feature which [when the
system was fully developed] would give it an unambiguous advantage
over any existing developed system',[87] and that its development was

fifteen years behind the LWR, the Board's decision had clearly been right – even if partly for the wrong reason. All the more so since the capital cost of the system was likely to exceed the LWR cost and the thermal efficiency would be less.[88] There could be no justification for a long costly development of a system which in the end would give no technical or economic gain.

Responsibility thus rested unmistakably on the AEA who had advanced the SGHWR as the best water reactor. It had rested on the AEA with Magnox and the AGR, but this had been obscured – now it could not be. The Authority's stance on the SGHWR had already been ambivalent in 1973–74. The Select Committee with its preoccupations had not observed this, and appeared shocked when the Authority down-graded its progeny in 1976; but what else could the AEA do? Benn said this was 'bound to impact on the credibility of other systems that the AEA had sponsored ... It must encourage us to scrutinise more carefully proposals that come from that source.'[89] How would it 'impact' on the credibility of ministers, following so devastatingly soon after their decision in 1974, a ministerial decision based on a much pondered rejection of the weight of engineering advice? The minister – and the Select Committee – left this question severely alone. Benn mentioned he had not been 'directly involved' in the choice (Varley had been the Minister), – which underlined the limited significance of ministerial responsibility.[90] Palmer stressed that his Select Committee had not *recommended* the SGHWR.[91] Both implied that the Minister would go on deciding and the Committee would go on expressing technical preferences. The Committee had advised the Minister to get better advice in 1967. It was probably the wrong message. How would he 'scrutinise' AEA proposals more carefully? Only one member of the Committee expressed any incredulity about ministers' capacity to make this kind of decision.[92] Would he obtain nuclear proposals from other sources and compare them? The indications were that he would *not* break the AEA monopoly, he spoke of competition in this context as 'absurd'.[93] Would 'scrutinising more carefully' just mean treating AEA proposals with more scepticism when weighing them against proposals for coal?

The crucial questions were not asked. Why should ministers and Parliament be expected to make this kind of choice well? Were two decades of mistakes just bad luck? Why did the countries – most other countries – who adopted different forms of organisation, where the major decisions were not by politicians, where reactor choice depended on utility/manufacturer negotiations, where there were usually several utilities and some competition between manufacturers, prove successful while Britain failed?[94] Why should an increasingly centralised system which provided no rationally conceived incentives, no mechanism for comparison, and no probability that good judgement would be rewarded, be expected to succeed? Why did no one acknowledge that the CEGB and the NNC, the manufacturer/utility axis, had been right in 1974?

## NOTES AND REFERENCES

1. The *Observer*, London, 25 May 1975, for example, had an article on 'Britain's Billion Pound Flop'. It estimated the cost much too low. Academic economists began to get on to the problem: I was asked to comment on two cost estimates in spring 1976.

2. Hill in evidence to the House of Commons Committee of Public Accounts, 22 January 1975, Q. 39. 'We are really proceeding much more cautiously than we did in the early days of nuclear power.'

3. *SC House of Lords, EEC Energy Policy, 1975*, Q. 787. 'I would guess we will have only two fast reactors operating at the turn of the century at the present rate of progress.'

4. Above, pp. 154–5.

5. See Hinton, *SC Science and Technology, 1974–5*, p. 183 and above, pp. 165–167.

6. AEC, *Civilian Nuclear Power*, 1962, p. 2.

7. Above, pp. 6–7, 118, 187 (n. 31). There was a useful summary by the AEC of the development programmes past and planned in JCAE, *Development, 1960*, pp. 125–8.

8. Above, p. 204.

9. Above, pp. 100–1, 103–4.

10. Above, pp. 97–8.

11. Above, p. 245.

12. Above, pp. 12, 169.

13. Above, p. 12.

14. Above, pp. 7, 12, 100, 103–4.

15. See, for example, *The Times*, London, 6 March 1957. 'Although there is complete assurance at the moment that no radical change in type of reactor will be economically desirable within the period it is well known that America and Canada are developing other types. Europe's inclination seems to be to sample all major types ... flexibility of approach will be necessary if the best use is to be made of these revolutionary new powers.'

16. *SC Estimates, 1959*, pp. 389–90. and the succeeding evidence of Harry West of AEI, especially Qs 2619–34. Also see the evidence of the other manufacturers.

17. US Government safety research could count as a subsidy, and enrichment was provided at a price lower than would have been charged if capital charges included not only interest but also profit and Federal Income Tax (FIT).

18. Gowing, *op. cit.*, vol. II, pp. 279–80. 'The argument [against] may have been covered in principle [I am not sure what this means!] but events in later years were to prove that other factors might affect the balance of argument.'

19. *Ibid.*, vol. II, p. 250.

20. *Ibid.*, vol. II, p. 265.

21. Charpie as quoted above (without attribution), p. 9.

22. Gowing, *op. cit.*, vol. II, pp. 283–4.

23. *Ibid.*, vol. II, p. 287.

24. *Ibid.*, vol. I, pp. 417–18.

25. *Ibid.*, vol, I, pp. 426–9 and vol. II, pp. 249–51.

26. It was found that at the value attached to plutonium the initial cost of the fuel rods for Pippa would be less than the value of the spent rods after irradiation. *Ibid.*, vol. II, p. 292. The value of plutonium was put at £3100 an ounce.

27. *Ibid.*, vol. II, pp. 293–5.

28. *Ibid.*, vol. I, p. 429.

29. K. E. B. Jay, *Britain's Atomic Factories* (London: HMSO, 1954) with a foreword by Duncan Sandys. The book was written in 1953 as publicity.

30. As historian and journalist I have come across so much laudatory description of industrial exploits, usually with skilfully selected statistics which cannot be subjected to any comparative check, that I take a cynical view of extreme claims. Government agencies are as given to this kind of exaggeration as private industry; the grotesque exaggeration of many claims by the AEA has been documented in this text. My interpretation of the data published by Margaret Gowing differs from hers in emphasis: I cannot speak for evidence, naturally, which she does not publish. Jay I regard as the first hagiographer of the atomic establishment. He showed no reservations over the reactor development programme though he knew and stated it had yet to be established by experiment. It was astonishing that in a technical appendix he remarked that few materials were available as moderators – graphite was a good one, heavy water better. He did not alongside these even mention light water, though by this time its use was well known. *Ibid.*, p. 96.

31. Lord Cherwell may have thought he had an inkling; and some of his colleagues may have thought it of him too.

32. Lord Waverley and Sir John Woods. I had great regard for Woods – 'John Henry' – with whom I worked a little during the war, and it was at his invitation (and Charles Wheeler's) that I became Director of the Economic Development Office set up for the Heavy Electrical Industry in 1962. Their object, though not avowed, was to encourage a more rational and efficient structure. He agreed that the nuclear industry should be covered, but AEI resisted. Woods died shortly after the Office was set up and he and I never discussed the basis of the 1953 recommendations.

33. Thomas Hobbes, *Leviathan* (Cambridge: Cambridge University Press, 1904) p. 45.

34. Edmund Dell, *Political Responsibility and Industry* (London: Allen & Unwin, 1973) p. 230. It is all in all a very curious passage. Starting with the statement that there are people who have good judgement and people who have bad judgement (not at all egalitarian), he chides Barbara Wootton for criticising the 'mystique of judgement' (to which he obviously subscribes) and then having given an extremely brief reference to aids to judgement, including perforce cost-benefit analysis, he concludes with a vague suggestion that most people are equally wise, and an implication that this wisdom is equivalent to judgement: a comforting discovery, 'the more so as it is a property so relevant to the ministerial conduct of industrial policy'. He quotes from Hobbes's *Leviathan* to support this democratic view (which is at odds with his starting-point), but he does not quote Hobbes on judge-

ment itself, which would be more apposite but less egalitarian. Hobbes, *op. cit.*, pp. 42–5 on judgement; pp. 81–2 on equality in wisdom. Hobbes here sets aside the 'acts founded upon words, and especially that skill of proceeding upon general and infallible rules, called science, which very few have, and but in few things'.

35. Above, p. 154.
36. Above, pp. 157–9.
37. Above, pp. 200, 269, 306 (n. 16).
38. Above, p. 118. The 250 MW stage was on export orders; GE did have an American order for this size (Bodega Bay, California) but the project was dropped because of earthquake risks at the site, as was a project for a Westinghouse 450 MW PWR at Malibu, California.
39. *Financial Times*, 23 June 1976, and below p. 298.
40. The Rasmussen results were foreshadowed earlier before publication. *Nucleonics Week*, 18 October 1973.
41. Above, p. 242.
42. The comparative growth of gross national product (in constant prices) in the countries since 1965 in annual rates per cent was:

|                  | *1965–70* | *1971* |
|------------------|-----------|--------|
| United Kingdom   | 1.8       | 2.4    |
| Germany          | 4.8       | 2.6    |
| Japan            | 12.1      | 6.7    |
| Sweden           | 3.8       | 0.1    |
| Switzerland      | 3.2       | 3.8    |

43. *Financial Times*, 17 March 1972.
44. *Structure of the Electricity Supply Industry of England and Wales*, Report of Committee of Inquiry, chairman Lord Plowden (London: HMSO, 1976). Hereafter cited as the *Plowden Report*.
45. Above, p. 12.
46. Above, p. 124.
47. Above, pp. 152–6.
48. Above, p. 172.
49. *Hansard*, House of Lords, 10 July 1963, cs 1387–90.
50. Hinton, Axel Ax:son Johnson Lecture, *op. cit.*, pp. 15–16. 'Firms ought to realise that much of their power plant business will have been swept into this field by the late sixties ... by the early seventies nearly 100 per cent of the public utility stations built each year will ... be nuclear.' This probably should have been so.
51. I have been informed by one of the abstainers.
52. Above, p. 169.
53. *SC Nationalised Industries, 1963*, vol. II, Qs 4091, 4093. *Wilson Report*, pp. 1, 23, 27–30. *Plowden Report, op. cit.*, p. 20. The faults in ordering in the early sixties were discussed in 1963 by some members of the Board, but they appear to have escaped Vinter.
54. *SC Science and Technology, 1973–4*, Q. 718; also *SC Science and Technology, 1972–3*, p. ix.
55. *Ibid.*, Q. 710A.

56. *SC Science and Technology*, *1967*, p. xlviii.
57. *The Times*, 19 December, 1946 (in a leader on the Transport Bill).
58. Burn, *The Steel Industry* (*1939–1959*), (Cambridge: CUP, 1964) pp. 366–9.
59. There was a debate in the House of Lords in 1963 in which Lords Coleraine, Stoneham, Aldington, Mills, Hawke, Carrington and Hailsham spoke, sustaining this view. The CEGB in particular must not be allowed to determine the size of the programme. *Hansard*, House of Lords, 16 July 1963, cs 1392–1403, 1430, 1442, 1446.
60. *Ibid.*, cs 1452–3.
61. Dell, *op. cit.*, p. 209.
62. *Ibid.*, pp. 38, 173.
63. *The Times*, London, 11 July 1974.
64. Dell, *op. cit.*, p. 226.
65. *Ibid.*, p. 225.
66. It was a view I did not share. For example I thought, and still think, that the promotion of aluminium smelting (dealt with in *ibid.*, pp. 38–9, 105–21) was a means of wasting resources. I remarked in a paper to the Society of Business Economists in April 1970 that 'particular instances of government aid which are in the bad old tradition are those to establish aluminium plants. I am always astonished that those who worry over the Concorde say nothing about bad steel or aluminium decisions.' *Changes in the Industrial Structure of the UK* (London: Society of Business Economists, 1970) p. 51.
67. Dell, *op. cit.*, p. 228.
68. *Ibid.*, pp. 196–7.
69. *Financial Times*, 23 June 1976.
70. It may seem hard to criticise the British arrangements for resulting in premature decisions and in delay in decisions; it is therefore perhaps useful to point out that the prematurity of decisions was a fault in the quality of the decisions, not in the time taken in making them.
71. *Hansard*, House of Lords, 10 July 1963, c. 1443.
72. Burn, *Political Economy of Nuclear Energy*, p. 114.
73. *The Times*, 19 September 1967 (in its obituary of Cockcroft).
74. Above, p. 174.
75. *United Kingdom Atomic Energy Authority Annual Report*, 1955–6, p. 29. The 'growth of interest and activity' in industry is dated from the Government's programme which made it clear that the power stations would not be built by the Authority. Firms also saw export prospects so there was a 'great increase in independent activity of industrial firms' in the last twelve to eighteen months.
76. Above, p. 172.
77. A perceptive view was expressed by Lord Champion, a Labour peer, formerly a signalman, in a House of Lords debate in 1964, which showed that light was spreading: but he was ahead of the field. 'A decision to buy the American system would be a considerable blow to our national prestige. But our attempt to keep one move ahead of any possible rival, both in aircraft and atomic energy, has been a very expensive business costing many millions of pounds which has certainly not paid off ... National pride will not help us in the export market if the American product is cheaper ... I do not think we shall ensure a market abroad by seeking some form of

nuclear energy which has been outstripped by somebody else.' *Hansard*, House of Lords, 10 June 1964, c. 913–14. The United Kingdom, nevertheless, did precisely this, not only in 1965, but in 1974.

78. Professor David Henderson analysed this together with the AGR episode in his inaugural lecture at University College London on 24 May 1976 entitled 'Two British Errors: Their Probable Size and some Possible Lessons'. *Oxford Economic Papers*, Oxford, July 1977.

79. The adoption of larger fossil-fuel generating units in the United Kingdom following the American lead provides an impressive instance of bad handling of an important development, which compares badly with the procedure followed in Germany. The steel strip mill decisions under Harold Macmillan provide an instance of specific intervention encouraging weak locations, over capacity and excessive subdivision (although promoting modern plant). I referred to the unsatisfactory pattern of investment resulting from government policies in 1945–49 in two articles on 'Policy for Industry' in *The Times*, 20 and 21 February 1950. I dealt with the early stages of the strip mill policy in *The Steel Industry, 1939–1959*, pp. 639–657. I returned to the general problem again in 'Investment, Innovation and Planning in the UK', *Progress*, London, September 1962, and in 'Why Investment has Fallen', *Lloyds Bank Review*, London, April 1963.

80. Thus in a recent article, 'What is Nationalisation For?' *Lloyds Bank Review*, July 1976, Professor Michael Lipton of Sussex University speaks of the 'voting test' as certainly not less just than a 'market test' and says that when 'voters mandate nationalisation, they direct a move from the market test, income-related rights, to the voting test, equal rights', and finally 'if the aims [of nationalisers] are wrong ... and the results serious enough, either MPs or voters will withdraw their support'.

This shows complete misconception of the present political process. It is well known that normally, and certainly at present, there is a majority against nationalisation. People moreover do not and *cannot* vote for specific policies but for groups of policies. Information to allow judgement of results is not provided or analysed by governments – it is often withheld. Governments do not fall on the failure of one element in policy. Parliament is quite inadequate as a check on this kind of thing (see Dell, *op. cit.*, pp. 198–201) and governments can control their party by the Whips (*ibid.*, pp. 200–4).

81. Above, p. 200.

82. Above, p. 232.

83. *The Times*, London, 28 June 1974.

84. Above, pp. 207–8.

85. Adam Smith, *Wealth of Nations* (London 1776) vol. I, p. 117.

86. David Henderson, 'Two British Errors: Their Probable Size and Some Possible Lessons', *op. cit.*

87. *SC Science and Technology, 1976*, Q. 68.

88. Above, pp. 229, 248. The Candu system also had the higher capital cost; but it used unenriched uranium fuel, which could be regarded as a distinguishing feature which appeared an advantage to some buyers. Net efficiencies were given as SGHWR 30 per cent, BWR 33 per cent. AEA *Annual Report*, 1974–5, p. 20.

89. *SC Science and Technology, 1976*, Qs 33, 55.

90. *Ibid.*, Q. 25. Wedgwood Benn was explaining he could not say how much weight was given to Cottrell when the decision was made.
91. *Ibid.*, Q. 1.
92. *Ibid.*, Q. 40.
93. *Ibid.*, Qs 26–7.
94. The nearest approach to this question came from Ian Lloyd (Conservative), who asked what weight the Minister attached to the judgement of the United States, Germany, France and the Soviet Union in committing themselves to the LWR, since they were – especially the Japanese – extremely safety-conscious. Wedgewood Benn said it represented a very successful commercial deal by the Americans – he evaded the real issues and no one pursued them.

# Glossary

*Most of the terms defined in this glossary relate to nuclear reactors and processes: it has seemed therefore most practical to preface it with a brief description of a reactor.*

A nuclear *reactor* is an apparatus in which a controlled nuclear *chain reaction* can take place. In the culminating stage of the exploration of atomic structure in the 1930s it was discovered that some isotopes of certain materials can be readily split by being hit by a *neutron* (q.v.); they are called '*fissile materials*' (see Uranium, Plutonium, Thorium). When *fission* takes place a very small part of the atom's mass is converted into a very large amount of energy, which provides heat, but there is also an emission of some neutrons, which can in turn produce fission by hitting neighbouring fissile atoms, with a further release of energy and emission of neutrons, so that a chain reaction can be set up. Fission can then become self-propagating, providing there is sufficient fissile material in a suitable disposition and concentration, and provided that capture of neutrons by non-fissile materials is kept at a low level (so that as many neutrons as possible are available to cause fission).

In the types of reactor used currently for power production the fissile material is contained in fuel elements which are surrounded by a material which does not readily absorb neutrons to act as *moderator* and slow down fast neutrons. The heat generated by fission is absorbed by a *coolant* circulated under considerable pressure through the reacting fuel elements and used to raise steam for turbo generators. The coolant, like the moderator, must not readily absorb neutrons. In order either to keep the rate of the nuclear reaction steady or to vary it, *control rods* which include materials which absorb neutrons strongly (like boron, cadmium or hafnium) are used; pushed further into the fuel assembly they reduce the rate of reaction and *vice versa*. Fission produces not only heat and neutrons but also radioactive waste materials and in varying proportions other fissile materials (in particular plutonium). These are formed within the fuel elements as the fissile materials in the elements are split. The ratio of fissile materials produced in the reaction to fissile materials used is called the *conversion ratio*.

To sum up, a reactor is an apparatus within which a controlled nuclear reaction can take place; it consists of a reactor vessel capable of withstanding the heat and pressure of the process, and containing fuel elements, coolant, control rods and, except in a fast breeder reactor, a moderator. Because of the high degree of radiation it is surrounded by *concrete shielding*.

*Advanced Gas Cooled Reactor* (AGR): a variant of the Magnox reactor (q.v.); with graphite as moderator and $CO_2$ as coolant, but using slightly enriched uranium oxide (2.25 per cent enrichment) as fuel in stainless steel cans and operating at higher temperatures than the Magnox reactor.

*Availability:* the proportion of a year (or other convenient period) during which a reactor (or power plant) is available for use at its rated capacity, though it may not be used continuously at its rated capacity.

*Boiling Water Reactor* (BWR): a reactor with ordinary (light) water both as moderator and coolant, using slightly enriched uranium oxide as fuel (2.3 to 2.4 per cent enrichment), in cans of a zirconium alloy (zircalloy). Steam is produced in the reactor vessel and used directly to drive a turbo generator; since the steam is radioactive the turbine is shielded.

*Breeder Reactor:* a reactor which produces more fissile atoms than it burns, where the conversion ratio, that is, is over 1:1. It would vary between 1.1 and 1.6:1. In a breeder reactor there would normally be no moderator and the neutrons are not slowed down. Hence it is also known as a *fast* reactor. *Fission* produced by fast neutrons results in the emission of more new neutrons than fission produced by slow neutrons.

*Breeding Ratio:* the ratio of the number of fissionable atoms produced in a breeder reactor to the number of fissionable atoms consumed in the reactor. *Breeding gain* is the breeding ratio minus one.

*Can(s):* the containers in which fuel rods are sealed.

*Centrifuge Enrichment:* process for enriching uranium based on the use of a centrifuge, using centrifugal force to separate the uranium isotopes.

*Chain Reaction* (above)

*Control Rods* (above)

*Converter Reactor:* a reactor which produces less fissionable material than it consumes. Current reactors (light water and AGR) used for power supply have a *conversion ratio* of 0.5 or 0.6:1. More advanced converter reactors are being developed with ratio of 0.7 to 1.0:1.

*Coolant* (above).

*Core:* the central portion of a reactor in which fission occurs, consisting of the fuel elements and moderator. (Control rods operate in the core, and coolant is blown or pumped through it.)

*Critical:* when a self-sustaining chain reaction is established a reactor is said to have become 'critical'.

*Decay Heat:* the heat produced by the decay of radioactive nuclides, which continues even after the reactor is shut down.

*Derating:* regulatory body action reducing the permitted maximum rate of output of a plant below its design capacity.

*Diffusion Plant:* a factory in which uranium enriched in U235 is pro-duced by the diffusion process (in which uranium atoms are passed through a long series of fine screens which retard the heavy isotopes).

*Doubling Time:* the time required for a breeder reactor to produce as much fissionable material as the amount usually contained in its core plus the amount tied up in its fuel cycle (fabrication, reprocessing, *et cetera*). It is estimated as ten to twenty years in typical reactors.

*Dry Well:* in a BWR, a steel vessel capable of withstanding high pressure which encloses the reactor pressure vessel, and is itself en-closed in concrete.

*Enriched Uranium:* natural uranium contains 0.7 per cent of the fissile isotopes U235. In enriched uranium this percentage has been made higher, so that it contains more fissile material.

*Enrichment:* the process of producing enriched uranium. See *Diffusion Process, Centrifuge Enrichment*.

*Fast Reactor, Fast Breeder Reactor:* see *Breeder Reactor*.

*Fertile Material:* see *Uranium*.

*Fuel Element:* a unit of nuclear fuel in the form in which it is used in a reactor. Normally the fuel – uranium either natural or enriched – is enclosed in a can which protects it from corrosion by the coolant and keeps in the fission products (waste and plutonium).

*Fissile Material* (above)

*Fission* (above)

*Half-Life:* the time in which half the atoms of a particular radioactive substance disintegrate to another nuclear form. Measured half-lives vary from millionths of a second to billions of years.

*Heat Sink:* anything that absorbs heat; usually part of the environment, such as the air, a river, or outer space.

*Heavy Water:* the oxide of deuterium, which is the heavy isotope of hydrogen. Used as moderator and sometimes as a coolant. It occurs normally as about 1 part in 60,000 in ordinary water.

*High Temperature (Gas-cooled) Reactor* (HTR or HTGR): a reactor in which helium is used as coolant, and the fuel spheres – of uranium, uranium dioxides (more highly enriched than in the AGRs or the LWRs) – or thorium or thorium carbide – are graphite-coated, and contained in a graphite sleeve instead of a metal can. Moderator and fuel are thus mixed. The gas is raised to much higher temperatures than in the AGR, and the conversion ratio is higher.

*Homogeneous Reactor:* a reactor in which the fuel is mixed with or dissolved in the moderator or coolant. Example: the HTR.

*Isotope:* atoms which are chemically the same may vary in weight: the different varieties are known as isotopes.

*Kilowatt:* a unit of power equal to 1.34 horse-power.

*Kilowatt-hour:* a unit of energy equal to the work done by 1 KW in an hour.

*Lead-time:* the expected time required from the placing of an order for a plant to the commercial operation of the plant.

*Light Water Reactor:* a reactor in which ordinary water is used as moderator and coolant. The type includes both boiling water reactors and pressurised water reactors.

*Load:* the amount of power needed to be supplied by an electric system, in total or at some specific point.

*Load Factor:* the ratio of the average load during a year (or any other selected period) to the maximum load occurring in the same period. The term can be applied to the experience of a whole system (system load factor) or that of an individual plant (plant load factor) or even that of an individual generating unit. Sometimes used for the ratio of total output of a generating unit in a period to its designed maximum capacity.

*Magnox Reactor:* a reactor using $CO_2$ as coolant, graphite as moderator, and natural uranium as fuel; the fuel is sheathed in a finned can of magnesium alloy known as Magnox, and this name was transferred to the reactor type. (Magnox was first used for the Calder Hall fuel.)

*Megawatt:* 1000 KW, equal to 1340 hp.

*Mills per KWh:* a mill is one-tenth of a cent; the price (and cost) of electric power in the United States is measured in terms of mills per KWh. With the exchange at $1.7=£1, 1 mill=0.059p.

*Moderator* (above)

*Neutron:* a nuclear particle having no electric charge which is found in the nuclei of atoms. Neutrons are released by fission, and they are the principal agents in initiating nuclear fission.

*Nuclide:* a general term applicable to all atomic forms of the elements. The term is often erroneously used as a synonym for 'isotope', which properly has a more limited definition. Whereas isotopes are the various forms of a single element (hence are a family of nuclides) and all have the same *atomic number* and number of protons, nuclides comprise *all* the isotopic forms of *all* the elements. Nuclides are distinguished by their *atomic number*, *atomic mass*, and energy state. Of 1700 nuclides 1400 are radioactive.

*On-Load Charging:* when the fissile material in a fuel element is burnt out the element must be replaced. In some reactors the reaction is stopped periodically and a large number of elements are changed at one time: in others elements are changed individually while the reactor continues to be 'on load'. Putting in elements is called 'charging' the reactor: hence 'on-load charging'.

*Outage:* the period when a generating unit is out of action whether for regular maintenance or due to breakdown.

*Pebble Bed Reactor:* a version of the HTR of German design in which fuel and moderator are in the form of spheres of about 6 cm. diameter, which circulate through the reactor, and are cooled by helium.

*Plutonium:* an element produced by neutron irradiation of the fertile (not fissile) isotope of uranium, U238. The isotope of plutonium Pn 239 is a fissile material, and an important potential nuclear fuel.

*Power Density:* the rate of heat generated per unit volume of a reactor core.

*Pump(ed) Storage:* an arrangement whereby electricity is used during off-peak periods to pump water into a storage reservoir; the water is used in peak periods through a hydroelectric plant to generate power.

*Rem:* (acronym for roentgen equivalent man). The unit of dose of any ionizing radiation which produces the same biological effect as a unit of absorbed dose of ordinary X-rays.

*Reprocessing:* the processing of reactor fuel to recover the unused fissionable material, the uranium, and plutonium, and to separate the waste fission products.

*Scram:* the sudden shutdown of a nuclear reactor, usually by rapid insertion of the safety rods. Emergencies or deviations from normal reactor operation cause the reactor operator or automatic control equipment to scram the reactor.

*Spent (depleted) Fuel:* nuclear reactor fuel that has been irradiated (used) to the extent that it can no longer effectively sustain a chain reaction.

*Steam Generating Heavy Water Reactor* (SGHWR): a type of reactor using heavy water as moderator and light steam as coolant, with a lightly enriched uranium fuel. Was originally called a steam *cooled* HW reactor; in Canada is called a boiling water HW reactor. The steam can be used directly in a turbine without a heat exchanger.

*Thermal Reactor:* a type of reactor with a moderator, which slows down fast neutrons.

*Thorium:* a metal mostly composed of its isotope 232 which is fertile and can be converted by neutron bombardment into the fissile material U233.

*Tritium:* a radioactive isotope of hydrogen with two neutrons and one proton in the nucleus. It is man-made and is heavier than deuterium (heavy hydrogen). Tritium is used in industrial thickness gauges, and as a label in experiments in chemistry and biology. Its nucleus is a *triton.*

*Turnkey Contract:* in this context, a contract to build a complete nuclear power plant in which responsibility for the whole project is undertaken by the firm which designs the nuclear system. As turnkey contractor the firm designs the whole station, manages the construction of the whole plant including the site development, provision of water, *et cetera,* and places all the sub-contracts.

*Uranium:* natural uranium includes 139 parts of U238, which is *fertile,* that is, can be converted into fissile material, and one part of U235, which is fissionable. Some U238 is converted in thermal reactors into plutonium, but larger proportions would be converted in intermediate and fast reactors.

# Selected Bibliography

This bibliography is divided into four parts. All the entries are publications cited in the notes and references at the end of each chapter. Where a publication is abbreviated in the text or notes, the abbreviation is listed first and then followed by the formal details.

## I BOOKS, ARTICLES, PAPERS, LECTURES

DUNCAN BURN, *The Economic History of Steelmaking 1867–1939* (Cambridge: Cambridge University Press, 1940). Out of print.

DUNCAN BURN, *The Steel Industry, 1939–1959* (Cambridge: Cambridge University Press, 1961).

DUNCAN BURN, *The Political Economy of Nuclear Energy* (London: Institute of Economic Affairs, 1967).

EDMUND DELL, *Political Responsibility and Industry* (London: Allen & Unwin, 1973).

MARGARET GOWING, *Britain and Atomic Energy, 1839–1945* (London: Macmillan, 1965).

MARGARET GOWING, *Independence and Deterrence: Britain and Atomic Energy, 1945–1952*, vol. I, *Policy Making*, vol. II, *Policy Execution* (London: Macmillan, 1974). These two volumes are the official history of Britain's atomic energy project.

R. F. W. Guard, 'The Year Since Geneva', pt. I, *Euro Nuclear*, London, September 1965.

ROLT HAMMOND, *British Nuclear Power Stations* (London: Macdonald, 1961).

DAVID HENDERSON, 'Two British Errors: Their Probable Size and Some Possible Lessons'. Inaugural lecture, University College, London, 24 May 1976, *Oxford Economic Papers*, Oxford, July 1977.

SIR CHRISTOPHER HINTON, *The Future for Nuclear Power*, The Axel Ax:son Johnson Lecture, Stockholm, 15 March 1957 (Stockholm: The Royal Swedish Academy of Engineering Sciences, 1957).

THOMAS HOBBES, *Leviathan* (London: Andrew Crooke, 1651). I have used the edition published by the Cambridge University Press in 1904.

K. E. B. JAY, *Britain's Atomic Factories*, with a preface by the Rt. Hon. Duncan Sandys (London: HMSO, 1954).

CECIL KING, *The Cecil King Diary, 1970–1974* (London: Jonathan Cape, 1975).

DAVID E. LILIENTHAL, *Change, Hope and the Bomb* (Princeton, NJ: Princeton University Press, 1963).

MICHAEL LIPTON, 'What is Nationalisation For?', *Lloyds Bank Review*, London, July 1976.

R. V. MOORE AND J. D. THORN. 'Advanced Gas Cooled Reactors – an Assessment', *Journal of the British Nuclear Energy Society*, London, April 1963.

L. B. NAMIER, *Diplomatic Prelude 1937–9* (London: Macmillan, 1948).

GLENN T. SEABORG, *Nuclear Milestones* (Washington: Atomic Energy Commission, 1971).

LORD SHERFIELD (ed.), *Economic and Social Consequences of Nuclear Energy* (Oxford: Oxford University Press, 1972).

K. L. STRETCH, 'Is Britain on the Right Track? A Critique of our Nuclear Power Programme', *Nuclear Power*, London, December 1958.

R. D. VAUGHAN Chief Engineer, TNPG) and J. O. Joss, Electrical Department, TNPG), 'The Current and Future Development of the Magnox Reactor', Paper No. 4 in the Anglo-Spanish Nuclear Power Symposium, Madrid, November 1964, issued at the symposium.

R. D. VAUGHAN, 'Experience with Integral Gas Cooled Reactors', a paper given at the Symposium on the Evolution of Proved Nuclear Power Reactor Concepts, Rome, 24–25 June 1965, issued at the symposium.

R. D. VAUGHAN and D. R. SMITH, 'Exploitation of the Advanced Gas-cooled Reactor for Hinkley Point B', a paper presented to a joint meeting of the Institute of Mechanical Engineers and the Société des Ingénieurs Civils de France, Paris, 6 December 1966. Subsequently published in London by the Institute of Mechanical Engineers.

G. H. WHITE, 'Boiling Water Reactor Progress', a paper presented at the Joint Japan-US Atomic Industrial Forum Meeting, Tokyo, Japan, 5–8 December 1965. G. H. White was General Manager, Atomic Power Equipment Department, General Electric.

H. E. WILLIAMS and D. T. DITMORE, *Current State of Knowledge High Performance BWR Zircaloy Clad $UO_2$ Fuel* (San José, California: General Electric, May 1970).

H. E. WILLIAMS and D. T. DITMORE, *Experience with BWR Fuel through September 1971* (San José, California: General Electric, May 1972).

HAROLD WILSON, *The Labour Government 1964–70* (London: Penguin Books, 1971).

## II REPORTS OF CONFERENCES, AD HOC COMMITTEES AND SYMPOSIA

*A Target for Euratom*, a report submitted by Louis Armand, Franz Etzel and Francesco Giordani at the request of the Governments of Belgium, France, West Germany, Italy, Luxembourg and the Netherlands (Brussels: Euratom, 4 May 1957).

*L'Industrie devant l'Energie Nucleaire*, papers given at the Second Conference on Nuclear Energy for Management, Amsterdam, June 1957 (Paris: OEEC, 1958).

*Nuclear Energy*, report of an FBI Conference held at Eastbourne, April 1958 (London: Federation of British Industries, 1958).

*Nuclear Power and the Environment*, Royal Commission on Environmental Pollution (London: HMSO, 1976).

*Nuclear Power in Japan*, report on a visit by a team from the British Nuclear Forum, May–June 1971, by H. Greenhalgh (London: British Nuclear Forum, 1971).

*Nuclear Supply Industries in Japan*, paper presented by Shiro Mochizuki on the visit of a team from the Japanese Atomic Industries Forum to Britain, September 1975 (London: British Nuclear Forum, 1971).

Paley Report: *Resources for Freedom*, report to the President by the President's Materials Policy Commission. Chairman William S. Paley. (Washington: US Government Printing Office, 1952).

Plowden Report: *The Structure of the Electricity Supply Industry in England and Wales*, report of the Committee of Inquiry. Chairman, Lord Plowden (London: HMSO, 1976) Cmnd. 6388.

*Public Health Risks of Thermal Power Stations*, report prepared for the Resources Agency of California by the School of Engineering and Applied Science, UCLA; principal investigators: Professors Chauncey Starr and M. A. Greenfield (Los Angeles: School of Engineering and Applied Science, UCLA, 1972).

*Réponses de la Commission d'Evaluation en Matière d'Enérgie Nucléaire* (Brussels: Ministère des Affaires Economiques, February 1976) p. 34.

*Report of the Committee of Inquiry into the Electricity Supply Industry*, Chairman, Sir Edwin Herbert (London: HMSO, 1956) Cmnd. 9672.

*The Industrial Challenge of Nuclear Energy*, paper given at the First Conference on Nuclear Energy for Management, Paris, April 1957 (Paris: OEEC, 1957).

*State, Science and Economy as Partners*, published in collaboration with the West German Federal Ministry for Scientific Research (Berlin: A. F. Koska, 1967).

Wilson Report: *Report of the Committee of Enquiry into Delays in the Commissioning CEGB Power Stations*, Chairman, Sir Alan Wilson (London: HMSO, 1969) Cmnd. 3960.

## FRANCE

Choix du Programme, 1970: *Rapport présenté par la Commission Consultative pour la Production d'Electricité d'Origine Nucléaire sur le Choix du Programme de Centrales Nucléo-Electriques pour le VI⁶ Plan*, November 1970, with a volume of Annexes (Paris: Ministère du Développement Industriel et Scientifique, 1970).

## GERMANY

*Fourth Nuclear Programme 1973–1976 of the Federal Republic of Germany* (Bonn: Federal Ministry for Research and Technology, 1974).

## UNITED KINGDOM

*British Nuclear Fuels Limited: Annual Report*, from 1972–73 (Risley, Lancashire: BNFL).

CEGB Appraisal: Central Electricity Generating Board, *An Appraisal of the Technical and Economic Aspects of Dungeness B. Nuclear Power Station* (London: CEGB, July 1965).

*Central Electricity Generating Board – Annual Report* and *Statistical Yearbook* (London: HMSO, until 1971–72, subsequently CEGB).

Report of NPAB, 1974: *Choice of Thermal Reactor Systems*, report of the Nuclear Power Advisory Board (London: HMSO, 1974) Cmnd. 5731.

*The Second Nuclear Power Programme*, report by the Ministry of Power, April 1964 (London: HMSO, 1964) Cmnd. 2335. This gives the references to the earlier collected papers on the programmes: 1955, 1957, 1960.

*United Kingdom Atomic Energy Authority – Annual Report*, from 1954–55 (London: HMSO, until 1971–72, subsequently UK AEA).

### COMMITTEES OF THE HOUSE OF COMMONS

*Committee of Public Accounts, 1975*
*Report of the House of Commons Committee on Public Accounts*, Minutes of Evidence (London: HMSO, 22 January 1974).

*SC Estimates, 1959*
Report of the House of Commons Select Committee on Estimates, Report and Minutes of Evidence and Appendices on *UK Atomic Energy Authority* (London: HMSO, 1959).

*SC Nationalised Industries, 1963*
Report of the House of Commons Select Committee on Nationalised Industries on *The Electricity Supply Industry*, Reports, Evidence and Appendices (London: HMSO, May 1963).

*SC Science and Technology, 1967*
Report of the House of Commons Select Committee On Science and Technology on *UK Nuclear Reactor Programme*, Report, Minutes of Evidence and Appendices (London: HMSO, October 1967).

*SC Science and Technology, 1969*
Report of the House of Commons Select Committee on Science and Technology on *UK Nuclear Power Industry*, Report, Minutes of Evidence and Appendices (London: HMSO, July 1969).

*SC Science and Technology, 1972–3*
Report of the House of Commons Select Committee on Science and Technology on *Nuclear Power Policy* (London: HMSO, June 1973).

*SC Science and Technology, 1972–3*
Report of the House of Commons Select Committee on Science and Technology on *Nuclear Power Policy*, Minutes of Evidence and Appendices (London: HMSO, January 1973).

*SC Science and Technology, 1973–4*
Report of the House of Commons Select Committee on Science and Technology on *The Choice of a Reactor System*, Report, Minutes of Evidence and Appendices (London: HMSO, December 1973, January 1974).

*SC Science and Technology, 1976*
Report of the House of Commons Select Committee on Science and Technology on *SGHWR Programme*, Minutes of Evidence (London: HMSO, August 1976).

COMMITTEES OF THE HOUSE OF LORDS

*SC, House of Lords, EEC Energy Policy, 1975*
Report of the House of Lords Select Committee on The European Communities R/3333/74 *EEC Energy Policy Strategy* (London: HMSO, 1974).

UNITED STATES

*Review of Methods of Mitigating Spread of Radio Activity from a Failed Containment System*, Natonal Safety Information Centre Staff, Oak Ridge National Laboratory, September 1968 (Oak Ridge, Tennessee: Oak Ridge National Laboratory, 1968).

AEC, ERDA AND NRC

Publications of the United States Atomic Energy Commission (AEC) whose functions were taken over by the Energy Research and Development Administration (ERDA) and the Nuclear Regulatory Commission (NRC) in 1975.

AEC, Civilian and Nuclear Power, 1962; *Civilian Nuclear Power*, a Report to the President (Washington: US AEC, 1962); 1967 Supplement (Washington: US Government Printing Office, 1967).

AEC, Concept, 1971: *Concept. A Computer Code for Conceptual Cost Estimates of Steam-Electric Power Plants*, Status Report, April 1971 (Washington: US Government Printing Office, 1971) Wash. 1180.

*An Evaluation of a Heavy-Water-Moderated Boiling-Light-Water-Cooled Reactor*, December 1969 (Washington: US Government Printing Office, 1969) Wash. 1086.

*Current Status and Future Technical and Economic Potential of Light Water Reactors*, March 1968 (Washington: US Government Printing Office, 1968) Wash. 1082.

*Growth of Foreign Nuclear Power*, a report to the AEC by Arthur D. Little Inc., April 1966 (Washington: US AEC, 1966).

*Nuclear Power, 1973–2000*, Forecasting Branch, Office of Planning and Analysis of the AEC (Washington: US Government Printing Office, 1972).

*Nuclear Reactors Built, being Built or Planned in the United States*, Annual Report (Washington: US AEC until 1973, subsequently ERDA).

Rasmussen Report: *Reactor Safety Study – An Assessment of Accident Risks in US Commercial Nuclear Power Plants*, report for the AEC under the independent direction of Professor Norman C. Rasmussen of the Massachusetts Institute of Technology. The Report Summary and Appendices (Draft) (Washington: US AEC, 1974).

*Reactor Safety: a Discussion by Officials of the Nuclear Regulatory Commission*, March 1976 (Washington: NRC, 1976).

*The Nuclear Industry*, produced by the US AEC Division of Industrial Participation (Washington: US Government Printing Office, annual until 1974).

*The Safety of Nuclear Power Reactors (Light Water Cooled) and Related Facilities*, a substantial study of almost 500 pages, final draft (Washington: US AEC, July 1973) Wash. 1250.

*1000 MWE Central Station Power Plants Investment Cost Study*, 4 volumes, study by United Engineers and Construction Inc. for the US AEC (Washington: US Government Printing Office, 1972) Wash. 1230.

*Trends in the Cost of Light Water Reactor Power Plants for Utilities* (Washington: US Government Printing Office, 1970) Wash. 1150.

*United States Nuclear Power Reactors: Operating History* (Washington: US AEC until 1973, subsequently ERDA, annual).
*US AEC Annual Report to Congress* (Washington: US AEC, until 1974, annual).

*National Power Survey*, a report by the FPC, in two parts (Washington: US Government Printing Office, 1964).
*Statistics of Electric Utilities in the United States, Privately Owned* (one volume), *Publicly Owned* (one volume) (Washington: US Government Printing Office, annual).
*Steam Electric Plant Construction Cost and Annual Production Expenses* (Washington: US Government Printing Office, annual).

The Joint Committee on Atomic Energy (JCAE) of the Congress of the United States has published a great volume of important material for over two decades. This includes collections of studies to form the background for discussions and reports of public hearings by the Joint Committee, including valuable documents in Appendices. Those listed are the volumes most used for the book. All were published in Washington for the JCAE by the US Government Printing Office.

JCAE, *Development, 1955*
   *Hearings on Development, Growth and State of the Atomic Energy Industry.* Parts 1 and 2 (there are 3 parts), February 1955.
*McKinney Report*
   *Report of Panel on the Impact of the Peaceful Uses of Atomic Energy.* Chairman of the Panel, Robert McKinney. Vol. 2 (Background material for Report) January 1956.
JCAE, *Development, 1960*
   Hearings on *Development, Growth and State of the Atomic Energy Industry.* February 1960.
   *Nuclear Power Economics 1962 through 1967*, February 1968.
JCAE, *Environmental*: pt 1 (1969); pt 2 (1970)
   *Hearings on Environmental Effects of Producing Electric Power*, Part 1 October–November 1969, Part 2 (vol. I) January–February 1970.
JCAE, *Antitrust:* pt 1 (1969); pt 2 (1970)
   *Hearings on Prelicensing Antitrust Review of Nuclear Power Plants*, Part 1 1969, Part 2 April 1970.
JCAE, *Licensing, 1971*
   *Hearings on AEC Licensing Procedure and Related Legislation*, June–July 1971.

JCAE, *Nuclear Power Problems, 1971*
 *Nuclear Power and Related Energy Problems,* December 1971.
JCAE, *Calvert Cliffs, 1972*
 *Selected Material on the Calvert Cliffs Decision, its Origin and Aftermath,* February 1972.
JCAE, *Authorizing*
 *Hearings on AEC Authorizing Legislation,* every fiscal year, usually in 4 parts.
JCAE, *To Amend Atomic Energy Act, 1972*
 *Hearings on HR 13731 and HR 13732 to Amend the Atomic Energy Act Regarding the Licensing of Nuclear Facilities,* March 1972.
House of Reps., *Interim Nuclear Licensing, 1972*
 House of Representatives Sub-Committee on Fisheries and Wildlife Conservation on *Interim Nuclear Licensing.* Printed for the Committee on Merchant Marine and Fisheries, March 1972.

## IV  SPECIALISED PERODICALS

*Atomic Energy Clearing House,* Washington, Congressional Information Bureau. Gives, for example, extended reports on JCAE proceedings; primarily United States coverage.
*Electric Light and Power,* Boston, May Calver Publishing Co.
*Info,* Washington, American Industrial Forum.
*Nuclear Engineering International,* London, IPC Electrical-Electronic Press Ltd.
*Nuclear News,* Hinsdale, Illinois, American Nuclear Society.
*Nucleonics Week,* New York, McGraw Hill. Well informed with broad international coverage.

# Index

Note: Items in the Notes and References at the end of each chapter are shown in the index by the page number followed by / and the reference number both in italics.

Acoustic tests, *see* Pressure vessels

Administrative Agencies (US) criticised, 49, 73–4

Advanced gas-cooled reactor (AGR), *see* Reactor types

'Adversary process', use in technical decisions analysed, 65, 74–6, 92, 236

Advisory Committee on Biological Effects of Ionising Radiation, *78/23*

Advisory Committee on Reactor Safety (ACRS), 56, 65, 66, 69, 71, 74, 75, 88–9, 243; independent statutory committee (1957), 46; approval of reactor safety record, 56, 66, 88–9; advice always taken, 72

AEI–John Thompson Nuclear Energy Co., *115/1*

Aldington, Lord, 213, *220/65, 221/78,* 231, 232, 239, 241, 251, *259/82,* 303, *309/59*

ALKEM, *109/20*

Allday, C., 107

Allgemeine-Elektrizitäts-Gesellschaft (AEG), 100, 101, 102, 106, *109/24, 191/106,* 203, 300

Allmänna Svenska Elektriska Aktiebolaget (ASEA) 103, 104, *109/30,* 200, 300

American Nuclear Society (ANS), *93/6*

Americans for Energy Independence Group, *95/38*

Anders, William C., 88, 90, *94/19, 25*

Anti-nuclear campaign in US: beginnings, *see* Environmentalist Groups, Intervenors: Phase I, focus on routine effluents, 62–4; Phase II, focus on risk of catastrophic accident, 64–8; Phase III, focus on fuel cycle and waste disposal problems, 15–16, 82, 85–92

Anti-nuclear campaign outside US, 106–7; in UK, 234–7

Anti-trust, 48, *53/41*

Anticipated transients without trip (ATWT), 84, *93/6*

*Appraisal of Technical and Economic Aspects of the Dungeness B Nuclear Power Station, An,* 10, 123–4, 128, *144/40–46 passim, 147/103 and 106, 148/115,* 150–1, 153, 161, 163–8, 172, 179, 181, *186/6–8, 187/31,* 216, 279–80, 286, 292, 299; *see also* Sketch Design for Dungeness B

Architect-engineers, 4, 5, 30, 32, 40, *52/25,* 56, *79/32,* 83, 89, *92/2,* 200, 203, *218/12,* 279

Argentina, 96

Argonne Laboratory (Ill.), *53/28,* 60, 118

Armand, Louis, *19/24*

Associated Electrical Industries (AEI), *115/1,* 162

Atkinson, Norman MP, 127, 201, *218/7*

Atomic Energy Authority Act, 1954, 176, 177

Atomic Energy Authority (AEA): antecedents, 5–6, 8, 116, 168, 177, 273; members, 6–7, 124, 155, 178, 226, 274; functions, 7–8, 159–60, 177; size, 274; monopoly, 6, 8, 12, 13, 159, 198, 202, 275, 277, 285, 288, 295, 297, 305; criticisms of in US, 150; conflict between advisory and development functions, 13, 160, 172, 178, 287; AEA/CEGB study SGHWR, 128; Ministers' powers, 6, 170, 177–9, 274; and BNX, 198; relations with Ministers, 170–80, 290–5; relations with CEGB, 15, 147–98 *passim*; relations with consortia, 8, 9, 11, 157,

Atomic Energy Authority – *cont.*
160, 170, 199, 207, 222, 289, 301; relations with APC, 121–67 *passim*, 197, 207; articulate body, 289; export activities, *220/60–63, 309/75*; seeks more power, 11, 181, 197, 301, 304; resists transfer of R and D, 214, 301; new powers acquired, 114, 207, 301; participation in D & C companies, 114, 206, *220/65*, 304; pro single D & C company, 198–9; in NNC, 195, 248; reactor policy, *see* Narrow front, Reactor types, Dungeness B; position in fuel industry, 114, 120, 208, 214; never commercial, 174; forecasting errors, 154–5, 157–8, 160, 267, 277, claims for AGR, 118, 121, 150, 267; failures in judgement and performance, 152–61 *passim*, 167, 182, 205, 286–7; bias in presenting policy, 280, 286–7, 301; credibility questioned, 304–5; Dungeness B, part played, 119, 151–61, 166, 266–7; recommends PWR in place of SGHWR, 249–50; export, 200, *220/60–63, 309/75*; FBR and SGHWR teams go to TNPG, 208

Atomic Energy Board (UK): under Cherwell (1952–4), 273; proposed by Benn, 206–7, 216; *see* Nuclear Power Advisory Board

Atomic Energy Clearing House (AECH), *93/4 and 10, 94/20–24 and 28–30, 95/34–5*

Atomic Energy Commission (AEC): executive agency, 4; Chairman, Presidential appointment – Seaborg, 60, Lilienthal, 61, Schlesinger, 64, Dixey Lee Ray, 91; duty to strengthen free competition in private enterprise, 4, 61; promotes participation in development by competing manufacturers and utilities, 4; by 1955 over twelve participation projects, 5; reactor experiments and experimental reactors, published lists, *17/4*; division of Reactor Development and Technology, 57, 66; estimates and forecasts of LWR costs, 30–3; comparison with fossil fuel plant costs, 33–6; bases of AEC cost figures, 35–44 *passim*; emphasis in 1968 Cost Study misjudged, 45;

Regulatory Licensing Division overloaded (1969–70), 45; organisation criticised by JCAE, 92; licensing responsibilities widened by NEPA and anti-trust responsibility imposed, 48; *see also* Regulations to ensure safety in US, *and* ASLB; tests to destruction, 235; *Safety of Nuclear Power Reactors* (1973), 57, 59; Rasmussen Report *Reactor Safety Study*, 59; *Tentative Staff Report*, 70; case for reorganisation, 76; replaced (1975) by ERDA and NRC, 16, 83; suppression of evidence alleged, 65, 70

Atomic Energy Committee (Sweden), 103

Atomic Energy Company (Sweden), 103–4

Atomic Energy of Canada Ltd (AECL), 239, *256/13*

Atomic Energy Office (*later* Atomic Energy Division, Ministry of Technology), 176–9

Atomic Industrial Forum (US), 83

Atomic Power Construction Co. Ltd (APC), *115/1*, 120–8 *passim*, *143/28–31, 144/54, 145/66, 73 and 91*, 156, 159, 161, 164–7, 170, 178, 181, 197, 203, 207, 266, 286; acquired by CEGB, *146/93*

Atomic Safety and Licensing Board (ASLB), 46–8; hearings called a charade, 46, 76; membership, 46; JCAE criticism, 92

Australia, 228, *257/40*

Automation and computerisation of controls, 42

Availability; statistics for early plants, 25–9; US efforts to raise, 85; misleading comparisons, 182–3, 229–230; *see also* Load factors

Avebury, Lord, *see* Lubbock, Eric, MP

Babcock & Wilcox: US – *52/22 and 25*, 92, *93/8*, 284; Germany – 98–103; UK – 5, *115/1*, 116, *143/15, 219/36*

Babcock English Electric Nuclear (BEEN), formerly NDC, *219/42*

Babcock-Brown Boveri GmbH (BBR), 102

Back fitting, 45, 84, 90

Badische Analin-Soda Fabrik A G

(BASF), *110/33*, 128, *145/76*, *220/67*

Batelle Memorial Institute, *53/28*, 77/5 and Rasmussen Report, 59

Bayernwerk, 100, 103

Belgium, 96, *191/106*, 250, *262/130*, 282

Benn, Anthony Wedgwood MP, 14, 127, *144/54*, *145/57*, 169, 171–9, *191/105*, 195, 206–10 *passim*, 214, 216, 224, 248–9, 251, 253, 255, 256, *262/132*, 264, 279, 298, 300, 301, 304, 305, *310/90*

Benson boilers, *146/95*

Berkeley Plant, 130, 184; advanced code review, *95/36*; generating cost, *22/36*; load factors, *192/118–119*

Berridge, D. R., 224, 227, *257/19*

Beryllium cans, *143/20*, 266

Bessborough, Lord, *22/40*

Bias: in *Appraisal*, 10, 151, 158, 161, 287; in AEA's presentation of policy, 156, 280, 286; seen as likely in AEA by Select Committee, 172, 301; in ministerial attitudes, 217; alleged in US atomic establishment, 69, 239

Biblis Plant – A, 97; B (1970) 60% dearer, 102

Big Rock Plant, 118, *187/31*

Bingham, Jonathan, Congressman, *80/51*

Blackett, Patrick Maynard Stuart (later Lord Blackett), 174, 300

Board of Trade, 264

Boardman, Tom MP, 212, 213–15 *passim*, 217, *221/79*, 264, 289

Boiler design, Dungeness B, Oldbury and Hinkley Point B (Benson), *146/95*

Boiling water reactor (BWR), *see* Reactor types

Bomb risk, 62, 90

Booth, Eric S., 125, 224, *257/19*

Bradwell Plant: corrosion, 130; generating cost, *20/36*; load factor, 184, *192/118–19*

British Electricity Authority (BEA), 161

British Insulated Callender's Cables (BICC), *143/15*

British Nuclear Design and Construction Co. (BNDC) (earlier NDC), 116, 121, 129, 131, *144/54*, *146/93*, 164, 207–8, 214, 215, *219/41 and 44, 220/48 and 55*, 224, 226, 228, 255, *258/44*

British Nuclear Export Executive (BNX), 197

British Nuclear Forum, 105–6, *110/32*, 195, 217, 226

British Nuclear Fuel Ltd (BNFL), 106, 195, 213, 255, 269, 289, 294

British Nuclear (Holdings) Associates, 212, 255

British Nuclear Policy: 1954–73 consortia changes, 115–16; 1955 First Nuclear Programme, 118; 1964 Second Programme, 120, 152, 154, 155, 167; 1973 National Nuclear Corporation (NNC) q.v.; one D and C company, 114, 126, 129, 173, 195–222 *passim*; need for marketing, 173–5; need for export, 195–8

British Safety Inspectorate, *see* Nuclear Installations Inspectorate (NII)

Brookhaven Laboratory, *53/28*, 60, 72, *79/38*

Brown, Dr Gordon, 119, 121–2, *146/93*, 287

Brown, Ronald MP, 121–2, 234–40 *passim*, 251, *259/82*, *261/113*

Brown, Sir Stanley, 121, 125, 131, 132, *143/28*, *144/54 and 56*, *145/66*, *148/111*, 162, 166, 181, 210, 224

Brown-Boveri (BBC), 100, 102, 103, *109/23*, 128, *146/91*, 203

Brown-Boveri-Krupp (BBC-Krupp), 100, *219/48*

Browns Ferry Plant, 31, 35, *52/21*, *93/16–17*; disastrous fire, 85–7; pressure vessel delivery delay, 44; came in at full power, *259/76*

Burch, James, *94/33*

Bush, Dr Spencer H., *53/28*, 66, 77/5, *79/38*

Business Men for the Public Interest, *53/30*

Calandria, The, 223, 231

Calder Hall Plant, 6, 8, 9, 11, 117–18, 130, *142/14*, *143/15*, 153, 154, 158, 182, 184, *192/118–19*, 271, 273

California: poll favours nuclear energy use, 16, 85, 91; resources, Agency of, Report for on *Public Health Risks of Thermal Power Plants*, 58, 67; projects dropped for earthquake risk, 62; 'Proposition 15', 85, 91–2

Callaghan, James MP, 278, 298
Calvert Cliffs judgement, 48–50, *53/44*, 85
Campbell, R., 255
Canada, 96, 115, 162, 227, 254, 271, 273; *see also* Reactor types, CANDU Capital costs: US, 31–43 *passim*; UK, 137–8, 227–8
Carbon dioxide-graphite system, 117–118, 169
Carl, Mrs William, *80/52*
Carrington, Lord, *218/6*, 264, 300, *309/59*
Central Electricity Authority, *see* British Electricity Authority
Central Electricity Generating Board (CEGB): 'all the best brains', 7, 285; ministerial powers over, 10, 165, 168, 179, 195, 215, 278, 291, 293; monopoly buyer, 161, 165; BNX, 198; external pressure on reactor choice, 10, 115, 163, 168–169, 215, 280; attitudes on: AGR (1969–76), 117–42; BWR (1964), 9, 155, 162, (1965) 10, 123, 161; CANDU (1962–3), 119, 162, 227, 278; HTR, 164, 222, 225, 226, 244; LWR (1964), 17, 113, 196, 225; PWR (1972), 17, 113, 194, 196, 240; SGHWR, 164, 215, 222, 244–245, 248–9, 281; relations with AEA, 13, 120–3, *147/98*, 150, 162–164, 167, 171–2, 202, 226; relations with APC, 121–7, *146/93*, 161–2, 164, 167, 266; *Appraisal, Dungeness B* (q.v.), 150; tender assessment, 123–4; Magnox generating costs cf. fossil fuel plants, 135; wants fewer and competitive suppliers, 120, 162, 202, 206; accepts single D & C Company, 114, 210; 'bond' or 'free', 161–4; oil-fired plant orders, 209; Select Committee and CEGB evidence, 125–128, 141, *145/66*, 201, *225/9*, 240; Vinter's praise, 289; Sizewell cancelled, *220/55*; Annual Reports, 132, *147/101, 256/3, 257/27*; Plowden Committee, 285; subdivision, 284–5
Centralisation, 6, 11, 17, 197–218 *passim*, 243–4, 268–9, 281, 285–6, 293, 301, 305–6
Centrifuge process, 103, 228
Champion, Lord, *309/77*

Chapelcross Plant, 130, 184, *192/118–119*, 273
Charpie, Dr Robert S., *18/8*, 55, 64, 182, *306/21*
Chemical reprocessing, *see* Nuclear fuel
Cherry, Myron N., 65, *79/40, 80/43*
Cherwell, Lord, 264, 271–4 *passim*, 276, *307/31*
Chrome steel problems, 130–1
Churchill, Winston, MP, 246
Citizens' Committee for Environmental Concern, *53/30*
Citizens for Survival, *53/30*
Civil servants, 10, 13, 121, 135–6, 152, 159, 170–1, 175–80, 185, *190/105*, 210, 233, 234–5, 242, 249, 264, 270, 273, 274, 278–80, 288, 289, 290, 296, 297, 302–4
Cladding, *see* Fuel rods
Clarke-Chapman Ltd, *115/1*, 116
Clarke, Sir Richard, 264
Coal: cheaper for US than UK, 160; prices – UK, 252; prices – US, 33–4, 84; sources of higher costs (US), 35–6
Coated particle fuel, 128, *186/4*
Cochrane, Dr Thomas, *79/32*
Cockcroft, Sir John, 5, 6, 164, 272, 273, 300
Coleraine, Lord, 287, *309/59*
Combustion Engineering Inc. (CE), *52/25, 93/8, 144/30, 189/66*, 284; Standard design PWRs, 84
Comey, David, 65, 67, 76, *93/16*
Commissariat à l'Energie Atomique (CEA), 97–9, 268
Commonwealth Edison, 36, *142/14*, 225
Competition: value, 216; from abroad, 159, 198, 205, *220/63*; from fossil-fuelled plants in US, 25–7; in UK and US, 299; between teams, 204; in home market, 205; scope, 295–304; CEGB wanted in plant supply (1967), 202, 206
Concorde, 255, 302, *309/66*
Conservative Party Policy on nuclear energy: 6–8, 114, 168, 181, *191/10*, 195, 203, 205, 209–11, 215, 216–18, *219/33*, 269, 272–5, 279, 281, 291–293, 295, 299, 301, 302
Consortia, The (UK), 8, 9, 11, 121, 126, 133, 151, 157, 160, 162–7 *passim*, 170–1, 194–5, 197–201, 207, 215, 222, 269, 279–80, 282,

287, 289–92 *passim*, 300, 301, 304; formation, *218/6*; lists, *115–16*

Construction costs, 31, 39–41, 252; bases, 136–9; LWR(US), *51/2 and 5*

Consumption of electricity, *see* Electricity

Containment: LWR, 42; integrity of GE BWR Mark I, 89; Mark III standard design (STRIDE) BWR, 89, *92/2, 94/34*; pressure suppression system, 118; concrete containment (UK), 119

Cooling Towers and Ponds, 31, 42, 49

Core cooling systems, *see* Emergency Core Cooling Systems (ECCS)

Core melts, degree of risk, 68

Corrosion, 124, 130, 131, 151, 153, 161, 166, 169, *190/105*, 235, 248, 280

Cost-benefit analysis: all environmental aspects by AEC, 48; too much cash spent for safety, 249; none *ex post facto*, 302

Cost-benefit balance, negative because effluent risks so low, 64

Costs: LWR costs in US, 30–51; rise in total costs of plants (1967–77), 30–31; changes in separate components (construction, IDC, escalation), 31, 38–9, 41; rise in fuel costs (1967–76), 32; changes in separate components (enrichment, fabrication, reprocessing, pre-credit), 32, 41; rise in generating costs, 32–3; LWR of fossil fuel plant costs, 33–6; sources of LWR cost increases, 36–51; general economic conditions (inflation, exceptional rise in construction costs, rising cost of money), 37–41; changes in design and specification, 41–3; delays and long lead times, 43–51, due to intense development and diffusion coinciding, 44–5, political administration and judicial infrastructure, 45–51; LWR costs cf. fossil-fuel plant costs in France and Germany, 97–8; AGR costs and prices in UK, 133–51; fall and rise in AGR prices (1965–70), 133; cost of AGR programme, 136–41; prospective costs of AGR and other reactor types in UK (1973–1974), 134, 228–30; generating

costs in Magnox and coal-fuelled plants in UK compared, 134–5; Department of Trade and Industry criticises accounting cost comparisons, 135; estimated cost of LWR in US cf. GHWR in UK (1976), 252

Cottrell, Sir Alan, 234–7, 240, 241, 246, 250, *259/76, 310/90*

Court decisions: Court of Appeal, D.C., US, 85, 92; *see also* Calvert Cliffs and Quad Cities

Cousins, Frank MP, 168–9, *189/70*, 264

Creative Initiative Foundation (CIF), 90, *94/33*

Critical Choices for America, Committee on, 60

Cunningham, Dr John MP, 185, 251


D & C company, *see* Single D & C company

Dalyell, Tam MP, 127, *145/64*

Davies, John E. H. MP, 195, 210, 264

Decentralisation, 216, *221/75*

Degussa, 101

Delays and long lead times: US, 43–51 *passim*, 252; outside US, 106–107; UK, 151, 251, 283, 298, 299

Delays in Commissioning CEGB Power Stations, Committee of Enquiry into – *see* Wilson Committee

Delays in delivery of conventional plants, 185, 283

Dell, Edmund MP, 276, 292, 294–6, 302, *307/34, 309/61–2, 64–5 and 67–8, 310/80: Political Responsibility and Industry*, 292

Demonstration Plant in UK discussions, 225, 226, 230, 231, 246, 265

Depreciation provisions, CEGB, 134–135, *148/112*

Dept of Energy, 251; Chief Scientist vice-chairman AEA, 303

Dept of Industry, 285

Dept of Trade and Industry, 135, 136, 210, 242; cost estimates, *148/116*

Derating of reactors: US, 71, 90, 130; UK, 153, 169, 184

Design and specification changes, 280; effect on costs, 41–4, 57

Development process, 2–8, 301–2; US and UK 'down-to-earth' engineers' philosophy, 272; US pattern, 4, 8, 280; UK, 5–14; Cockcroft and

Development process – *cont.*
Hinton approach, 7, 272; US programme plant and system variety, 4–6; UK one system, 7; LWR planning, 153, 280; AEA's misjudgement of US procedure, 158; AEA lacks experienced development engineers, 158; criticised for 'drawing board' development and no feedback, 158–9; Hill (1974) wants demonstration plant for new systems, 230; but Varley opts for SGHWR programme not demonstration plant, 281
Diffusion Plant, 117
Dignam, T. J., 76
Direct cycle systems, 9, 118 (BWR), 223 (SGHWR), 254 (HTR)
Ditmore, D. T., *187/31*
Doggett, F. J., 227, 231
Dounreay, 154, 207; DFR *145/60*
Dragon Project, 253
Drax Plant, 135–6, 185, *193/120*
Dresden Plant, *52/21*, 118, 141, *142/14*, 158; zirconium can, *187/31*, *188/38*
Dungeness A, 135; load factors, 184, *192/118–19*
Dungeness B, 9, 132, 133, 150–5, 158, 159, 161, 165–70, 279; problems, 122–31; 'greatest breakthrough', 10; inadequate prototype, 119, 123, 153, 160; extent of extrapolation, 123; tenders and contract, 119–24, *140/28*, 163, 168; AEA's part, 120–4, *143/28*; favourable assessments, 123, 163; APC tender not 'in depth', 122–3; initial time scale, 124, 150, 153; design deficiencies and delays, 124–5, 130, 151; price too low, 129; APC's weakness, 120–7; management transferred to BNDC, 131; estimate of cost (1965) *148/119*, 132–3, (1975) 136–137; *see also Appraisal*, APC, 36-rod cluster, sketch design for
Dunster, H. J., *78/25*, 249
Düsseldorf, 100

Earthquake risks, 30, 62; Bodega Bay site, *308/38*
Ebben, Stephen, *81/59*
Eccles, Sir David (later Lord Eccles), 7–8, *18/21*, 117, 264, 274, 275
Economic environment changes and effect on costs, 36–41

Eden, John, 264
EDF 4 Plant (France), *108/5*, 164
Edison Electric Institute, 83
Edwards, Sir Ronald, 7, *221/75*
Effluents, radioactive: from LWRs, 42, 47, 62–4, 72, *79/30*; from oil-fired plants, *77/10*
Egypt, 90
Electrical, Electronic, Telecommunications and Plumbing Union, 246–7
Electrical power, demand for: fall (1972), 224; forecast, 224–5; slow growth, 300–1
Electricité de France (EDF), 97, 98, *108/5*
Electricity, consumption of: US, 25, 83; UK, 283
Electricity costs, 140
Electricity Council, *188/40*, 215–16, 225, 247, 256, 285
Emergency Core Cooling Systems (ECCS), 42, 47, 65, 71, 72, 90, 231–7 *passim*: criticism discussed, *258/71*; computer system not sophisticated enough, 66; interim criteria, 65, and revised, 74; rule-making hearings, 65–77; semi-scale test, 67, 76, *79/40*: public 'bored' by, 85; LOFT results, 90, *259/81*
Energy Research and Development Administration (ERDA), 35, 83, 86, 243
Engineered safeguards, 42, *52/25*, 56–8
Engineering standards: upgraded for nuclear plants (US), 42–3, 56–7; UK, 235; universal problem, 266, 280
England, Glyn, 256
English Electric Company (EE), *115/1*, 162, 207, *219/36 and 41*
English Electric–Babcock and Wilcox–Taylor Woodrow Group (BEEN), 207
Enrichment, *see* Uranium, enriched
Enrichment costs in US, 32, 41, 84, *306/17*
Environment Control Administration, *79/28*
Environment Protection Agency (EPA): AEC must accept standards, 48, *53/39*; said Rasmussen understated hazards, 86
Environmental Impact Statement, 48, *53/39*
Environmentalist Groups: US, 15, 16,

23, 49, 55, *53/27 and 30, 79/40*, 281; influence outside US, 16, 106–7; US groups quoted in UK, 236–7; UK, 195, 247, 251, 294; *see also* Intervenors, Friends of the Earth

Escalation: in US, 31, 36–8, 40, *51/2*, 83, 134, 137; not included in UK forecasts, *258/44, 262/121*

Establishment, Atomic: in US, 57, 61, 67, 69–73 *passim*; in UK, 9, 11, 16, 267, 278, 288

Estimates, *see* House of Commons Select Committee on

Etzel, Franz, *19/24*

Euphoria over AGR, 10, 14, *20/42*, 113, 152–3, 163, 224, 267

Euratom (European Atomic Energy Commission), *19/24*, 100; *Target for Euratom, A*, *19/24*

European Communities Energy Policy Strategy, *see* House of Lords Select Committee on

European Economic Community (EEC), fanciful target, 107

'Event trees', 59

Experts, lay judgements on, 68–9

Exportability of reactors: LWR 'the only exportable reactor', 301; British reactors in general, 8, 113, 128–9, 175, 205, 279, 301; AGR, 10, 126, 128–9, *191/105*, 194, 197, 209; Magnox, 9; SGHWR, 128; 196, 209, 226, 227, 228, 265

Extrapolation, 11, 122, 151, 158, 165, 166, 230, 286

Fairey Engineering Ltd, *115/1*, 226

Farmer, F. R., *77/7*

Fast Breeder Reactor (FBR), *see* Reactor Types

'Fault trees', 59

FBI Conference (1958), 182, *187/30, 191/111, 192/113*

Federal Radiation Council, 62, *78/23*

Fermi, Enrico, 60

Finland, 104, 228, *257/40*

Fishlock, David, *145/63, 187/27, 192/114*

Flowers Report, The, 17, 86

Forecasts and forecasting: Paley and McKinney Reports (1952–5), 1–3, *17/1, 18/5*; Hinton's Magnox costs, 12, *19/34, 142/9*; sources of over-optimism in these, 155, 158;

date for competitive UK nuclear power periodically deferred, 155, 267; FBR successive forecasts *186/18*; Hinton on the best, 155; LWR supporters', 253; Penney on prospective AGR costs (1967), *20/42*; unrealistic basis for AGR cost estimates (1964), *143/24*, (1967), 14, *20/42*; need for marketing men's forecasts (Benn), 173; electricity demand and forecasts, 288; NPAB, 244–5

Fort St Vrain Plant, HTR, 253

Fortescue, Peter, *19/20*

Fossil-fuelled plant: US costs, 33–4, 82; UK manufacturing and design failures, 185, 252; steam plant working costs, *263/138*

Fragmentation of production of nuclear plant: US, 45, *52/25*; UK, 11, 14, 166–7, 199, 266; West outside US, 200

Framatome (France), 98, *257/31*

France, 15, 96–8, 114, 118, 155, 164, 183, 216; cf. German organisation, 97–103, 268, 271

Francis, Owen, 224, *257/19*

Franklin, Dr N. L., 107, 255, 294

Friends of the Earth (FOE), *53/30*, 184, 234, 236–7, *257/44–258/46 and 67, 259/71, 260/88*

Fuel and power costs: US, 33–4, 36–7, *51/4 and 5*; Germany, 97; UK, 27, 135–41 *passim, 147/106–7, 148/119–20, 149/121*, 252–3

Fuel fabrication US: fall in cost (1971), 32, 41

Fuel manufacture UK, 106, 204, 213: cf. US, 213; *see also* British Nuclear Fuels Ltd

Fuel reprocessing US, 32, 41, 64, 84

Fuel rods: LWR – Diameter, 42; Temperature, 74; PWR-internally pressurised, 42; BWR – Hot vacuum dried, 42

Fuel rod clusters – AGR, 120, 121, *143/20*, 151, *187/31, 188/41*

Gag behaviour, 132, *147/100*

Gas-graphite reactor, *see* Reactor types

Gaseous effluents, *see* Effluents

Gaulle, General Charles de, 97

GE anti-nuclear engineers, 85, 87–90, *94/32 and 34*

General Atomic, *21/20, 219/48*

General Dynamics, later General Atomic q.v.

General Electric Company UK (GEC), 114, 194–5, 211–12, *219/41, 220/60*, 227, 254, 255, 269, 283, 284; absorbed AEI, *115/1*; absorbed English Electric, 116, 207; in BNDC, 116; *see also* Weinstock

GEC–Simon Carves, *115/1*, 121

General Electric Company US (GE): *modus operandi*, 211; BWR–6 (1972–3), 43, 82; gaseous release, 63; three separate identical systems for emergency core cooling, 72; Dresden, *142/14*; standard nuclear designs agreed PDA, 84; Reed Report, 85; resignations, 85, 87–9; TNPG, 161; CEGB, 162; link with AEG, 203; Select Committee visit, 239; fuel manufacture, *187/31*; Bodega Bay, *308/38*

Generating costs, *see* Costs

Generating plants, large in US, 24–5

Geneva Conference on Peaceful Uses of Atomic Energy (1964), 127, 156

German Atomic Forum, *109/19*

Germany: Atomic Energy Advisory Committee, 99; Federal Republic, 11–15 *passim*, 28, 97–103 *passim*, 106, 114, 216, 217, 229, 232, 268, 282, 284; Government nuclear expenditure, *109/14*; 4th Nuclear Programme, *110/34*; safety, *110/33–4*; utilities' autonomy, 101, 206

Ghalib, S. A., 151, *186/3–4*, 223, 231, 247

Gibb, Sir Claude, 182

Ginna Plant, *260/89*

Ginsburg, David MP, 203, 211, 212

Giordani, Francesco, *19/24*

Gofman, Dr John W., 62–4, 67, *78/25, 79/26*, 87, 249

Goldsmith, Sir Henry D'Avigdor, *186/15*

Gowing, Margaret, 6, *18/9–13*, 118, *142/2–6, 12 and 14, 186/22*, 272, *306/18–20, 307/22–8 and 30*

Grain (oil plant), 136

Grainger, Leslie, 126, 134, *145/60*, 170, *188/31*

Graphite, 7, 10, 117, 130, 131, 154, *30E/30*; *see also* Reactor types, gas-graphite

Greece, 228, *257/40*

Green, Professor Harry P., 75–6

Greenfield, Professor M. A., *77/9*

Greenhalgh, G. H., *110/32*

Ground rules: US, 34–5, 41; US 'standard' used in UK controversy, *190/105*; UK, 133–5

Growth rates, comparative, 282–3

Guard, R. F. W., *19/29*, 128, 134, *144/48*, 166

Gundremmingen Plant, 100, 103

Haddam Neck Plant, *260/89*

Hailsham, Lord (formerly Quintin Hogg MP), 14, *20/37 and 40*, 162, 264, 291, *309/59*

Hammond, Rolt, *143/15*

Harris Opinion Poll, 16, 91

Hartlepool Plant, 131, 133, *146/87 and 91*

Harwell: UK atomic research centre, 5–6; Cockcroft, first director, 5; envisages reactor development by Harwell and industry, 5–6, 272, 291, 300; but Harwell's only development outlet Risley, 274; misses PWR significance (1951–2), 272; opposition to AGR, 7–8, 117, 266, 277, 280; prestige, 182, 288; research on SGHWR, 247, *262/120*; early support for LWR, 7, 12, 275, 277; *see also* Cottrell, Grainger, A. J. Little, Marshall and Merrison

Hawke, Lord, *309/59*

Hawkins, Arthur (later Sir Arthur), 131, *147/97 and 99*, 150, 151, 161, 163, 166, 181–3, *221/78*, 224–6, 230, 231, 234, 240, 255, 256, *258/44*, 280

Head Wrightson & Company, *115/1*, 116

Health and Safety Executive, 249

Heat exchangers outside Magnox pressure vessel (1957–8), 119

Heath, Edward MP, 299

Heavy water, 117, 162, 223, *307/30*

Helium, 7, 117, 128, *186/4*

Henderson, Professor David, 86, *310/78*

Hendrie, Dr Joseph M., 66, 74–5, *79/38*

Hennings, John R., *95/41*

Herbert Committee Report, 7, *18/18*

Heysham Plant, 129, 133, 136, *146/87*, 207, *219/44*

High temperature (gas-cooled) reactor

(HTGR and HTR), *see* Reactor Types

Hill, Sir John, 124, 141, 154, 158, 163–167 *passim*, 181–2, 185, 210, *221/78*, 226–7, 231, 234–5, 238, 240, 244, 247, 249–50, 255, *257/42*, *259/76*, 267, 278, 286, 294, *306/2*

Hinkley Point A, *192/118 and 119*, *193/120*

Hinkley Point B, 122, 125, 127, 130, 131, 133, 135, 136, *146/89*, *147/106*, *148/119 and 120*, 185: *Exploitations of the Advanced Gas-Cooled Reactor*, *189/58*

Hinton, Sir Christopher (later Lord), 5, 12, 13, 104, 105, 117–18, 154, 155, 157, 160–3 *passim*, 166, *186/12*, *187/28*, 196, 198, 224, 228, 234–6, 272, 275, 285, 287, 300, *306/5*: *The Future for Nuclear Power*, *19/34*, *142/9*, *308/50*

Hitachi (Tokyo Atomic Industrial Consortium), 105

Hobbes, Thomas, 276, *307/33 and 34*

Hochtemperatur Reaktorbau (HRB), Germany, 253

Hoechst Chemical Company, 99, 103

Hogg, Quintin, *see* Lord Hailsham

Hosmer, Representative Craig, 74–76

House of Commons, 130: Committee of Public Accounts, 167, *189/57*, *59, 63 and 69*, *306/2*; *see also* Select Committees on Estimates, National Expenditure, Nationalised Industries, and on Science and Technology

House of Lords, *see* Select Committee on European Communities Energy Policy Strategy

How, Sir Friston, 177, 179

Humboldt Bay Plant (Calif.), 118, *187/31*

Hunterston Plant A, 130, *192/118 and 119*

Hunterston Plant B, 131, 133, 136, *145/66*, *148/114 and 120*

Idaho, 60, 90, 107, 236, 238, 242; *see also* Semi-scale experiment

Imperial Chemical Industries Ltd, 5, 273, 300

Import of nuclear plants, potential, by Generating Boards, 203

In-service monitoring, *77/3*

Incentives to speed and efficiency: US,

297–8; lacking in UK, 13–14, 298–9

India: atomic bomb, 90; CANDU order, 96

Indian Point II plant (Westinghouse), 85: 'An accident waiting to happen' 87

Industrial Reorganisation Corporation (IRC), 115, 200, 206, 207, *219/36 and 37*, 300

Inspector/Inspectorate of Nuclear Installations, *see* Nuclear Installations Inspectorate

Institution of Professional Civil Servants (IPCS), 121, *144/31*, 195, 196, 209, 227, 246, 294

Insulation, 130, *146/89*, 151, 161

Integration – backward or forward, 200

Interatom, 100, 103, *109/27*

Interest – self-interest and public interest, 72, 304

Interest during construction (IDC) US, 31, 39–40, 43, *51/2*; UK, 132, 134, 136, 137–40, 248, *258/44*, *262/121*

Interest rates, *see* Money, cost of

International Combustion Ltd (ICL), *115/1*, 120–5 *passim*, *219/36*

International Commission on Radiological Protection (ICRP), *78/22 and 23*, 249

International Companies, 98, 100, 101, 105–6, *219/36*, 232, 284, 300

International Energy Agency (OECD), *110/36*

International safety standards, 107

Intervenors US: and ASLB, 46–7; increased activities late 1960s, 62; part in ECCS rule-making hearings, 65–77; methods of formulating and presenting case, 65–9, 75–6; assert all experts biased, 69, 239; lack access to information, 69; seek public funding to make case, 69; persuasive though technically weak, 73; favoured by media, 73; political 'winds' favourable, 73; utilities concede to save time, 73; 'too academic' 72; assessment of impact, 74–7; evidence used in UK, 236; could contest US Attorney General's licensing decisions, 48

Irradiation facilities, 172, 174, 208

Israel, 90

Italy, 96, 104, 216, 268, 282, 284

Jackson, H. W., 122, 155, 166
Jackson, Professor Willis (later Lord Jackson), 300
Jaffé, Professor L., quoted *54/45*, *80/54*
Japan, 97, 98, 105–6, *110/36*, 216, 229, 232, 268, 282, 284: LWR order, 96; Magnox, 105; supplies pressure vessel to US, 44; groups making PWRs and BWRs, 105; Japanese Atomic Industries Forum (JAIF), 105, *110/32*
Jenkin, Patrick, 242, 264
Jet pumps for recirculation, *188/38*
Joint (Congressional) Committee on Atomic Energy (JCAE), *see* US Congress
Jones, Aubrey, 264
Joseph, Sir Keith, 14, *20/40*
Joslin, Murray, *142/14*
Joss, J. O., *20/36*, *188/52*
Judgement, Ministerial, 276–7, 289, 291, 294–7
Jukes, J. A., *142/10*
Julich Research Institute, 99

Karlsruhe Atomic Research Institute, 99
Kasper, Raphael, *80/53*, *81/59*
Kearton, Sir Frank (later Lord Kearton), 206
Kennedy, Edward, 69
Kernreaktorteile (KRT), 101
King, Cecil, 211, *220/57*, 255
Korea, 96
Kouts, H. J. C., *79/41*, 241, 250
Kraftwerk Union (KWU), 102–3, *109/22 and 24*
Kronberger, Dr H., *145/69 and 73*, 177
Krupp, 100; *see also* Brown-Boveri-Krupp (BBC-Krupp)
Krypton (87 and 88), 63

Labour Committee Against Proposition 13 (Calif.), 91
Labour Party policy on nuclear energy, 168, 170–5, 196, 203, 205–209, 245–51 *passim*, 256, 279, 282, 283, 292, 293–6 *passim*, 298–302 *passim*; *see also* Liberal-Labour majority on Select Committee
Lawrence Radiation Laboratory, 62
Lawyers US: and ASLB, 47; muscle

in for political influence, 74; usefulness in technical judgement questioned, 75–6; Cherry and Roisman, 65; *see also* Judges
Lead in development: US, 8–9, 14; UK claims, 9–10, 13–14, 157, 168, 194, 241, 265; UK 'firsts', 9, 114; Germany – HTR, 15; France – FBR, 15, 194
Lead times: in US: big increase in Phase 2, 30, 40; average time from order to commercial operation (1966–74), 43–4; sources of increase, 43–51; impact of longer lead times, 43–4; hopes of reduction, 51, 82, 84; building required only 4½–5½ years, 50; lead time taken as 9 years (1975–6), 51; in Germany and Japan 4½–5½ years proved possible, 106, 232; in UK 5½ years (1973–4) claim for LWR but disputed, 230–1
Lead times and repeatability in UK, 230–2
Leadbitter, Ted MP, 234, 239
Learning: expected source of cost reduction (US), 42–3; standard designs as means, 82–4; learning curve for LWR probably 90%, 42; slowed down by volume of orders, 43; Prospects improved 1973–76, 84
Le Creusot, *257/31*
Lee, Fred (later Lord Lee), 10, 163, 168, 264
Legge-Bourke, Sir Harry MP, 121, 122, *143/28*, 162, *188/41*
Liberal-Labour majority on Select Committee, 197, 204, 205, 210
Liberal MP on Select Committee, *see* Lubbock, Eric
Licensing US, 238, 282, 300: increased time taken, 45–51 *passim*, 83–4; sources of delay, 45, 74–6; public hearings, 21–2, 24, 31–4, 36, 47, *52/20*, 74–6; impact of NEPA, 47–48, 92; agencies involved in addition to AEC, 46–8; rule-making hearings, 46, 65–77 *passim*; antitrust requirements, 48; conflict on State rights, 49; judges' impact on licensing, 49; impact of lawyers, 65, 74–6 *passim*
Light water reactor (LWR), *see* Reactor types
Lilienthal, David, 61, 69, *78/13 and 21*

Lipton, Professor Michael, *310/80*
Liquid Metal Fast Breeder Reactor (LMFBR), *see* Reactor types
Little, Dr A. J., *187/30*
Little, Arthur D., Inc., *108/5*
Lloyd, Geoffrey MP, 264, 274
Lloyd Harbour Study Group, *80/52*
Lloyd, Ian MP, 238–9, *311/94*
Load factors: in cost estimates, *29/1*, 84, 134; comparison of Magnox and LWR records, 153, 183–4, *190/105*, *192/118 and 119*, 229, *258/44*
Los Alamos Plant, *53/28*
Loss of coolant accident (LOCA), 71, 74
Loss of fluid tests (LOFT), 85, 238; first results, 90; OECD, *110/36*; became international, *259/81*
Lubbock, Eric MP (later Lord Avebury), 126, 129, *144/57*, 183, *192/ 115 and 117*, 201, 203
Ludwigshafen, *110/33*
Lyons, J., *144/31*
Lyttelton, Oliver (later Lord Chandos), 291

McAlpine, Sir Edwin, 122, *146/86*, *189/55*, 223, 247
McAlpine & Son, Sir Robert, *115/1*, 116
McKinney Report, *18/5 and 8*, *77/1*, 182
McMahon Act, 156
Macmillan, Harold MP, 264, *310/79*
Magnox, *see* Reactor types
Maine Yankee Plant, *260/89*
Mandel, Professor Heinrich, 100, *109/ 16 and 19*, *145/75*
Manufacturer/utility axis, 4–5, 267–9, 282, 305, 306
Market for nuclear plants: US, 25, 26, 82, 85, 92; UK, 107, 216, 224–5, 250, 283–4, 300–1, 303; other countries, 96
Market orientation, 173–5, 195–8, 205, 208
Marsh, Richard MP, 14, 169, 170–1, 179, 264
Marshall Report – *An Assessment of the Integrity of PWR Pressure Vessels*, 242, 249, 250
Marshall, Dr Walter (deputy chairman AEA), 242, 246, 249, 250, *262/120*, 265
Marsham, Dr N., 255

Marviken Plant, 104
Maschinenfabrik Augsburg – Nürnberg AG (MAN), 102, *109/27*
Mason, Roy MP, 130, 264
Matthews, R. R., 234, 240, *261/92*
Maudling, Reginald MP, *19/34*, 264
Media: support for environmentalist cause in US, 73; in UK (World in Action), 241; susceptible to AEA appeal, 288; *Multi-Media Confrontation* (US), *80/44*
Menzies, Sir Peter (Chairman, Electricity Council), *221/78*, 244
Merrison, Sir A. W., 216, 244
Metal fatigue and fracture, 124, 236, 245, 248
Metropolitan-Vickers (Metro-Vick), 5, *143/15*, 300
Michaels, M. I., 176, *190/105*
Middle East: US and West's dependence for fuel, 23; prices, 253; OPEC policies, 107
Midland Plant (US), 92
Miller, Keith, *95/36*
Mills, Lord, *218/6*, 264, *309/59*
Minister of Energy, 264, 298
Minister for Industry, 212, 289
Minister of Power, 130, 163, 168–9, 171, 176, 179, *218/19*, 233, *257/19*, 264, 289
Minister for Science, 179, 291
Ministerial intervention – non-nuclear misjudgements, 302
Ministers responsible for atomic energy UK, 178, 264
Ministers UK: decision-making powers, 195, 255, 269, 274, 281, 293, 295; restrictive, 6, 214; in AEA, 6–8, 170, 171, 274, 288, 303; in NPAB, 216–17; in CEGB, 10, 168, 170–1, 215, 281; in NNC, 195, 213; in NII, 282; qualifications for job, 170, 270, 274, 275–7, 291–305 *passim*; Select Committee view of qualifications, 169, 176, 180, 290; compared with those of Company chairmen, 275, 276, 295, 297; need for objective advice, 178, 215, 277, 286, 290; options, 270–1; mistakes and misjudgements, 13–14, 179, 251, 264–266, 270, 278, 279, 283, 284, 290, 291, 294–5, 302; ministerial responsibility, 150, 168, 170–1, 177, 178, 216, 242, 264–72, 269–306 *passim*; emotive element, 280

Ministry of Fuel & Power, *218/6*, 273
Ministry of Supply, 264, 274, 275; Atomic Energy Dept. Industrial Division, 7, 8, *155–6*, 168, 176, 178, 273
Ministry of Technology, 129, 168–9, 173, 176, 178, 179, 206, *218/19*, 264, 289
Mitsubishi, 105
Mochizuki, Shiro, *110/32*
Moeller, Dr Dade, 88
Molten Salt Reactor, *see* Reactor types
Money, cost of: US: interest rate on short-term loans, 38–41; return on utility bonds, 39–40; UK: discount rates assumed by CEGB, 136–7, *148/109*; rates used by Dept of Technology & Industry in capital cost estimates, *148/116*; Treasury Test rate of discount, 138; *see also* IDC
Monopoly, 13, 114, 165, 200, 202, 203, 213, *257/31*, 285, 295, 297; dual (buyer and developer), 288; 'Atomic establishment', 288; experience of monopoly public bodies 290; 'Minister + monopoly', 290; *see also* AEA, CEGB and Single D & C Company
Moore, R. V., *79/28*, *142/1*, *143/19*, *145/69*, 155, 183, *186/10*, *192/117*, *221/78*, 223, 227, 231, 244, 273
Morris, J. R. S., 216, *221/78*, 244
Mortgage bonds, use by US utilities to fund new investment, 40
Morton, R. A., 210, *219/37 and 44*
Muntzing, L. Manning, 51, 82, 84, 86

Nader, Ralph, 23–4, 45, 73, 91
Namier, L. B., *191/107*
Narrow front policy, 7, 12, 13, 155, 159, 269, 274, 277, 291
National Coal Board (NCB), 126, 158, 170, *186/13*
National Council of Radiation Protection and Measurement, US (NCRP), *78/22*
National Economic Development Council (Neddy) – First plans, 224
National Environment Protection Act (NEPA), 47–8, 92; judicial attitudes to, 49–50
Nationalised industries, *see* House of Commons Select Committee on
National expenditure, *see* House of

Commons Select Committee on
National interest, *see* Public interest
National Nuclear Corporation (NNC), 113, 116, 195, 213, 226, 229, 231, 232, 247, 248, 251, 254, 255, 282, 293, 303: right in 1974, 306; operating subsidiary Nuclear Power Co. q.v.
National Power Survey (US), *17/3*
National Regulatory Commission (NRC) (US), 83–92 *passim*: accused of allowing commercial considerations to prevail over safety, 89; and of failing to examine energy conservation, 92
National Resources Defence Council (US), *79/32*
National Union of Mineworkers (NUM), 246
Nationalisation, *310/80*
Nationalised industries, *see* House of Commons Select Committee on
Nationalism, 113–14, 156, 168, 182, 200, 205, 233, 241, 246, 279, 284, 293, *309/77*
NDC, *see* British Nuclear Design and Construction (BNDC)
Neave, Airey MP, 181, 183, *191/110*, 227, 234, 237
New York bans street movement of nuclear fuel, 91
Nicolaides, Professor Leander, *108/9*
Nicolin, Curt, 200
Nine Mile Point Plant, 120
Nixon, President Richard: first energy message, 44; appoints Schlesinger Chairman of AEC, 64
Non-radiological environmental impacts, 47
North of Scotland Hydro-electric Board (NSHEB), 215, *258/47*
North Sea Gas, 283, 301
North Sea Oil, 196, 252, 253
Nuclear Design and Construction Co., *115/1*
Nuclear electric power plant production schedules US, 50
*Nuclear Energy* (FBI), *187/30*
Nuclear Installations Inspectorate (NII) (British Safety Inspectorate), 131, 210, 226, 229, 231, 233–5, 241–2, 246, 250, 255, 281, 282; Contrast with US, 242–3
*Nuclear Milestones* (US 1971), *78/20*
Nuclear plant construction industry in UK: engineering firms want wide

scope as in US, 7, 300; consortia's limited role, 8, 11, 164, 266; consortia want to choose and develop UK and foreign systems, 164, 290, 300; large export of nuclear plants forecast, 8, 197; export hopes disappointed, 10, 173, 175, 196; failure attributed wrongly to consortia, 11, 165, 197; reduction in number of consortia and pressure for single D and C company, 197; AEA plan for such a company organised round itself (1967), 197–202; Select Committee (1967) different plan for single company, 203–6; Benn decides against single company (1968), 206; first reorganisation (1968–9), 113–14, 206–9; fuel manufacture remains separate from D & C company, 206; some AEA teams transferred, 207; but little AEA R and D work transferred, 208; no export orders – nothing to export, 209; second reorganisation (1972–1973), established single D and C company, 114, 210; former opponents accept, 210; proposed functions and form contrasted with those of US companies, 213; wide Ministerial control established, 215; the UK industry virtually destroyed, 17, 254, 279, 290

Nuclear plant makers in US: scope of activities, 4, 156–7, 182, 213, 267–269; licenses to foreign firms, 98, 100–1, 105

Nuclear Power Advisory Board (NPAB), *148/118*, 195, 214, 216–218, *221/77*, 235, 240, 244–5, 248, 249, *262/121*, 281, 293, 294

Nuclear Power Collaboration Committee (AEA), 120

Nuclear Power Company, The (NPC), 158, 255: Whetstone and Risley, 116, 158, 254, 255

Nuclear Power Costs (UK), Working Party on, 170

Nuclear Power Group, The (TNPG), *108/5*, *115/1*, 116, 120–1, 126–8 *passim*, *143/20*, *145/73 and 85*, *146/89*, 151, 161, 164, 182, *186/3 and 10*, *189/54 and 55*, *191/106*, *110 and 112*, 197, 208, *220/48 and 55*, 224, 239, 304; *see also* Vaughan

Nuclear Power Plant Co., *115/1*, *192/ 112*

Nuclear Regulatory Commission (NRC) (US), 83–92 *passim*, *93/6, 7 and 17*, *95/34 and 36*; *Reactor Safety*, *94/19, 25, 26 and 29*

Nuclear Steam Supply System (NSSS), *52/25*, 84, *92/2*

Nuclear weapons, France, 98; UK, 5; US, 4, 11

NUKEM, 101–2, 103, *109/20*: *see also* Uhde, Lurgi consortium

Oak Ridge National Research Laboratory, 55, 59, 238, 241–3 *passim*, *258/71*, *259/80*

Office of Nuclear Regulation (NRC), 88

Ohio Edison Co., 92

Oil-fired plants UK: costs, 96, 133, 135, *147/102*; orders, 209; attractions over AGR, 283; attractions of LWR over, 283

Oil imports UK, *108/2*

Oldbury Plant, 119, 130, *145/85*, *146/ 89 and 95*, 153, 165, *186/10*, 222; load factors, 184, *192/118 and 119*; concrete pressure vessels, *189/55*

Onload fuel charging, 130–1

OPEC oil policies, 107

Open tender system, 213, 214

Oppenheimer, John, 60, *78/20*

Orders, *see* Market for nuclear plants

Organisation for European Economic Cooperation (OEEC), 98, *108/9*; later Organisation for European Cooperation and Development (OECD), *110/36*

Organisation of civil nuclear development: US: initial, 1–3; aims, competition and private enterprise, 267; executive agency AEC, 4; role of Congress (JCAE), 2; of maufacturing firms, 4, 201, 213, 297; of utilities, 5, 206; of architect engineers, 4, 200–1; nine reactors to be developed (1955), 5; safety organisation, 1, 15–16, 46–51, 55–6; organisation criticised by Nixon (1971), 44; alleged conflict between AEC's interests as promoters and regulators, 70; organisation change, 82; AEC, replaced by ERDA and NRC, 83; criticism not eliminated, 89; Select

Organisation of civil nuclear development – *cont.*
Committee on US organisation, 204; *see also* ACRS, ASLB and NEPA
UK: initial 5–8; AEA set up and monopoly, 6–7; utilities are government agencies, 6–7; role of Ministers and civil servants, 7–8, 13, 168–80; role of plant manufacturers, 6; consortia, 8, 164–8; government programme of commercial plants of one system (1955), 6, 8; centralisation, 11; advocacy of single D & C company, 197–202; established by Conservatives (1972–1973), 210; its constitution contrasted with that of US companies, 210–16; Conservatives set up NPAB, 216–18; safety organisation contrasted with US, 242–3; role of Select Committee, 180, 203–6; operation of US and UK systems contrasted, 298–9
France, 96–8 *passim*, 268
Germany, 11, 98–103, 114
Sweden, 103–4
Japan, 105–6, 199, 268
Orr-Ewing, Sir Ian (later Lord Orr-Ewing), 180
Oskarsham Power Group, 104, *110/31*
Oyster Creek Plant, 30, 31, *52/21*, 96, 120, 141, 158, 161, 183, *188/33 and 38*, 290

Pacific Gas & Electric Co., 118
Paley Report, The, 1–3, *17/1*, 61, 182
Palisades Plant, *260/89*
Palmer, Arthur MP, 150, 163, 167, 180–2 *passim*, *188/47*, 195, 198, 211, 214, 215, *221/75*, 225, 227, 228, 234–41 *passim*, 245–6, 249, 251, *261/113*, 269, 289, 301, 302, 305
Parsons, C. A. & Co., 5, 6, *115/1*, *143/15*, 182, *191/112*
Participation in US development strategy, 4
Pastore, Senator John A., 91
Pebble bed reactor, 100; *see also* Reactor types, HTR
Penney, Sir William (later Lord Penney), 124, 125, 155, 160, 171, *186/18*, *188/35*, *190/95*, 216, *218/7*,

*219/41*, 244; Report to Select Committee (1967), 14, 152–8 *passim*, 197–202, 213, *265*, 275
Phénix Plant (FBR), 15
Pippa, 6, *18/13*, *143/14*, 272–4; uranium in initial costs, *307/26*
Plowden, Sir Edwin (later Lord Plowden), 7, *187/30*, 285: Plowden Committee, Plowden Report, *Structure of the Electricity Supply Industry in England and Wales*, 216, 285, *308/44 and 53*
Plutonium (Pu): US manufacturers build and operate Pu-making reactors during war, 4; UK military demand for Pu doubles 1952, 6, 117; order for Pippa (Calder Hall) to make Pu and electricity, 6, 273; by-product value attached to Pu in fuel cycle (US), 41; value exaggerated in early UK cost estimates, 154, *307/26*; French start making Pu, 98; focus in anti-nuclear campaign, in US, 84, 90, 91, 304; in UK, 304
Point Beach Plants I and II, *260/89*
Political, legal and administration factors increasing delays and lead times, 44–51
Powell Committee (1964), 171, 210, 264
Powell, Sir Richard, 264
Preliminary Design Approvals (PDA), *see* Standard plants
Pressure tubes, 223
Pressure vessels: late delivery in US, 44; sub-contracted to European and Japanese firms, 44; steel, *77/3*, 119, 153, 232–5, 237, 241, 245, 250, 282; concrete, 119, 122, 124, 125, 130, 132, *146/95*, 151, 153, 165, 184, *189/55*; *see also* Marshall Report
Pressures, operating: Magnox, 153, 184; AGR, 121, 124, 130, 153; BWR and PWR, 11
Pressurised Water Reactor (PWR), *see* Reactor types
Price, David MP, 167, 180, 200, *218/12*
Prices: US: retail and wholesale, 37; coal, 3, 33–5; oil, 3, 33–4; construction, 36–9; electricity, 33–4; gas, 34
UK: coal, 252–3, *263/138*, 283; oil (1967–70), *108/2*, 283
Proctor, Sir Dennis, 264

Productivity factor in construction costs, 37, *52/10*

Programmes, UK nuclear plants: projected (1950), 277; First 10-year Programme (1955, 1957), 8, 9, 12, 118, 274, 277; Second Programme (1965), 120, 152, 154, 155–6, 290–291; 1969 Programme, 129; 1974 Programme, 281, 293; 'two cardinal errors', 267–8; attractions of 10-year programme, 156, 271–2; Hill advocates development type programme, 226; NPAB thinks Tombs' plan for SGHWR programme 'practicable', 244–5

Proliferation US, 16, 77

Proposition, 15; *see* California

Prototype Fast Breeder Reactor (PFR), *see* Reactor types

Public Accountability in UK, 299, 302, 303

Public Accounts Committee, *see* House of Commons

Public hearings US, 242–3: when required, 45, 46, *53/27*; for rule-making, 47, 65, 75; criticised, 74–77, 92; Tombs' view on value in UK, 234; as sources of delay, *see* Licensing

Public interest, 155, 185, 214, 276, 281, 288, 290, 291, 295, 298, 302


Quad Cities Plant, *53/39*

Quality control, 166, 185


Raby, Colonel G. W., 122, 126, 127, *145/64*, 164

Radioactivity: Effect on steel, *77/2*, 235, 237–8; Monitoring, 42; *see also* Effluents

Rasmussen Report, 56, 59, 61, 67–8, *77/2*, *78/11–12 and 17*, *80/42*, 82, 85–7, 282

Rasmussen, Professor Norman, 59, *94/19 and 29*

Ray, Dixey Lee, 91 *95/38*

Reactor types:
AGR: adopted by AEA, 7, 9; opposition at Harwell, 7, 117; rejected by French, 96, 118, 269; prototype, 11, 117–49; operates, 113, 119; change to concrete pressure vessel, 119; AEA attitude to, 14, 119, 120, 125, 153, 286–7; chosen against

BWR (1965), 10, 123; problems in fuel development, 126; 36-ring fuel clusters, 120–1, *187/31*; orders, 127, 128–9; development troubles, 124–5, 129–32; exportability, 10, 113, 129, 151, 209; re-design needed, 227; inherent high costs, 132–4, 228–9, 256–7; HTR and SGHWR to be cheaper (1969), 222; CEGB attitude, 131–2, 222; favourable comment, 255; cost of programme, 137; long-term comparative costs, 132–9; a failure, 194

Air-cooled graphite moderated reactor, 272

BWR: System, 9; stages in development, 71, 118, 158; rapid increase in size (1963–7), 43; BWR, 6 and Mark III containment, 42, 82, 89; standardised design, 82, 85; early development not familiar in UK, 161; disparaged in UK, 182; ordered in US as competitive (1963) Oyster Creek, 9; ordered by TVA (1966–7), 35; CEGB attracted (1964), 9, 161–2; design improvements rejected by CEGB (1965), 10, 123, 161; hot vacuum dried fuel rods, 42; Rasmussen Report, 59; AGR/BWR comparison by CEGB and AEA (1965), *see* Appraisal; Oyster Creek completion delayed, 183; gaseous effluent, 63; Mark I containment problems, 89; Mark III, 4, 42, 82, 89; vibration in fuel channels, *95/34*; fire at Brown's Ferry, 86–7; costs, 12, *21/28*, 82, *188/35*; and *see* LWR; French reckon BWR and PWR costs equal, 97; BWR in Germany, 100; in Sweden, 104; in Japan, 105; in Finland, Italy, Spain, Switzerland, 104; France ordered 2 but cancelled 98

CANDU: System, 223; CEGB (Hinton) attracted (1962–3), 119, 162; could be used as stopgap (Hill), 227; exports, first orders India, Argentina and Korea, 96; AECL favour for UK, 239,

CANDU – *cont.*
*256/13*; drawbacks and attractions, *310/88*

FBR: Early US assessment, 4; and development (from 1945), 272; Early UK programme, 7–8, 272; 'Universally agreed *terminus ad quem*' (Hill), 226; UK claims lead, 11, 14, 155, 225; Dounreay PFR (Prototype fast reactor), 14, 154, 207; optimistic AEA forecasts belied, 154; Phénix, French prototype, unlikely to operate till 4 years after PFR (Penney), 155, but operates before PFR, 15, 155; UK at best level pegging (1972), 9, 265; no 'commercial' plants before 1990, 155; Dr Marshall on prospects, 265; German FBR government sponsored development by Interatom, 103; prospects seen as remote, 103; in Sweden no funds for work on FBR, 104; in Japan research treats as remote, 105; US and Germany important work on components and fuel, 15; a UK programme of 'commercial' FBRs could be premature, 15; TNPG takes in AEA FBR team, 208; Working Party, 210; alternative coolants, 254

Gas-Graphite-CO₂ Cooling: Initial choice for UK, 7–9; in France, 97; Ministers choose for UK, 7, 117, 269, 271; US experts criticise for 'low material economy', 269; Harwell scientists judge poor system for enriched uranium, 7, 117; French agree and turn to LWR, 118, 268; but AEA and Government choose CO₂ cooling with enriched uranium, 117–18; Rotherham defends choice, *142/10*; Germany no experiments with CO₂ cooling system, 11, 99; AEA calls 'most advanced system available' (1964), 152; Select Committee says choice may have been mistake (1967), 169, but attitude ambivalent, 203; Weinstock considers

choice a main source of export failure, 215; inherent high costs, 126, 132–3, 227; gas-graphite distinguishing feature of UK development, 253; Harwell Pippa feasibility study, 272

Gas-graphite-helium cooled reactor, *see* HTR

HTR: attraction of system, 7; early UK work, 7, 11; UK claim lead, 11; early German work (pebble bed), 100; leading AEA expert joins US firm, *19/20*; Select Committee urges rapid development (1967), 169; calls 'leading large reactor for export' (1969), 129; large plants ordered 1970 but cancelled 1974–5, 25; TNPG expects quick development, as economical, 129; CEGB calls for tenders (1970), 222; technical difficulties delay, 222–3; US and Germany take lead in development, 169, 194, 253; operating prototypes larger than Dragon (international unit on AEA site), 253; industrial uses, 17; ERDA cautious optimism, *79/32*, 86; AEA rates SGHWR more developed than HTR (1972), 227; CEGB wants large demonstration plant (1973), 225; Hill's preference for HTR (1974), *257/42*; UK development stopped (1974), 246, 253; significance, 253; orders for 800–1000 MW plants in Germany and US, 253; HTR loses ground (1975–6), 253; large plants cancelled, 253

LMFBR Liquid Metal Fast Breeder Reactor, *see* FBR

LWR: two types BWR and PWR q.v. Competitive plants ordered (1963), 10–11; competitiveness accepted (1965–76) in US, 25–7, 36, 83–4; outside US, 10, 96–108; orders in US, 26, 82, 92; outside US, 25–7, 96; early Harwell support, 7, 12, 275; used in all industrial countries save UK and Canada, 23; early recognition in Europe, 96;

Euratom (Germany), 11, *19/24*, 100–1; Sweden, 104; NNC attracted, 226; CEGB attracted (1964), 120, 163; but turned down in favour of AGR, 163; components, UK export prospects, 226; CEGB against (1969–70), 222; sensational rise in costs US, 16, 30, 83; sources analysed, 30–51; environmentalist opposition US, 16, 23–4, 45, 51, 60–4, 68–77, 85–7; licensing delays US, 232; prospects for fall in real costs US, 84–5; standardised components, 293; exportability, 301; CEGB wishes to order (1973–4), 196, 225, 246; not investigated for safety in UK prior to SGHWR choice, 242, 281; AEA says LWR can be stopgap in 1974 and recommends in place of SGHWR, 225–6; 20-year safety record, 56; after SGHWR choice, LWR supporters asked to keep silent, 246, 288

Magnox: constraints determining choice (1953), 114–15; Pippa to supply plutonium with power as by-product (1953), 6, 117; US criticism of system for power production (1955), 9; basis of first UK programme, 8–9, 12; two exported (Italy and Japan), 9, 105; no subsequent exports, 9, 96; French use same system for early plants, 97; but abandon it for LWR, 97, 118; costs in UK above forecasts, 12, 154, 274, 277; concrete pressure vessels for Oldbury and Wylfa plants, 130; vibration and insulation problems, 130; corrosion problems, 130; all CEGB Magnox save one derated, 153; load factors fall as capacity and pressure increase, 184, *191/118–19*; Select Committee's extravagant claims for Magnox achievement (1964), 158, 170; TNPG suggest new fuel design could make Magnox cheaper than AGR, 164; CEGB claims Magnox gives cheapest power, 135; but figures neglect

inflation and interest costs, 135; move from Magnox to AGR, 7; not chosen for test by Germany, 12; not success in Japan, 105; FOE figures, *258/44*

Molten salt reactor, 243

PFR, *see* FBR Dounreay

PWR: used first for submarines, 11, 98, 272; System, 9; Rasmussen Report, 59; Harwell failed to see scope for large reactors, 272; early US prototypes for power stations, 118; shipping port Plant ridiculed in UK, 182; English Electric offer PWR for Dungeness B, 162; rise in costs US, *see* LWR; estimated costs in UK (1973–6), 229, 252; orders in US (1972–4), 82, 85; Fuel rods internally pressurised, 42; ATWT, 84; controversy on safety in US, 55–77, 83; UK discussion, 182, 236–45; Marshall Report on pressure vessel, 242, 246, 249, 250, 265; English Electric and Rolls Royce approach Westinghouse re building in UK, 162; CEGB wants to buy (1973), 113, 194; NNC wishes to make (1973–4), 229, 303; and envisages link with French Westinghouse, 232; standard design PDAs in US, 84; Select Committee opposes PWR for CEGB, 230; and Government refuses CEGB, 17, 246; use and manufacture in Germany, 100; in Japan, 105; in Switzerland, Italy and Spain, 104; French Westinghouse orders, 98, 268; used in Finland and Sweden, 104; NNI instructed to make safety study, 246; Idaho test, 236, 238

SGHWR: cross between HTR and BWR, 223; 'The best BWR' AEA claim, 11, 223; 100 MW prototype built (1962–7), 12, 223; Hinton (CEGB) opposed (1962), 162; lower costs than AGR-AEA and CEGB (1969), 128, 222; modular system as source of low costs, 227, 231; export potential claimed, 227;

SGHWR – *cont.*
said to depend on British utility, 227; AEA attempts special inducements, 228; NSHEB secures tender, *258/47*; called 'proved reactor', 223; reasons for CEGB opposition, 215, 225, 230, *257/43*; TNPG absorbs SGHWR team, 224; SSEB and NSHEB support SGHWR (Tombs), 215, 227, 228, 244; AEA attitudes, 128; (Moore 1972), 223; Hill (1973–1974), 227, (1976), 249; CEGB and NNC, 230; Select Committee attitudes, 127–8, 228, 245; Government decides utilities should order, 196, 245, 281, 288, 293; development problems, 230, 231, 247–8; safety problems, 241, 245, 247, 249; AEA recommends abandonment, 249; research curtailed, 250; development suspended, 304; estimates of costs, 229, 248–9; Canada wanted UK to adopt, 239

Reed Report (GE), 85, 88–90

Regulation to ensure safety: in Germany, 106

in UK: Nuclear Installations Inspectorate, 233, 241–2; Chief Inspector appointed by Minister, 233; size of staff (1973), 242; allocation of staff, 242; a few to study US published evidence, but not design and manufacture, 242; calls on AEA for research, 243; reports on individual plant proposals not published, 243; Marshall Report on PWR vessel safety, 242, 249, 250; Health and Safety Executive responsible for NII from 1975, 249

In US: AEC responsibility, 15–16, 243; widened by NEPA, 47–48; other agencies involved, 15, 47, 243; openness, checks and balances, 243–4; AEC licensing procedure, 45–51, 56–7; ACRS and ASLB, 46–8, 56, 74–5; public hearings by ASLB, 46–7, 76, 244; rule-making hearings, 46, 65, 75; AEC criteria on effluents changed, 64; on ECCS, 65–7, 74; standardised designs, 55, 82, 84; delays in licensing, 45, 83; AEC regulation criticised: for bureaucracy (JCAE), 45, for excessive caution, 65, 71–2; for subordinating regulation to promotion, 61, 70; regulators expect to cut licensing time (1973), 82, 84, but no success by 1976, 83, 84; functions passed to NRC (1975), 83, 84, and ERDA, 83; NRC criticised, 89; UK views on US regulation, 234, 238, 239, 241–4; Openness of US system, 243; *see also* ACRS, ASLB, Intervenors, State rights, Licensing and Public hearings

Reload cores, *52/25*

Replacement power costs, 126, 138–40 *passim*, *147/106 and 118*

Replication, 127, 136, 181, 204, 304

Research and development, UK: Control, 6, 13, *145/60*, 198, 203, 204, 207–8, 213–14, 241, 248, 277, 292; costs, 137, 140–1, *191/106*, 204, 248; on carbon deposition on fuel, *147/98*

Responsibility, Ministerial, *see under* Ministers–UK

Reyrolle, A. & Co., *115/1*

Reyrolle-Parsons, 227

Rheinisch Westfälisches Electrizitätswerk AG (RWE), 99, 100, 102, *109/22, 145/75*

Richardsons, Westgarth, *115/1*

Rio Tinto Zinc (RTZ), 101

Rippon, Geoffrey MP, 264

Risks: 'scram' failure, 234; 'abnormal' incidents, 66; 'common mode' accidents, 86; 'steam binding', 90; vibration, 124, 130, *147/100*, 151, 161; corrosion, 124, 130, 131, 151, 153, 161, 169, 235; static adhesion, 124; cracks, fretting, etc., *77/3*, 124, 235, 236; metal/water reactions, 236; metal deformation, 236; copper embrittlement, *77/3*; plutonium toxicity, 77, 90, 91; 'cold spots', 151; loss of coolant, 71, 74, 235; pipe or vessel rupture, 58, *94/29*; radioactivity, 42, 55, 58; cancer, 58, 62, 68; gaseous and radioactive effluents, 42, 58, 62, 64; bombs and bomb tests, 62, 86, 90; core melts,

67, 68; earthquake, 31, 42, 62; catastrophe, 64–8; sabotage, 16; tornado, 31, 42; tsunamis, 42; missiles or aircraft, 42; control room mishaps, 42; 'routine' effluents, 58, 62; weld faults, 77/3; *Public Health Risks of the Thermal Power Stations, Report on,* for the Resources Agency of California, 77/9

Risley, 5, 6, *18/13*, 117, *145/60*, 198, 207, *219/41*, 224, 273–5 *passim*, 288, 291

Rittenhouse, P., *258/71*

Robinson Plant, *260/89*

Roisman, Anthony Z., 65, *79/37*, *80/43*

Rolls Royce Ltd., 162, 211, 280

Rotherham, Dr L., 118, *142/10*, 224, 226, 228, 234, 235, *257/19*

Royal Commission on Environmental Pollution, *93/15*

Rule-making hearings, 47, 65, 75

Rusche, Bernard, 88, *94/26 and 29*

Russia, 156

Sabotage, Risk of, US, 16, 77, 91

'Safer than safe enough', 72, 89, 153, 249

Safety and environment, 232–44, 245, 248: types of risk, *see* Risks: Degree of Risk: Probability assessments – development of methodology, 57–62, 85–6; initial work by F. R. Farmer (AEA), 77/7; risks from nuclear and other sources compared (AEC report 1973), 57, (Rasmussen Report), 67–8, 82, 86, (California Report) 58, *77/9 and 10*: estimates in Brookhaven Report (1957), 60–61; probability assessments, 57–62; probability methodology dismissed by three GE engineers, 87–8; uncertainties in measuring radiation risks, 62–3; bases of assurance: confidence of experts, 56, 58, 59, 66, 89, 90, 96, 233, 241; the safety record, 56, 58, 63–4, 66, *79/36*, 233; exceptional research, 56, *77/3*, 90–1, 107, 233; conservatism in design, operation and maintenance, 56, 63; exceptional quality control and engineered safeguards, 56, 63, 249; stringent regulation and surveillance, 56, 237; safe 'to any degree of con-

cern', 249; *see also* Rasmussen Report

Safety Inspectorate – UK, *see* Nuclear Installations Inspectorate

Salisbury, Lord, 264

San José, 91, *94/33*

San Onofre Plant, *260/89*

Sandys, Duncan (later Lord), *142/7*, 264, 274, 291, *307/29*

Sargent and Lundy, 30, 32, 33

Saturated steam turbines, 223

Sax, Professor J. L., *54/45*

Schlesinger, James, 64

Schoeters, Ted, *187/27*

Schubert, A. E., *77/4*

Schulten, Professor R., 100

Scientists and Engineers for Secure Energy (SESE), 91

'Scram' failure, 234

Seaborg, Glenn D., 60, 64, *78/20*

Secretary of State for Energy, 196

Secretary of State for Trade and Industry, 233

Select Committee (House of Commons) on Estimates (1959), 157, 177, 275: Report on UK Atomic Energy Authority, *143/16 and 17*

Select Committee (House of Commons) on National Expenditure, 224

Select Committee (House of Commons) on Nationalised Industries – *The Electric Supply Industry*, *143/18*, 164–5, *186/14*, *188/36 and 42*, 288–9, *308/53*

Select Committee (House of Commons) on Science & Technology (The Select Committee) established, 10; members, 180–1, 234, 239, 240; sub-committee procedure, 180–1; contrast with US JCAE, 181; on UK nuclear policy, 12; treatment of evidence, 125–8, 181–2, 184–5, 239–41; criticism of Ministers, 169, 305; recommends Technical Assessment Unit, 216–217, 286, 290–5; criticises CEGB, 125–7; respect for AEA, 181, 209; recommends more power for AEA, 209; AEA criticised for abandoning SGHWR, 249; ambivalence on $CO_2$/graphite reactor systems, 170; attitude on AGR, 121, 125–8; on HTR, 129, 169, 208, 244; on LWR, 129, 181, 183, 225, 232–3, 241–2, 250; on

Select Committee – *cont.*
SGHWR, 129, 181, 228, 249–50, 304–5; advocates single D & C company, 173, 197, 203–4, 210, but Party division on this, 203, 205, 210; criticism of NNC shareholding, 195–6; attitudes on competition, 203–9, 213; misjudges export prospects, 128–9, 209; against nationalism (1967), 203; for nationalism (1973–4), 210, 228, 241; on safety, 232–44 *passim*; no study of US Safety organisation, 242–4; 'cost was a secondary matter to coal and oil shortage', 12; failed to probe AGR failure, 17, 125–8, 141, 181–185, 286; questions on international competition, 171–2; ministers need more expert advice, 175, 176, 180; sub-committee's visit to GE and Westinghouse (1967), 204, 239

Select Committee (House of Lords) on European Communities Energy Policy Strategy (1975), 107, *110/ 37, 306/3*

Semi-scale test, 67, 76, *79/41, 80/52,* 236, *259/76*

Shaw, Milton, 57, 64, 66–7, *79/39 and 41*

Shippingport Plant, 118, 182

Shoreham (L.I.) Plant, *80/52*

Siemens AG, 100–1, 107, 129, 203, 300

'Significant event' statistics, *79/34*

Single D & C Company, 11, 113, 114, 126, 129, 173, 195–221 *passim, 220/65,* 269, 279, 281, 283, 284, 289, 293, 299, 304

Sizewell Plants, 129, 135, 209, *220/55;* load factor, 184, *192/118–19*

Sketch design for Dungeness B, 123, 132, 150, 153, 159, 163, 178, 266, 280, 286

Smith, Adam, 304, *310/85*

Smith, D. R., *189/58*

Sodium-cooled thermal system, 100

Solar heat, 2

South Africa, 90

South of Scotland Electricity Board (SSEB), 136, 137, *145/66,* 151, 183, 196, 215, 216, 227–9, 244, *262/121*

Spain, 96, 104, 284

Sporn, Philip, 32, *51/5, 78/20*

Springfields Laboratory, *145/60*

Standard Plants in US: Licensing authority offers shorter review for, 51; manufacturers offer designs, 84; first preliminary (PDA) approvals, 84

Stanford Research Institute, 59

Starr, Chauncey, *77/9*

State Power Board, Sweden, 104

State rights US – claims to control radioactivity and thermal pollution, 49

Static adhesion, 124

Steam-generating heavy-water reactor (SGHWR), *see* Reactor Types

Steel, effect of radiation on, *77/3; see also* Pressure vessels and Marshall Report

Steel membrane distortion at Dungeness B, 125, 166

Sternglass, Dr Ernest V., 87

Stewart, J. C. C., 158, *219/41,* 234, 255, *258/44*

Stoneham, Lord, *309/59*

Strachan & Henshaw, *115/1,* 116, *143/ 15*

Stretch, K. L., 154, 157–8, *186/13, 187/29*

STRIDE (Standardised Reactor Island Design) for BWR, *92/2*

Sub-contractors and sub-contracts, 199–200, 213

Subdivision of Production, *see* Fragmentation

Submarines, nuclear: France, 98; US, 11, 272; behaviour of steel in PWRs, *77/3*

Suez, impact on UK nuclear policy, 12, *19/33,* 269

Sulphur removal, Cost of, 36, 83

Surry I Plant, *260/89*

Sweden, 96, 98, 103–4, *109/29,* 216, 232, 245, 268, 282, 284; success of private enterprise development, 104; State Power Board chose LWR, 104; Social Democrats' large nuclear programme, 107

Switzerland, 96, 104, 216, 232, 282, 284


Tamplin, Dr Arthur R., 62–4, *79/32,* 87, 249

Taylor, Dr Lawriston, *79/26*

Technical Assessment Unit, 216–17, 286, 290–5

Teller, Edward, 60, *78/14*

Temperatures, operating, 42, 153
Tender Assessment procedure (CEGB), 123–4
Tennessee Valley Authority (TVA), 4–5, 31, 33, 35: Significant design changes, 42; capital charges, *190/ 105*; *and see* Brown's Ferry Plant
Test rate of discount (TRD), 138
Thermal pollution standards US, 49, 73
Thirty-six-rod cluster fuel element (for AGR) – new fuel assembly, 120–1, *143/20*, 151, 156, 158
Thorium as fuel, 254
Thorn, J. D., *142/1*, *143/19*, *186/10*
Thomson, Sir George, 182
Tokai Mura Plant, 105
Tombs, F. L., 183, *221/78*, 227, 230, 231, 233–4, 240, 244, 245, 248, 251, 253, 256, *257/42*, *263/143*, 294
Tornado risks, 30
Toshiba (Tokyoshibaura Electric Co.), 105
Trade Unions, 207, 254, 274, 279; *and see* IPCS EETPU NUM
Transfer of AEA teams, 206–8
Trawsfynydd Plant, 126, 127, 139; Load factors, 184, *192/118 and 119*
Treasury, The, 176, 177, 179, 185, 210, 289
Tremmel, Ernest, 222
Tugendhat, Christopher MP, 241
Turkey Point 3 & 4 Plants, *260/89*
Turnkey system, 203
TVA, *see* Tennessee Valley Authority

Uhde, Lurgi, NUKEM & St Gobain (private consortium), *109/26*
UK government centralised policy, 6, 8: ten-year programme, world's first, 11; safety research subsidy and cheap enrichment, *306/17*; *see also* AEA
Ultrasonic testing techniques, *77/3*, 235
'Unconscious levity' of UK government, *191/107*, 302
Underground nuclear plants proposed, 60
Union of Concerned Scientists, *53/30*, 237
United Power Co. (UPC), *115/1*, 120–1
University of California, Berkeley, 62
University of California, Los Angeles (UCLA), 59, *77/9*

Upstairs-downstairs view of British industrial economy, 289, 301
Upton, Arvin E., *81/58*
Uranium enrichment (diffusion plant): UK, 7, 272; Urenco, 103; US, 32, 41, 84, *306/17*
Uranium mining, milling and conversion, 32, *256/13*
Urenco, 103
US Congress – Joint Committee on Atomic Energy (JCAE): contrast to Select Committee, 181; experts' assessments for (1954–6), 2, 4, 182; powerful watchdog over AEC, 4; discusses rising coal prices, 35; criticises AEC licensing as badly organised, 45; initiative in setting up ACRS, 46; staff criticism of EIS, 48; evidence to JCAE on NCRP and ICRP, *78/ 23*; JCAE on high effluent limits, 64; evidence to JCAE on semi-scale tests, 67, *79/41*; JCAE attacked as too pro-nuclear and pro-AEC, 61, 70; all AEC safety research data sent to JCAE, 71; criticism in JCAE hearings of ASLB public hearings, 76–7, 92; discussions on waste disposal, 92; members pressurised by Nader groups among constituents, 91; hearings on GE resigned engineers, 88–9; Calvert Cliffs decision, *53/44*; McKinney Report, *18/5*
US nuclear achievements disparaged in UK and discrimination against US products; 14, 156, 158, 161, 182–3, 201, 210, 224, 227, 228, 232, 233, 237, 239, 284, 293
Utilities (Electricity supply companies): US: numerous, most large ones private enterprise, 4; investment and finance sources, *52/12–14*; *but see* TVA; participation in early development of nuclear plant, 4–5, 270–1; keen on cost-cutting innovations, 11; left to buy nuclear plant on commercial criteria, 4–5, 8, 206, 267–8, 271; free to choose system, 267–8, 271; no government direction, 8, 267–8; no subsidy, 8, 271; 200 MW utility-manufacturer plants planned 1955, 9, 25; (*see also* Manufacturer-Utility axis);

Utilities: US – *cont.*

many consider LWR competitive (1965), 25; factors determining utility orders for LWRs, 24–8, 35–6, 82; many orders cancelled (1974–5), 83, but LWRs still deemed competitive by most utilities (1976), 19, 83; financial difficulties with rising costs and delays, 36–45, 83; influence of higher cost of money, 39–40; and of public regulation, 42, 45–51 *passim*; and slow market growth, 83; *see also* individual utilities e.g. TVA, Commonwealth Edison

UK: few, all nationally owned, 6; dominated by CEGB q.v.; advantages of sub-division discussed, 215–16, 284; members of Boards appointed by Ministers and subject to directives, 6–7, 275; Ministers must approve investment plans, 7, 275, 281; utilities and first ten-year plan, 8, 161; CEGB and AGR choice, 161–3; uncertainty about Ministerial part, 117–18, 271–9; Ministers say utilities must buy SGHWR, 17, 114, 245; France, 97, 98, 268; Germany, 97, 100–1, 268; Italy, Spain, Switzerland, Finland, 104, 268; Japan, 105–6, 268; Sweden, 104

Vallecitos Laboratory, Calif., 118, *142/14*, *187/31*
Varley, Eric, MP, 196, 218, 241, 242, 245–7, 249, 253, *260/89*, 264, 281–282, 285, 288, 293–4, 296, 299, 305
Vaughan, Roger D., *20/36*, *146/85*, *95 and 96*, *186/10*, *188/52*, *189/55 and 58*
Vermont Yankee Plant, 92
Vibration, *95/34*, 130, *147/100*, 151, 161
Vinter Committee, 185, 209–10, 224
Vinter, F. R. P., 179, 210, 224, *256/16*, 279, 288

Wage rise in US, 35, 38
Walker, Peter MP, 195, 216, 264
Waste, US problems, 16, 77

Water, pure, standards in US, 49
Waterford Plant (US), *53/41*
Waverley, Lord, *307/32*
Weapons, nuclear: France, 98; UK, 5; US, 4, 11
Weinstock, Sir Arnold, *188/47*, 194, 207–8, 211, 215, 224, 226, 234, 237, 240, 245, 255, *257/31*, 293
Welding, *77/3*, 183, *192/117*
West, Harry (AEI), *306/16*
Westinghouse Electric Corporation, *29/1*, *52/125*, 67, *80/41*, 104, 120, 162, 211, 239, *260/89*, 268, 284, 289, *308/38*: Framatome (France), 98, 232; Fuel economy, *92/3*; Germany, 100; Standard PWRs, 84, *93/9*; Zero release plants, 63
Whessoe Co., *115/1*, 116
Whetstone Works, 207, 255
White, G. H. (GE), *142/14*, *187/31*
Whitman, Merrill, *79/39*
Williams, E. C., Chief Inspector of Nuclear Installations, 210, 233–40 *passim*, 242, *259/76*, *260/89*, 281
Williams, H. E., *187/131*
Wilson Committee (1969), 166–7, *189/ 61 and 64*
Wilson, Sir Alan, *189/61*
Wilson, Sir Harold, 168, *190/79*
Windscale, 118, *186/9*, *187/29 and 31–32*; to produce plutonium, 272
Windscale prototype AGR, 11, 119, 122, 123, *142/14*, *143/24*, 152, 155, 158, 161, 165–6, 178, 222
Winfrith Heath Prototype (SGHWR), 230, 231, 248
Winnaker, Dr Karl, 99
Wolman, Professor Leo, *54/45*, *80/54*
Woods, Sir John, *307/32*
Wootton, Barbara, *307/34*
Working Party on Nuclear Power Costs, Ministry of Power, 126, 170
World in Action – TV feature on nuclear risk, 241
Wylfa Plant, 119, 130, *146/85*, 153, *188/52*; Load factors, 184, *192/ 118–19*

'Xenon 135–138', 63

Zero release plants, 63
Zion Plant, 44, 225, *259/76*, *260/89*
Zirconium cans and tubes, *181/31*, 201